Monte Verità Proceedings of the Centro Stefano Franscini
Ascona

Edited by K. Osterwalder, ETH Zürich

Soil Monitoring

Early Detection and Surveying of Soil Contamination and Degradation

Edited by
R. Schulin
A. Desaules
R. Webster
B. von Steiger

1993

Springer Basel AG

Editors:

Professor Dr. Rainer Schulin
Institut für terrestrische Ökologie
Fachbereich Bodenschutz
ETH - Eidg. Technische Hochschule
Grabenstrasse 3/11A
CH-8952 Schlieren, Schweiz

Dr. André Desaules
Swiss federal research station for
agricultural chemistry and
hygiene of environment
CH-3097 Liebefeld-Bern, Switzerland

Professor Dr. Richard Webster
Institut für terrestrische Ökologie
Fachbereich Bodenschutz
ETH - Eidg. Technische Hochschule
Grabenstrasse 3/11A
CH-8952 Schlieren, Schweiz

Berchtold von Steiger
Institut für terrestrische Ökologie
Fachbereich Bodenschutz
ETH - Eidg. Technische Hochschule
Grabenstrasse 3/11A
CH-8952 Schlieren, Schweiz

Library of Congress Cataloging-in-Publication Data

Soil monitoring : early detection and surveying of soil contamination and degradation / edited by R. Schulin ... [et al.].
-- (Monte Verità : proceedings of the Centro Stefano Franscini, Ascona)
Proceedings of the workshop ... held at the ETH seminar centre, Stefano Franscini of Monte Verità, Ascona (Switzerland) from October 18–23, 1992"--Pref.
ISBN 978-3-0348-7544-8 ISBN 978-3-0348-7542-4 (eBook)
DOI 10.1007/978-3-0348-7542-4
1. Soil pollution--Measurement--Congresses. I. Schulin, R. (Rainer) II. Series: Monte Verità (Series)
TD878.S665 1993
628.5'5'0287--dc20

Die Deutsche Bibliothek – CIP-Einheitsaufnahme

Soil monitoring : early detection and surveying of soil contamination and degradation / ed. by R. Schulin ...
– Basel ; Boston ; Berlin : Birkhäuser, 1993
(Monte Verità)

NE: Schulin, Rainer [Hrsg.]

Originally published by Birkhäuser Verlag Basel, Switzerland in 1993

Camera-ready copy prepared by the editors
Printed on acid-free paper produced of chlorine-free pulp

9 8 7 6 5 4 3 2 1

PREFACE

This volume contains the proceedings of the workshop on "Soil Monitoring: Methods for Early Detection and Surveying of Soil Contamination and Degradation", held at the ETH seminar centre "Stefano Franscini" of Monte Verità, Ascona (Switzerland) from October 18 - 23, 1992. Seventy participants, representing a variety of institutions, nations, and disciplines, discussed the concepts, approaches, status, gaps, problems, and perspectives of soil pollution monitoring.

The idea for this workshop came from A. Desaules when he was installing the Swiss National Soil Monitoring Network (NABO) as his doubts about the philosophy of soil monitoring prevailing at that time increased. This philosophy essentially equated soil monitoring with repetitive surveys of soil pollutant concentrations at permanent observation sites. He sought others interested in discussing alternatives, and he found a ready partner in the ETH-Institute for Terrestrial Ecology. Soon it was realized that a discussion of the NABO would immediately raise general questions with respect to the conceptual basis of soil monitoring and that a minimum agreement on this basis was indispensable to discuss more specific problems related to the realization of the NABO.

As a result, a workshop was organized whose objectives were in particular (i) an assessment of current knowledge on soil monitoring by permanent networks, (ii) a synthesis of the experience from different disciplines related to soil monitoring, (iii) the identification of research gaps with respect to long-term and large-scale soil monitoring, and (iv) the design of a platform for the development of soil monitoring strategies and methodology.

The workshop consisted of a series of sessions on specific topics followed by a final synoptic session. Each topic session was started with formal presentations of invited speakers to introduce and review the most important aspects. Then open questions on each topic were discussed within small groups of mixed composition. Specialists in the relevant fields were invited to prepare questions and to summarize the discussions. Their reports which include their personal interpretation as well as the views of the groups follow the papers on the particular topics in this volume.

The papers have been reviewed for publication, and we thank the referees for their advice and constructive criticism.

Many participants helped behind the scenes in the organisation and management of the workshop and thereby contributed to its success. We mention in particular: P. Blaser, F. Crameri, R. Dahinden, U. Eggenberger, I. Forrer, P. Fry, G. Geiger, U. Hoins, R. Meuli, J. Obrist, A. Pepels, K. Studer, S. Zimmermann, and M. Zysset. The team of the Centro Stefano Franscini, led by Mrs K. Bastianelli, provided a very pleasant and productive environment. We are also indebted to Mrs M. Webster for reading and editing the scripts.

The workshop was supported and generously funded by the ETH Zürich through Prof. Dr K. Osterwalder and his staff, whom we thank. We also thank the Swiss Federal Research Station for Agricultural Chemistry and Hygiene of Environment (FAC), Liebefeld-Bern, the Swiss Federal Office of Environment, Forests, and Landscape (BUWAL), Bern, the Ciba Geigy AG, Basel, and the Metallwerke AG, Dornach, for their sponsorship.

Schlieren, July 1993
Bern, July 1993

Rainer Schulin
André Desaules
Richard Webster
Berchtold von Steiger

TABLE OF CONTENTS

SECTION 4 Survey and Data Analysis

PAPERS

REPORT

Soil Monitoring, Monte Verità, © Birkhäuser Verlag Basel

INTRODUCTION

The soil has a limited capacity to immobilize and degrade undesirable chemicals. In recent years this fact has become of increasing concern, both because of the burdens of such chemicals that the soil is now known to carry and because people have become aware of the damage being done to their environment and the hazards to health that they represent. In response some countries have enacted laws that oblige their governments and other agencies to monitor soil pollution so that they can diminish it or at least prevent its becoming worse. Long term monitoring began in the early 1980s after people began to notice forest decline and attributed it to pollution. In many respects, however, we lack the scientific basis for designing and implementing schemes for monitoring. Pollutants in the soil vary greatly in both space and time, and this variation depends on a host of factors specific to the particular sites of interest including the nature of the soil, the behaviour of the pollutants, and the climate.

Different types of pollution demand different strategies for measuring and assessing them and for their remediation. The scientific and technical problems posed are legion, and solving them all is taxing hundreds if not thousands of scientists throughout the world. The workshop had to concentrate, and it focused on soil contaminants that threaten the long-term soil fertility and related functions of the soil on a regional scale; i.e. on a scale larger than the individual field. The pollutants of most concern were the heavy metals and persistent organic poisons. Controlling these involves taking into account the unique nature of the pedosphere.

Although the soil is accessible in the sense that we can get to any place, determining its contents is difficult with current technology. We have to remove material from the field, take it to the laboratory where the pollutants can be extracted and measured in the extractant. This is expensive and takes time. It also means that the quantities actually recorded depend on the procedure used for extraction. Perhaps more seriously the soil on which the measurements are made is no longer there to measure on a second or subsequent occasions, and later measurements can be made only at sites elsewhere. Inevitably therefore measurement of change with time is confounded with spatial variation. This means that a monitoring programme must be designed in such a way as to separate these two sources of variation.

The soil at any one place has a history, which can be inferred and distinguished from that elsewhere by the soil's characteristics. This information can be valuable, either because it can tell us what fate to expect for certain pollutants, or because by stratification we can economize on sampling. Much of it exists in the form of maps of soil and geology, and it should be possible to marshall it more effectively than in the past using automated geographic information systems (GIS). This application to soil monitoring should be investigated further.

The soil can store large quantities of most pollutants, and by comparison the amounts received annually from contaminated fertilizers and sewage and in fallout from the atmosphere are small. Accumulation may appear to be slow, therefore, and it might go undetected for a long time against the background spatial variation. Nevertheless, the long-term result can be serious, and if it is detected only at this late stage then remediation is likely to be expensive or even impossible. To prevent this from happening, to maintain the proper functioning of the soil, and to protect the quality of food grown on it and water draining from it any increase in potentially dangerous substances should be detected as soon as possible so that deterioration can be halted. Solutions based on inventory are not the only ones: we do not have to wait till accumulation becomes 'significant'. There are other ways of assessing the likely impact of pollution on the soil. One in particular is to estimate the fluxes of pollutants into and out of the soil: it is the mass balance approach, and it has had considerable success in studying cycling of elements in forest ecosystems. Its value for soil monitoring needs to be explored more.

Some contaminants sorb so strongly on the solid phase of the soil that they are for practical purposes immobile: they might accumulate but have no deleterious effects. In fact because they are so persistent they might serve as markers against which to judge temporal fluctations in other pollutants. They constitute the soil's 'memory' of pollution. The soil is also a complex reactor, and a pollutant that might be quite immobile and innocuous in one state of the soil could become lethal if the soil changes. For example, some of the heavy metals are virtually insoluble in a neutral or calcareous soil and pose no threat to life. However, a decade or two of acid rain could bring the soil's pH down to less than 5, thereby mobilizing the metals, making them available to plants and releasing them into drainage water. It is a 'chemical time bomb'. Simple measures or predictions of contaminant burdens are not enough, therefore. It is necessary to know and understand the processes in the soil that can lead to such changes, and to anticipate them.

These were the topics chosen for discussion at the Monte Verità workshop and which are now the subjects of this book. The papers cover the principles of soil monitoring, current technology for it, and practical applications. As in the workshop itself, they are grouped by topic, each of which is followed by one or two reports of the group discussions.

The first three papers describe the objectives, experiences, and problems of soil monitoring, with emphasis on Switzerland. They give examples of new concepts, in which soil contamination on a regional scale is estimated by the mass balance approach.

The second section assesses specific fluxes for contaminant mass balances in soil. It opens with a paper discussing the general problems of the mass balance approach to soil monitoring, in particular that of measuring the fluxes in and below the soil and the relevance of spatial scale. Two papers cover the input of contaminants from the atmosphere by deposition and filtering by vegetation canopies. These are followed by four papers on the processes generating fluxes out of the soil by volatilization, leaching, erosion, and plant uptake. This section shows clearly that the 'state of the art' is not yet good enough for calculating mass balances as routine in long-term monitoring, and that it must be improved to be effective. In discussion the evidence was assessed, and participants attempted to set priorities for further research and for developing the practical application of measurements.

A section on assessing soil contamination begins with a description of problems in analysing for inorganic and organic soil contaminants and aspects of sampling efficiency for calculating mass recovery. The difficulties were pursued in the group discussion, and the topic report places them in the context of long-term monitoring. Microbiological characteristics of the soil appear to have much potential as indicators for monitoring in soil pollution, and this section includes an extensive review of the topic. Some soil contaminants are ferromagnetic or correlated with other ferromagnetic materials, and another technique with potential for monitoring soil pollution is based on measuring the magnetic susceptibility of the soil.

Finally, there is a section on the representation and statistical analysis of *data*. One paper lists the sources of error and how to control them by sound sampling, quality control, and attention to detail. Spatial variation in the soil is a fact now widely recognized, and two papers describe the sampling strategies to be adopted to estimate change of contamination in time and local fluctuation against this background. As above, GIS are destined to play ever increasing roles in soil monitoring, and they are the subject of another paper. The discussion distinguished actual knowledge from inference, stressing the need to design survey and analyse the resulting data specifically for the objectives of the investigation.

The workshop closed with a synthesis of the week's proceedings by two research scientists and the view of an administrator. Superficially the two seemed contradictory, yet at depth they agreed. The primary objective of soil monitoring is to provide sound and relevant information on which political decisions, which may be far reaching, can be based for managing and protecting our environment.

Soil monitoring combines science and technology. The functional relations between what is observed and the underlying processes must be understood. Equally, the techniques used for observation must be soundly based, using well-tried methods and equipment, and reproducible by different observers. But technology does not stand still, and new technique should be brought into play when its superiority is established. Our knowledge of ecosystems and the the way they work is never complete, and so there will always be scope for research. The synthesis identifies gaps in that knowledge. And, although soil monitoring should not be equated with soil research, the scientists responsible for it must be *au fait* with the latest research and willing to incorporate its conclusions into their programmes.

Thus, soil monitoring draws on research, implements the current concepts with today's expectations, and communicates its findings to decision-makers and administrators. Clearly, the people involved must understand one another, and they must be open to communication. The scientists must be able to explain in lay terms what they do and why they do it, they must explain their need for instruments and monitoring sites, and why standardization is important. The politicians and executives must appreciate the rigour of scientific enquiry and the personnel, equipment and time that it demands. Together they must agree clear pertinent objectives that will win public support and commitment for as long as is necessary to obtain the information and ensure that the soil is maintained in good heart.

SECTION 1

Objectives, Concepts and Experiences

SOIL MONITORING IN SWITZERLAND BY THE NABO-NETWORK: OBJECTIVES, EXPERIENCES AND PROBLEMS

A. Desaules

This contribution identifies major problems and gaps of current long term soil monitoring concepts related to pollution as seen from the experience of the NABO-network in the light of three main objectives, as follows.

a) The assessment of soil pollution and related ecotoxicological risks raises the problems of separating anthropogenic pollution from natural background for inorganic compounds. The different rating standards (e.g. guide levels, frequent contents, assumed natural background levels) give different results of the pollution in the same soil. The assessment of soil pollution depends therefore very much on the definiton of adequate standards of reference and will therefore always remain more or less an approximation.

b) The areal assessment of soil pollution must cope with spatial representativity of observation points in a heterogeneous medium. The 102 NABO observation sites spread over Switzerland are too few to be treated geostatistically. Grouping the laboratory results by functional criteria, especially land use and parent material, may possibly lead to statements related to space.

c) The temporal evolution of topsoil concentrations of pollutants was studied so far on 20 selected sites for 5 years. The fact that there were both increases and decreases suggests that the temporal changes are a complex result not only of inputs and outputs, but also of processes of residual enrichments and dilution by pedoturbation, especially ploughing and earthworm activity.

Further problems of long term soil monitoring are the restrictions of biomonitoring due to the poor reproducibility and the difficulty of general risk assessment because of the heterogeneity and buffer capacity in soils compared to air and water.

Introduction and objectives

The Ordinance Relating to Pollutants in Soil (VSBo, 1986) explicitly charges the Federal Research Station for Agricultural Chemistry and Hygiene of Environment to run a national measuring network for monitoring soil pollution known as NABO (NAtionales BOdenbeobachtungsnetz). According to the conceptual framework NABO should consist of about 100 observation sites from which soil samples are taken every 5 years. The sites should cover agricultural land (50 %), forests (30 %) and extensively used open land (20 %). The location of the observation sites should be representative with respect to land use, climate, geology, soil type and geographic distribution. The analytical programme is given by the 10 heavy metals and fluoride for which guide levels exist in the ordinance (VSBo, 1986).

The sampling for NABO started in 1985, and at present 102 observation sites are operational. Among them 20 have been sampled for a second time after 5 years. An observation site consists of a reference area of 100 m^2 from which soil samples are periodically taken from the ground surface to a depth of 20 cm by 4 replicate composite samples. Additionally, soil samples are taken only once from a soil profile pit nearby (approximately 2 to 3 m distant). All samples are stored. So far analytical results of so called "total contents" for Pb, Cu, Cd, Zn, Ni, Cr, Co, Hg and F have been produced. The heavy metals are extracted in a boiling 2 M HNO_3 solution. The elements Mo and Tl, for which guide levels exist, were omitted because of poor analytical comparability. The same is true for the soluble contents of 0.1 M $NaNO_3$ extractions which are mostly less than the limit of detection. Furthermore the following soil characteristics were determined: pH, $CaCO_3$, organic C, clay and silt, Fe- and Al-oxides, cation exchange capacity, soluble P and bulk density. The actual input of pollutants is approximated on the one hand by annual data of existing networks measuring atmospheric deposition or concentrations in moss and on the other by protocols of agricultural inputs on the observation sites themselves. The results under the given conditions will be published in a first report (BUWAL, 1993).

The objectives of soil monitoring related to pollutants are three:

a) to assess soil pollution and related ecotoxicological risks,

b) to seize soil pollution through space, and

c) to follow soil pollution through time.

Achieving the objectives is complex because of the nature of soils and the processes of pollution. The soil cover is heterogeneous and so is the pattern of its contamination. Moreover there are great differences of the toxicity of pollutants and of the buffering and degrading capacities of soils.The aim of this contribution is to identify and describe major problems and gaps of current soil monitoring concepts through the experience of the NABO-network in relation to the three objectives mentioned above.

The assessment of soil pollution and related ecotoxicological risks

According to the Ordinance (VSBo, 1986) soil pollutants are natural or man made substances which can damage fertility of the soils. They include heavy metals and organic compounds containing chlorine. Soil pollution has first to be assessed on the basis of the existing guide levels which are, however, applicable only to soils with humus contents less than 15 %. If guide levels are lacking, the assessment has to be made on the basis of criteria of the following definition of soil fertility.

A soil is fertile if it:

a) is characterized by diverse and biologically active animal and plant community, possessing a typical structure and undisturbed capacity of decomposition;

b) makes possible undisturbed growth and development of plants and plant communities (both natural and influenced by Man) and does not affect their characteristic properties;

c) assures that plant products are of good quality and wholesome for Man and animals.

To make the definition operational a guideline to assess soil fertility (BUWAL/FAC, 1991) has been published. It is mainly a compilation of tests with rating keys.

The current state of NABO is restricted to the assessment of soil pollution as far as it is possible by the "total contents" of 9 chemical elements which are primarily of natural origin in soils. Methods of biomonitoring have therefore to be excluded. Different chemical methods are presented and discussed below. There are basically two methodological approaches to separate anthropogenic pollution from the natural background in soils: sequential extractions and more or less specific standard comparisons. Sequential extractions cannot be considered here, since only "total contents" are available for NABO so far.

The guide levels of the Ordinance (VSBo, 1986) are countrywide standards and do not take into account spatial variations of natural background concentrations from parent material and pedogenic processes. Guide levels may already be exceeded by natural contents. An indication for this is given in Table 1 by the number of exceeded guide levels in subsoils compared to topsoils and the examination of the parent rocks. Soils on ultrabasic rocks for example contain more Ni, Cr and Co than other soils on other parent material. Guide levels are therefore only rough standards and not accurate scales for effective soil pollution.

Table 1. Number of exceeded guide levels out of 102 observation sites

element	Pb [50]	Cu [50]	Cd [0.8]	Zn [200]	Ni [50]	Cr [75]	Co [25]	Hg [0.8]	F [400]
topsoil (0-20 cm)	6	6	5	0	5	1	1	0	57
subsoil (BC- or C horizon)	0	3	2	0	8	2	1	3	62

The values in brackets are guide levels in mg/kg of "total contents" (VSBo, 1986)

The results in Table 2 show how differently soil pollution is assessed in the same soil according to what standard is used. The assessment by guide levels is negative for all three elements. The same holds for the comparison with "frequent contents" (Meyer, 1991) of the topsoil. Related to frequent subsoilcontents, however, a slight but significant pollution of Cu appears. The comparison of the topsoil with the subsoil reveals a soil pollution for Pb only by the weight related procedures (a, c, d, and e). There is no pollution of Pb detected by the volume based procedures (b and f), because the bulk density of the organic topsoil is much less than for the mineral subsoil.

In Table 3 soil profile budgets (after Ruppert, 1991) for Pb, Cu and Zn of the same soil as in Table 2 are shown. The soil profile is somewhat heterogeneous in character and is podzolised. Therefore the definition of natural background contents remains somewhat speculative despite of soil scientific considerations. For Pb the (B)C-horizon and for Cu the Of-horizon have been assumed to reflect best the natural background for the overlaying horizons. Both elements are known to be almost immobile in soils. For Zn, however, there is a distinct translocation from the E- into the I-horizon. The natural background of Zn for the overlying soil horizons has thus been approximated by the graphically determined mean of Zn out of the E- and I-horizon. The results yield significant soil pollutions for all three elements down to the reference horizons for natural background concentrations.

By comparing the methods of soil pollution assessment for the same soil it becomes obvious that, besides the different used standards (guide levels, "frequent contents", subsoil, natural background), differences in soil characteristics are at the source of the different results, too. Therefore soil samples with (very) different soil characteristics, especially bulk density, organic matter content, stone content and water content should be taken into account first, before comparing pollution concentrations.

The methods are ranked by order of increasing specificity, complexity and accuracy. General standards such as guide levels or "frequent contents" will never assess real soil pollution for a particular site. For this site specific approaches are required, as comparison with the corresponding subsoils which are assumed to be unpolluted, or even complete soil profile budgets. The weakness of site specific approaches is that the topsoil differs from the subsoil, and thus unequal materials are compared.

Finally, the accuracy of soil pollution assessment depends on the definition of adequate standards of reference. The best case are certainly site specific natural background concentrations. But even these are based on assumptions and remain approximations. It will (probably) never be possible to assess the accurate actual soil pollution *a posteriori*. In this case the question is not to find the "right" method to assess soil pollution but the most suitable one.

Table 2. Assessment of soil pollution of Pb, Cu and Zn by three methods of standard-comparison in a podzol under forest

Basic data:	Pb	Cu	Zn	dimension
- measured values				
topsoil (0 - 20 cm)	20	7	26	mg/kg
subsoil (C-horizon)	(8)	18	43	mg/kg
- guide levels (VSBo 1986)	50	50	200	mg/kg
- median of frequent contents[1]				
topsoil (forest)	21	12	47	mg/kg
subsoil	14	14	41	mg/kg
- analytical reproducibility	(± 15)	(± 10)	(± 10)	%
- bulk density				
topsoil	0.5	0.5	0.5	kg/dm^3
subsoil	(1.4)	(1.4)	(1.4)	kg/dm^3
- water content (dried at 40^0 C)				
topsoil	4	4	4	% (w/w)
subsoil	1	1	1	% (w/w)
- organic matter				
topsoil	33	33	33	% (w/w)
subsoil	0.5	0.5	0.5	% (w/w)
- stone content (>2mm)				
topsoil	(0)	(0)	(0)	% (w/w)
subsoil	(25)	(25)	(25)	% (w/w)
Standard comparisons:	Pb	Cu	Zn	
1 comparison with guide levels (topsoil)	**-30**	**-43**	**-174**	mg/kg
2 comparison with frequent contents				
topsoil	-1	**-5**	**-21**	mg/kg
subsoil	**-6**	+4	+2	mg/kg
3 comparison of top- / subsoil related to:				
a) weight	**+12.0**	**-11.0**	**-17.0**	mg/kg
b) volume	-1.2	**-21.7**	**-47.2**	kg/dm^3
c) oven dry matter	**+12.7**	**-10.9**	**-16.3**	mg/kg
d) mineral fine earth	**+21.8**	-7.7	**-4.4**	mg/kg
e) bulk soil (with stones)	**+10.0**	**-15.5**	**-8.4**	mg/kg
f) b,c and e	**-3.7**	**-28.3**	**-62.5**	kg/dm^3

bold = significant difference to the value of comparison (> 2x analytical reproductibility)

( ) = approximated value

[1] (MEYER, 1991)

Table 3. Assessment of soil pollution of Pb, Cu and Zn by the method of soil profile budget (after Ruppert, 1991) in a Podzol under forest

Basic data

horizon	(cm)	bulk density (kg/dm^3)	org. matter (%)	stones (%)	bulk-soil (t/ha)	stones (t/ha)	org. matter (t/ha)	min. fine earth (t/ha)
Ol	8	0.2	70	0	160	0	112	48
Of	4	(0.4)	(65)	0	160	0	104	56
E(Ah)	5	(0.5)	10	(5)	250	12.5	25	213
Ih	7	(0.6)	2.5	(10)	420	42	10	368
Ife	13	0.6	2	(20)	780	156	16	608
(B)C	43	(1.0)	0.5	(25)	4300	1075	21	3204
				sum:	6070	1285	288	4497

Assumed natural background:

	Pb	Cu (mg/kg)	Zn
Ol, Of, E(Ah), Ih = Ife = (B)C	(8)		
Ol = Of		8	
Ol, Of = graphic mean out of E- and I- horizon			43

Soil profile budgets:

horizon	fine earth total	fine earth mineral	element quantity: total	natural		fine earth **anthropogene**
(mg/kg)	(mg/kg)	(g/ha)	(g/ha)	(g/ha)	(g/ha)	(mg/kg)
Pb						
Ol	62	207	9920	384	9536	**60**
Of	45	129	7200	448	6752	**42**
E(Ah)	12	13	2856	1704	1152	**5**
Ih	10	10	3780	2944	836	**2**
Ife	(8)	8.2	4992	4864	128	0.2
(B)C	(8)	8.0	25800	25800	0	0
		sum:	54548	36144	18404	
Cu						
Ol	11	37	1760	384	1376	**9**
Of	8	23	1280	448	832	**5**
E(Ah)	10	11	2380	2380	0	0
Ih	12	12	4536	4536	0	0
Ife	14	14	8736	8736	0	0
(B)C	19	19	61275	61275	0	0
		sum:	79967	77759	2208	
Zn						
Ol	43	143	6880	2064	4816	**30**
Of	47	134	7520	2408	5112	**32**
E(Ah)	30	33	7140	7140	0	0
Ih	53	54	20034	20034	0	0
Ife	52	53	32448	32448	0	0
(B)C	42	42	135450	135450	0	0
		Sum:	209472	199544	9928	

bold = significant difference to the natural background (> 2x analytical reproductibility)
( ) = approximated value

Monitoring soil pollution through space

There are basically two methodological approaches of spatial representation which may be combined: one is functional and the other geostatistical. The functional approach is based on plausible and empirical functional specifications which can be related to space (Desaules, 1991). An example are soils in vineyards which contain much Cu, because Cu is regularly applied to vineyards. The approach is by nature deterministic, and it is very important that the spatially located data are accompanied by complementary functionally relevant information. A problem is that the functionally assessed spatial variability is an empirical result and not under statistical control. The geostatistical approach, however, is stochastic and of probabilistic value related to space.

A major question for NABO is how far the 102 observation sites can represent spatial soil pollution of Switzerland. The results for Pb in Figure 1 are given as an example of geographic site distribution.

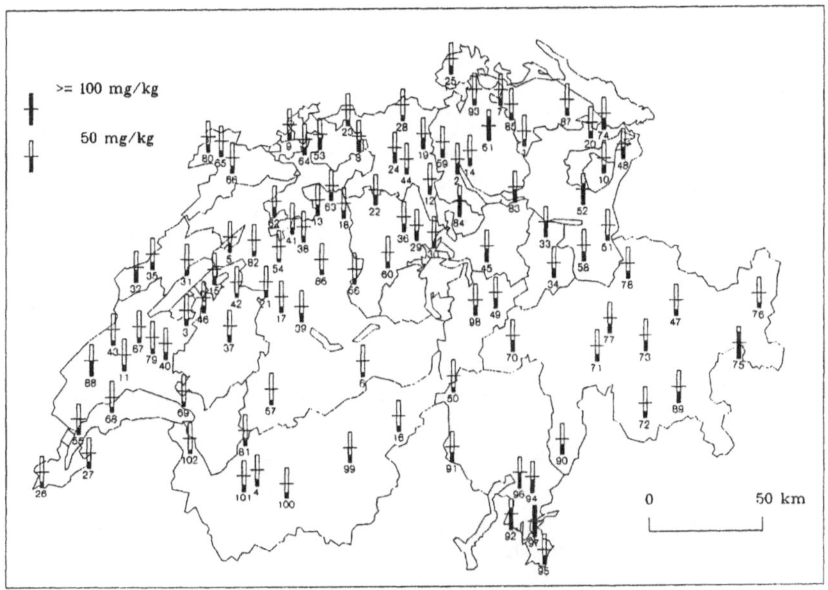

Figure 1. Observation sites with contents of Pb in topsoils (0-20 cm) in mg/kg

There are too few sites to be treated by geostatistical procedures, and they have been selected with respect to specific critieria, not randomly. Distributed in a regular grid, the sites would have been about 14 km apart from each other. From this, and given the heterogeneity of soils and processes of soil pollution, it is clear that the 102 observation sites are individual cases and are of little spatial representativity. The grouping of laboratory results by functional criteria as land use, parent material and others, however, may lead to possible spatial statements.
The comparison of "frequent total contents" (range of the central 80 % of all values) in Table 4 shows that despite the different areas and number of measuring points the ranges are similar and that, with the exception of one case, less than 10 % of the values exceed guide levels.
If the "frequent total contents" are grouped by categories of land use or parent material as in Table 5, or by order categories, the ranges become more or less specific and can thus be indirect indicators of spatial differences. The most important problem is to define significant and distinctive categories which may easily be related to space. The sampling pattern will then be of secondary importance. An open question is: what is the potential of geostatistics for small-scale soil monitoring?

Table 4. "Frequent total contents" of Pb, Cu and Cd in topsoils (0-20 cm) of Switzerland and two subareas.

area	Pb [50] mg/kg	Cu [50] mg/kg	Cd [0.8] mg/kg	n	source
Switzerland ($41\ 290\ km^2$)	16 - 38	6 - 35	0.11 - 0.49	102	BUWAL (1993)
Area below 700 m a.s.l. ($13\ 630\ km^2$)	14 - 41	9 - **61**	0.11 - 0.45	237	Vogel *et al.* (1989)
Canton of Zurich ($1\ 730 km^2$)	19 - 54	9 - 44	0.10 - 0.62	425	AGW (1989)

Lower value = 10 % percentile - higher value = 90 % percentile
The values in brackets are guide levels (VSBo, 1986) and exceeding values are bold

Table 5. "Frequent total contents" of Pb, Cu and Cd in soils of Switzerland grouped by land use and parent material (adapted from NABO in prep.)

grouping criteria	Pb [50] mg/kg	Cu [50] mg/kg	Cd [0.8] mg/kg	n
topsoils (0-20 cm)	16 - 38	6 - 35	0.11 - 0.49	102
Land use:				
- arable land	16 - 28	13 - 37	0.16 - 0.36	27
- vegetables, orchards, vineyards	18 - 77	23 - **465**	0.20 - 0.62	13
- permanent grassland	18 - 35	9 - 31	0.20 - 0.50	13
- pasture land	14 - 41	4 - 21	0.10 - 0.56	11
- spruce forest	20 - 41	5 - 19	0.08 - 0.35	16
- deciduous forest	14 - 37	3 - 20	0.09 - **0.96**	12
subsoils	10 - 19	6 - 25	0.05 - 0.24	102
Parent material:				
- acid cristalline rock	10 - 12	13 - 21	0.03 - 0.07	7
- carbonate rocks	10 - 19	6 - 17	0.09 - 0.80	12
- mixed unconsolidated sediments	10 - 21	8 - 22	0.06 - 0.20	32

Lower value = 10 % percentile - higher value = 90 % percentile
The values in brackets are guide levels (VSBo, 1986) and exceeding values are bold

Monitoring soil pollution through time

Can significant differences of pollutant concentrations be measured through time in soils? Retrospective analysis of archived soil samples from long term field trials do partly confirm this (e.g. Jones *et al.*, 1987).
For NABO the question was whether 5 years are long enough to show significant differences of pollutant contents in soils by using the following procedure. The first sampling in 1985 consisted of only one simple composite sample from a depth of 20 cm, whereas for the second sampling in 1990 quadruple replicate composite samples have been taken to estimate the variability of the observation sites. The corresponding soil samples of both sampling times have been analysed simultaneously to minimize the analytical variation. The examples for Pb and Cu in Figure 2 show some significant increase but also decrease of the contents. No distinct relationship is apparent

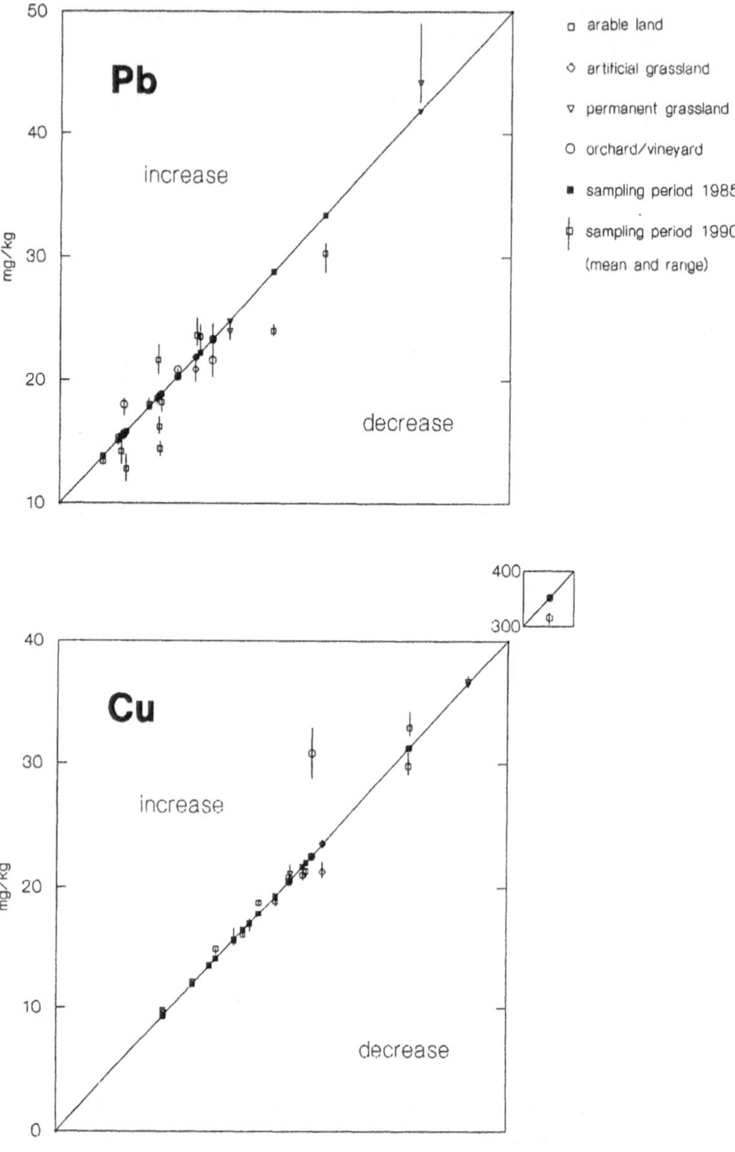

Figure 2. Changes of soil contents of Pb and Cu on 20 observation sites after 5 years

either from element concentration or from land use of the 20 observation sites so far. The increases can be explained by atmospheric depositons, agricultural inputs and residual concentration mainly due to mineralization of humus. Possible reasons for decreases are dilution by pedoturbation (ploughing and earth worm activity among others), erosion, and vertical translocation of pollutants.

The results of Table 6 give further evidence that anthropogenic pollution, which is normally brought into soils through the surface, is not maintained strongly stratified in the topsoil, but is more or less mixed into the soil. This makes soil a difficult environmental archive of anthropogenic pollution to "read".

Table 6. Number of observation sites without mechanical soil treatement where topsoil samples (0-5 cm) exceed contents of corresponding reference samples (0-20 cm)

	(n)	Pb	Cu	Cd	Zn	Ni	Cr	Co	Hg	F
all	(56)	34	29	42	34	12	18	11	41	12
forest	(28)	13	13	21	18	4	11	3	18	5
open land	(28)	21	16	21	16	8	7	8	23	7

Nevertheless, the figures in Table 6 are open to the following possible interpretations: The prevailing larger concentrations in the samples from 0-5 cm soil indicate anthropogenic pollution. This is especially so for the elements Pb, Cd, Zn and Hg, which are transported mainly through the atmosphere. But under forest and grassland also biogenic enrichment in the organic matter especially of Cu, Zn, and Co are possible. Smaller concentrations are attributed to pedoturbation and translocation. Soil under forest is likely to be more acid and the soil biology is less active (e.g. fewer earthworms) than in soil of agricultural land. This calls for smaller concentrations in the soil layer of 0-5 cm under forest, especially for the more mobile elements such as Cd, Zn and Ni, but also for soils of open land, due to dilution by biological pedoturbation. The results are not coherent throughout, because the input, residual concentration, dilution by pedoturbation and output by translocation overlap and vary in intensity and efficiency at the individual sites.

In Table 7 the results for 20 sites on open land are summarized by indicating the significance of the change of soil concentrations for 8 elements after 5 years. All sites show clear changes (4 out of 4 samples) for some elements. Clear increases have been observed on 16 of the 20 sites. Striking are the increases on both orchard sites (4 and 20) but also on one arable site (15). The increases, especially for Cu and Zn, on the orchard sites are attributed primarily to the application of biocides. The general increase for all 8 elements on the arable site (15), which has an organic topsoil, is explained by residual concentration as an effect of humus mineralization. Very surprising is the general decrease even for Cu on the vineyard site (5). The reason seems to be dilution by deep ploughing between the two samplings.

Table 7. Changes of soil concentrations for 8 elements of 20 sites on open land after 5 years

site nr.	land use	Pb	Cu	Cd	Zn	Ni	Cr	Co	F	total: 8 elements +	−	+3	−3
3	AKO	−3	+3	−3	+1	−3	−2	−3	+3	3	5	2	4
4	IOI	+3	+3	+3	+3	+3	+3	+2	−3	7	1	6	1
5	IRE	−2	−3	−3	−3	−3	−3	−3	−3	0	8	0	7
9	AKO	−3	−2	+3	−3	+3	0	+3	−1	3	4	3	2
11	AKO	0	−3	−2	−3	−3	−3	−3	−2	0	7	0	5
13	AKO	−3	−3	−3	+1	+2	+1	−3	−1	3	5	0	4
14	AKW	−2	−3	−2	−2	−3	0	−2	+1	1	6	0	2
15	AKO	+3	+3	+2	+3	+3	+3	+3	+3	8	0	7	0
17	AKO	−3	+2	−3	+3	+3	+2	−2	+2	5	3	2	2
20	IOI	+3	+3	+2	+3	+3	+3	+3	+3	8	0	7	0
21	AKO	−3	−1	−3	−3	+3	−3	−3	−2	1	7	1	5
23	AKO	+2	+3	+2	+1	+3	+2	−2	−3	6	2	2	1
25	AKO	+3	−3	−2	−3	−1	−3	+2	+2	3	5	1	3
26	AKO	+1	+2	+3	+3	+3	−1	−2	+3	6	2	4	0
28	AKO	−3	+3	+1	+3	+1	+3	+2	−2	6	2	3	1
29	DWI	−3	+3	−1	+3	+2	0	+3	−2	4	3	3	1
31	AKO	−2	+3	+3	+2	−3	−3	−2	−2	3	5	2	2
33	DWI	+3	+2	+3	+3	−1	+3	+2	+3	7	1	5	0
36	AKO	+3	−3	+2	−3	+2	0	−3	−3	3	4	1	4
38	AKO	−3	−3	+2	−3	0	+3	−3	+2	3	4	1	4
total:	+	8	11	11	12	12	9	8	9	80			
20 sites	−	11	9	9	8	7	7	12	11		74		
		+3	6	8	5	8	8	6	4	5		50	
		−3	8	7	5	7	5	5	7	4			48

AKO = arable land; IRE = vineyard; + = increase; 1 = median
AKW = ley grassland; DWI = permanent grassland; − = decrease; 2 = 3 out of 4 samples
IOI = orchard; 0 = no change; 3 = 4 out of 4 samples

Wherever possible the total inputs of soil pollutants have been assessed for the NABO sites. The basis are the annual mean values of atmospheric depositions from existing measuring points nearby and calculated agricultural inputs on NABO sites by annual protocols of soil treatments filled out by the farmers. In Figure 3 the results of the total mean annual inputs are compared to the corresponding changes of topsoil contents (0-20 cm) for 14 observation sites on open land of different land use. Bulk densities as well as stone contents have been taken into account to transform annual change of soil contents from mg kg^{-1} a^{-1} into g ha^{-1} a^{-1}. Despite this, there is no concordance. The differences are one or two orders of magnitude, and except for the two orchard sites, no trend is apparent. There are several reasons for this: First, change of topsoil contents are governed not only by anthropogenic inputs, as explained above.

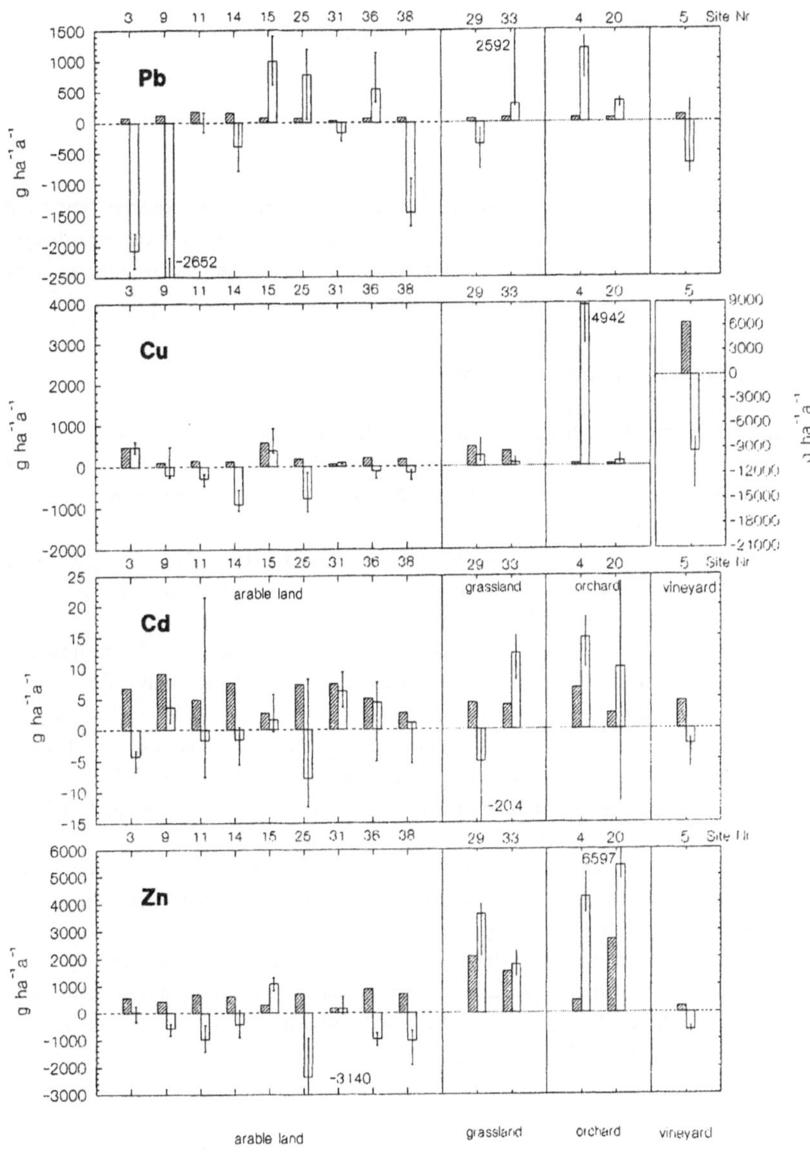

Figure 3. Comparison of mean annual inputs of heavy metals and corresponding changes of topsoil contents (0-20 cm)

Second, the relationship of input (g $ha^{-1}a^{-1}$) on change of topsoil contents (mg kg^{-1} a^{-1}) of 20 cm soil depth is very sensitive to analytical accuracy. To illustrate this an input of 2.4 kg/ha in a topsoil free of stones with a bulk density of 1.2 kg/dm^3 increases the concentration by only 1 mg/kg. Third, the procedures to assess atmospheric and agricultural inputs are approximations of unknown accuracy. The only way to check the input balance would be a profit and loss account of the soil concentration.

Jones et al. (1987) found discrepancies between input estimates and observed time trend pattern of heavy metal contents in soils, too. They emphasize the need for elemental budget studies under closely monitored conditions. It is crucial to know whether the required accuracy can be reached to make the approach successful.

Gaps and outlook for future research

In the light of the three basic objectives mentioned in the introduction NABO is only a beginning, and many gaps are already apparent. NABO is still far from being a routine soil monitoring network; it needs further substantial research in several fields to provide more satisfactory answers to questions of soil pollution.

NABO is actually only a partial approach of soil monitoring. The analytical programme is restricted to "total contents" of 9 chemical elements. So far no organic pollutants or physical and biological investigations have been considered.

The Ordinance Relating to Pollutants in Soil (VSBo, 1986) sets guide levels for 11 chemical elements which apply only to topsoils containing less than 15 % humus. Organic soils and compound pollution are thus excluded of soil pollution assessment by the existing guide levels. The compilation of concentration limits in soils and surrounding media by the Swiss administration in Table 8, is a striking example of ecological inconsistency concerning administrative limits.

Where limits have not been set by the administration, hardly any data exist. According to Table 8 Cd is the only element where administrative limits exist for every medium. Therefore complete sets of data for elemental budget studies are probably available only for Cd. This, however, leaves still the question of how reliable and comparable the data are. In these circumstances it is often impossible to apply rigorously the principle of causality stated by the Swiss Federal Law Relating to the Protection of the Environment (USG, 1983). Further questions concern the methods of sampling and analysis. Are the methods compatible, and are the analytical accuracy and comparability in soils (e.g. Desaules *et al.*, 1992) and the other media sufficient to assess specific fluxes for contaminant mass balances in soils?

Table 8. Administrative concentration limits of Switzerland in Soils and surrounding input- and output- media

Medium	Element										Source
	Pb	Cu	Cd	Zn	Ni	Cr	Co	Mo	Tl	F	
input											
Atmosphere											
suspendet dust	+	–	+	–	–	–	–	–	–	–	a
deposition	+	–	+	+	–	–	–	–	+	–	a
Agrosubstances [1)]											
min. fertilizers	–	–	+	–	–	+	–	–	–	–	b
sewage sludge	+	+	+	+	+	+	+	+	–	–	c
compost	+	+	+	+	+	+	(+)	(+)	–	–	b
Soil											
"total contents"	+	+	+	+	+	+	+	+	+	+	d
"soluble contents"	+	+	+	+		+	–	–	–	+	d
output											
Water											
sewage inlet	+	+	+	+	+	+	+	–	–	+	e
Plants											
foodstuffs	+	+	+	+	+	–	–	–	–	–	f
fruits a. vegetables	+	–	+	–	–	–	–	–	–	–	g

[1)] for biocides no restrictions are given

administrative limit: + existent, (+) in elimination, – inexistent (irrelevant?)

Sources: a = LRV 1985, b = StoV 1986, c = KSV 1981, d = VSBo 1986, e = VoA 1975, f = FIV 1986, g = BAG 1980

The small-scale monitoring of soil pollution by NABO has been related indirectly to space by selected and visible criteria of land use and geology. There are still too few point observations to map concentrations of soil pollutants of known variability. There is a need for unbiased and space related, statistically controlled procedures (geostatistics) of practical potential for small-scale soil monitoring.

The evaluation of biomonitoring to assess soil pollution and related risks is another wide field for future research. There are two main topics. One is to improve the reproducibility of biological observations for long term soil monitoring. The second is to find methods of interpretation and to control interpretations for the often very specific biological properties in soil.

Risk assessment of soil pollution is at its very beginning compared to pollution of water and air. Guide levels are still very general and do not really assess actual toxic risks of soil pollution to specific organisms. The approach of dose "effect-levels" is not as successful for soil as for water or air because of the heterogeneity and buffer capacity of soil. Research for alternative concepts of risk assessment in soil must consider not only the chemical load but also chemical speciation and buffering soil properties. The concept of "chemical time bombs" defined as "sudden effects of chemical loads in soils and sediments, triggered by slow alterations in the environment" (Stigliani, 1991) seems promising.

Acknowledgement

I am grateful to my colleagues R. Dahinden, S. Geering, E. Meier and K. Studer working for NABO for their substantial contributions and support and to R. Frossard and P. Lischer for reading the paper. I am also indebted to the authorities of our Research Station, in charge of NABO, as well as to the Swiss Federal Office of Environment, Forest and Landscape, responsible for the Ordinance Relating to Pollutants in Soil.

References

AGW 1989. Schadstoffbelastung des Bodens im Kanton Zürich - Resultate des kantonalen Bodenrasternetzes. Amt für Gewässerschutz und Wasserbau, Zürich.

BAG 1980: Richtwertempfehlungen für Blei- und Cadmiumgehalte von frischem Obst und Gemüse. *Kreisschreiben vom 19. Mai 1980.* Bundesamt für Gesundheit (BAG), Bern.

BUWAL (ed.) 1993. NABO-Nationales Bodenbeobachtungsnetz: *Schriftenreihe Umwelt Nr. 200.* Bundesamt für Umwelt, Wald und Landschaft (BUWAL), Bern.

BUWAL/FAC 1991. Wegleitung zur Beurteilung der Bodenfruchtbarkeit. Bundesamt für Umwelt, Wald und Landschaft, Bern und Eidgenössische Forschungsanstalt für Agrikulturchemie und Umwelthygiene, Liebefeld-Bern.

Desaules, A. 1991. The soil vulnerability mapping project for Europe (SOVEUR): Methodological considerations with reference to conditions in Switzerland. In: Batjies, N.H. and Bridges, E.M. (eds). Mapping of soil and terrain vulnerability to specified chemical compounds in Europe at a scale of 1:5'000'000. *Proceedings of the International Workshop (20-23 March 1991) held at the International Soil Reference and Information Centre (ISRIC)*, Wageningen.

Desaules, A., Lischer, P., Dahinden, R. & Bachmann, H.J. 1992. Comparability of chemical analysis of heavy metals and fluorine in soils: Results of an interlaboratory study. *Communications of Soil Science and Plant Analysis* **23**, 363-377.

FIV 1986: Verordnung über Fremd- und Inhaltsstoffe in Lebensmitteln vom 27. Februar 1986 (Stand am 1. Januar 1989). *AS 1986, 647.* Das Eidgenössische Departement des Innern, Bern.

Jones, K.C., Symon, C.J. & Johnston, A.E. 1987. Retrospective analysis of an archived soil collection: I. Metals. *The Science of the Total Environment* **61**, 131-144.

KSV 1981: Klärschlammverordnung vom 8. April 1981 (Stand am 1. Juli 1985). *AS 1981,408.* Der Schweizerische Bundesrat, Bern.

LRV 1985: Luftreinhalte-Verordnung vom 16. Dezember 1985 (Stand am 1. Juli 1988). *AS 1986,208.* Der Schweizerische Bundesrat, Bern.

Meyer, K. 1991. Bodenverschmutzung in der Schweiz. *Themenbericht des Nationalen Forschungsprogrammes "Boden"*, Liebefeld-Bern.

Ruppert, H. 1991. Zur Problematik der Abschätzung anthropogener Stoffgehalte in Böden am Beispiel von Schwermetallen. *GLA-Fachberichte* **6**, 36-61, Geologisches Landesamt Bayern.

Stigliani, W.M. (ed.) 1991. Chemical time bombs: Definition, concepts, and examples. *International Institute for Applied Systems Analysis (IIASA) Executive Report* **16**, Laxenberg.

StoV 1986: Verordnung über umweltgefährdende Stoffe vom 9. Juni 1986. *SR 814.013.* Der Schweizerische Bundesrat, Bern.

USG 1983: Bundesgesetz über den Umweltschutz vom 7. Oktober 1983 (Stand am 1. Oktober 1989). *AS 1984,1122.* Die Bundesversammlung der Schweizerischen Eidgenossenschaft, Bern.

VoA 1975: Verordnung über Abwassereinleitungen vom 8. Dezember 1975 (Stand am 1. Januar 1983). *AS 1975,2403.* Der Schweizerische Bundesrat, Bern.

Vogel, H., Desaules, A. and Häni, H. 1989. Schwermetallgehalte in den Böden der Schweiz. *Bericht Nr. **40**. Nationales Forschungsprogramm "Boden"*, Liebefeld-Bern.

VSBo 1986. Ordinance Relating to Pollutants in Soil of June 9, 1986. *SR 814.12.* Government of Switzerland. English version issued by the Swiss Federal Office of Environment, Forest and Landscape, Bern.

A. Desaules, Swiss Federal Research Station for Agricultural Chemistry and Hygiene of Environment, Schwarzenburgstrasse 155, CH-3097 Liebefeld-Bern, Switzerland

Discussion

Question by H. Puchelt

a) Is there any lead-mineralization in the south of Switzerland where large lead concentrations have been found in the soil?

b) Is there an area of heavy automobile traffic?

c) Was a discrimination made between natural (geogenic) and anthropogenic lead by isotope analysis?

Answer by A. Desaules

a) As far as we know none of our observation sites in southern Switzerland is affected by Pb-mineralization, since the subsoil contents do not exceed the country-wide average.

b) The main valleys of southern Switzerland are effectively suffering from heavy road traffic. However, the fact that we find increased contents of Pb and other heavy metals such as Cd and Zn in topsoils far from any traffic at high altitudes might hint at pollution from the industrial area of Milano. This idea is supported by the particular climatic situation of southern Switzerland with important air currents from South to North and heavy precipitation.

c) There was no discrimination made by isotope-analysis. The only indications for the anthropogenic origin we have so far are increased contents in topsoil compared to corresponding subsoils.

Soil Monitoring, Monte Verità, © Birkhäuser Verlag Basel

MONITORING AND CONTROL OF REGIONAL MATERIAL FLUXES

Hans-Peter Bader and Peter Baccini

Environmental protection aims to prevent damage by early recognition of possible hazards. The potential of accounting techniques for materials (similar to financial accounting) to forewarn of critical loadings on water, air and soil is assessed. The increase of the annual mean concentration of the various substances introduced by man are very small and directly measurable only after many years. However, analysis of material flux can quantify such changes. This is illustrated for zinc.

Introduction

A regional economy can be defined as an open system, comprising the major compartments "anthroposphere" (where human activities such as nourishment, residence, and transport take place) and "environment" (Baccini & Brunner 1991). The latter can be subdivided into soil, water and air. The study of metabolic processes in regions densely populated by affluent societies has shown (Brunner *et al.*, 1990) that aquatic and terrestrial ecosystems are important sinks for various substances introduced by man, e.g. nitrate in groundwater, and zinc in agricultural soil. Their metabolism have been unstable for decades. The increases in annual mean concentration, however, are small (< 1%) and not directly observable by monitoring the environmental compartments. In agricultural soil only mass balances permit us to quantify changes in annual mean concentration on a regional scale (Baccini & von Steiger, 1993).

The present strategy in environmental protection utilizes data mainly from field observations (physical, chemical and biological) to evaluate the metabolic state of the environment. For certain noxious substances *threshold values* are set. These are based on toxicological and political arguments, and if they are exceeded then people can intervene to stop further increases. In these circumstances the ecosystems themselves are main indicators of their own metabolic states. Such a strategy allows slow but steady increases in the concentration of the materials up to these threshold values. This contradicts the *principle of precaution*, a basic principle of all environmental protection laws.

Early recognition of potentially damaging changes in the metabolism of ecosystems is possible only by monitoring and controlling the overall material fluxes on a regional scale. More precisely, an overview over all material fluxes enables us to detect possible hazards earlier than by measuring the environmental compartments. In most regions many data exist to serve compartmental interests, e.g. preservation of water, air or soil quality. So far it has been possible neither to combine such information in to a whole nor to elucidate the relevant interdependencies. The aim of this paper is to present one possible method to quantify and qualify regional material fluxes on the basis of a flux analysis. Section 2 contains a brief discussion of some phenomenological aspects, section 3 presents the mathematical approach, which is illustrated in section 4.

Phenomenology

A first assessment of the material fluxes through the anthroposphere reveals that about 90 t of goods are used to fulfil the needs of modern man in his household (Table 1). The main fluxes consist of water for cleaning and removal of faeces, followed by fossil fuels and construction materials such as gravel and stone. The large reservoirs of the activities residence and transport contain the goods with long lifes such as buildings, roads and vehicles.

As in many other countries, the flux of goods through Swiss households increases each year with a rate determined mainly by the national income. The input and output of the households are not in equilibrium: the reservoir of the anthroposphere increases at a rate of about 2 % each year. The enrichment is due mainly to the increase in the infrastructure for residence and transport. In affluent societies the supply of goods is organized efficiently. The main criteria for the flux of input goods into a household are the availability of products, the need for them and their prices. Product quality is mostly a function of the acceptance by the consumer; in rare cases it is defined by public or private authorities (e.g. safety requirements). The output from households is not of a predictable quantity nor quality.

Table 1. Flux of goods through a Swiss household, in tonnes per capita per year.

Activity	Input		Output			Reservoir
	Total	Solids	Sewage	Off-gas	Solid waste	tonnes per capita
eating	1.7	0.2	0.9	0.7	0.1	0
breathing	4	0	0	4	0	0
cleaning	60	<0.1	60	<0.1	<0.1	0.1
residence	11	2.4	0	8.6	0.4	100+2*
transport	12	6	0	6	2	160+4*
Total	89	9	60	20	2.5	260+6*

* Annual increase of the reservoir, in tonnes per capita per year

Mathematical description

Our aim is to describe mathematically the material and energy flow through anthropogenic systems. Since energy and mass are physical quantities we apply the ordinary procedure for describing physical systems, as follows.

1) *Find a set of variables describing the system completely:*

Let us define:

$\rho_i(t,\mathbf{x})$:	density of element i	kg m^{-3}
$e(t,\mathbf{x})$:	density of the energy	J m^{-3}
$\mathbf{v}_i(t,\mathbf{x})$:	flow velocity	m s^{-1}
$\mathbf{s}\ (t,\mathbf{x})$:	flux of the energy	W m^{-2}
$\rho_i\mathbf{v}_i(t,\mathbf{x})$:	flux of the element i	kg m^{-2} s^{-1}

where t and $\mathbf{x}$ denote the time and space point, respectively.
It is obvious that these quantities, as functions of time and space, describe the system completely. Indeed, the two densities describe the actual state of the system and the two fluxes the actual change of the system.

2) *Find the governing equations for these variables:*

As is known from physics, conservation laws or principles of balance apply for energy and mass. We consider the integral form which is easy to interpret. For simplicity we restrict it to the mass balance equation:

$$\frac{\mathrm{d}}{\mathrm{d}t}\int_V \rho_i \mathrm{d}\tau = -\oint_{\partial V} (\rho_i\,\mathbf{v}_i, \mathrm{d}\mathbf{f}) + \int_V r_i\,\mathrm{d}\tau \qquad (1)$$

Here V is an arbitrary (balance) volume, $\mathrm{d}\tau$ the infinitesimal volume element and $\mathrm{d}\mathbf{f}$ the infinitesimal surface element. (...,...) is the euclidean scalar product and r_i the production rate of element i. The left hand side is the change of the mass of element i per unit time in V. The first term on the right hand side represents the total net flow of element i through the surface ∂V of V and the second term the net production rate of element i in V.

In the systems considered here $r_i \equiv 0$, and the balance equation (1) reduces to the conservation law. Of course a similar law also applies to the total mass and the energy respectively. Since equation (1) is a first order differential equation in time, the density $\rho_i(t,\mathbf{x})$ at time t follows from the initial density $\rho_i\,(0,\mathbf{x})$ and the flow velocity $\mathbf{v}_i\,(t,\mathbf{x})$.This is simply the mathematical formulation of the fact that the density describes the actual state and the velocity its change as mentioned above. This demonstrates how governing equations reduce the set of variables to a set of independent variables.

So far the approach is general and applies to many different fields, as for example to the continuum mechanics and hydrodynamics. These fields differ in the form of the momentum balance equation, which is the analogue equation for the flow velocity as equation.(1) for the density.

In the following section we specialize to anthropogenic systems.

3) *Apply the general formalism to anthropogenic systems:*

For man-made systems it is generally not possible to know all the element densities and element flows at each space point for each time, nor is it necessary. It is sufficient to know the material flows for selected so called "processes", as, for example, waste treatment plants, households, the soil. Mathematically these "processes" are just the balance volumes described above. Of course, it is always possible to refine the system by considering more and smaller balance volumes, the so called "subprocesses". For instance, as a first approximation the soil as a whole may be considered as one single balance volume. In further steps the different soil layers or different lots could be considered as balance volumes. The material flow analysis for anthropogenic systems as described by Baccini & Brunner (1991), can be formulated mathematically as follows.

i) Define the balance volumes V_j to be considered ("processes").
ii) Define the element (and energy) flows $\Phi_i^{(r\rightarrow j)}$ of element i from balance volume V_r to balance volume Vj.

From the balance equation (1) follows directly:

$$\frac{dM_i^{(j)}}{dt} = \sum_r \Phi_i^{(r\rightarrow j)} - \sum_s \Phi_i^{(j\rightarrow s)} \tag{2}$$

where:

$M_i^{(j)} = \int_{V_j} \rho_i \, d\tau$ is the mass of element i in V_j

$\Phi_i^{(r\rightarrow j)}$ is the element flow from balance volume V_r to balance volume V_j.

As for equation (1) the left hand side of equation (2) is the change of the mass of element i in the balance volume V_j per unit time. The first term on the right hand side represents the total input to V_j. Therefore the sum extends over all external input fluxes as well as all internal fluxes originating from other balance volumes. The analogue holds for the second term, which represents the total output of V_j.

The element flows $\Phi_i^{(r\to j)}$ form a matrix, the input-output matrix **A** according to :

$$a_{rj}^{(i)} = \Phi_i^{(r\to j)} \tag{3}$$

The columns of the matrix **A** represent the inputs and the rows the outputs of the balance volumes. From equation (2) and (3) it follows immediately that

$$\frac{\mathrm{d}M_i^{(j)}}{\mathrm{d}t} = \sum_r a_{rj}^{(i)} - \sum_s a_{js}^{(i)} \tag{4}$$

Equations (2) and (4) respectively are the basis for all material flow calculations, e.g. for the stationary case $\left(\frac{\mathrm{d}}{\mathrm{d}t} = 0\right)$ as well as the dynamic case including a sensitivity analysis.

Clearly the $a_{rj}^{(i)}$ contain the whole dynamic of the system. As an illustration we discuss two important cases.

a) *Experimental case.* Here all the $a_{rj}^{(i)}$ are known from experiments. The solution of equation (4) follows directly by simple integration. In principle, everything is fixed; there are no parameters to optimize or to vary. However this procedure neglects the errors of the measurements. A careful analysis should include also the propagation of the errors to investigate the confidence interval of the calculated variables. Of special interest is the case where, in addition to all fluxes, also the mass-change-rates with errors are known. Then the system is overdefined, and a least square fit with the balance equation (4) as constraint gives the best estimates for the variables with their confidence intervals.

b) *Input-output case.* For the stationary case the input-output relation in each balance volume can be characterized by transfer coefficients, namely:

$$a_{rs}^{(i)} = I_r^{(i)} k_{rs}^{(i)} \tag{5}$$

where :

$I_r^{(i)} = \sum_l a_{lr}^{(i)}$ is the total input to V_r

$k_{rs}^{(i)}$ is the transfer coefficient from V_r to V_s and describes the relative amount of the total input to V_r transferred to V_s

The transfer coefficients represent properties of the process. The system is described completely if all input fluxes and transfer coefficients are known. In particular the internal and output fluxes are functions of the input fluxes and the transfer coefficients.

In that case two important questions can be answered by simulations.

(1) Which are the most important input fluxes? (Requires variations of the input fluxes).

(2) What are the effects of a change of the internal properties of the processes, e.g. the process technology? (Requires variations of the transfer coefficients)
Both of these sensitivity studies are necessary for managing or optimizing the system.

The concept of transfer coefficients can be extended to time dependent systems. The generalization of equation (5) is

$$a^{(i)}_{rs}(t) = \int_0^t dt'\, k^{(i)}_{rs}(t-t')\, I^{(i)}_r(t') \tag{6}$$

All quantities are depend on time. The interpretation of the integrand is straightforward : $k^{(i)}_{rs}(t-t')\, I^{(i)}_r(t')$ is the amount of the input to V_r at time t' transferred to V_s at time t. Or in terms of a probabilistic approach : $k^{(i)}_{rs}(t-t')$ is the probability that the input to V_r at time t' flows to V_s at time t. A typical example for $k^{(i)}_{rs}(\tau)$ is a Gauss-function :

$$k^{(i)}_{rs}(\tau) \sim e^{-\frac{(\tau-\tau_0)^2}{2\sigma_0^2}}$$

where τ_0 is the average time delay between input and output and σ_0 is the standard deviation. Of course τ_0 and σ_0 depend on the balance volumes V_r and V_s.

For simulations of such systems this box model formalism is implemented in the computer program SIMBOX.
SIMBOX contains three modules. These are
1) Preprocessor : there the system (balance volumes and fluxes including their data) can be defined.
2) Processor : calculates the desired quantities. The following procedures are implemented :
- input-output case for flow equilibrium including a sensitivity analysis
- least square fit for the best estimates for experimental data and their errors.
In progress is the dynamic case.
3) Postprocessor : there are several possibilities to present the results graphically.

In the following the term flux instead of flow is applied to define a quantity per volume and time.

Case study: zinc flux

We illustrate the material flow analysis for the element zinc for a standard region with the following characteristics:

Area	2500 km^2
Inhabitants	1000000
Industrialisation	average for Swiss urban region
Agriculture	" " "
Waste treatment plants	" " "

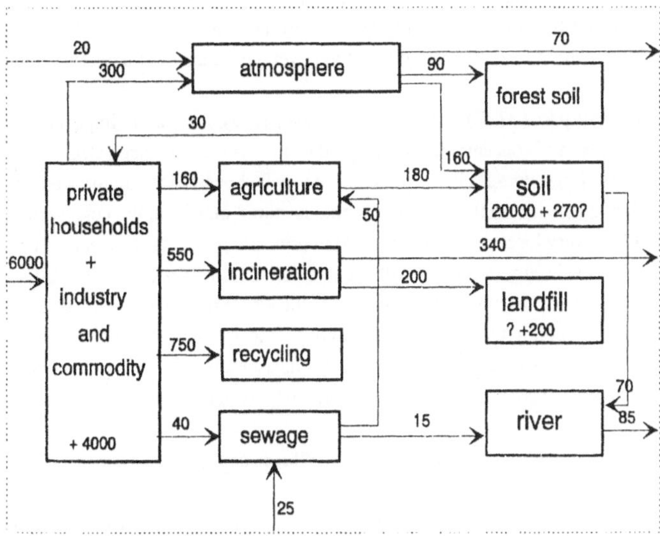

Figure 1. Regional zinc flux in g per capita per year.

Since we first of all are interested in the large element fluxes, we concentrate on total zinc and we consider a simplified system, shown in Fig. 1.

Zinc flow data: The inputs to industry and commodity, agriculture via fertilizer and the output from the soil by erosion were estimated, the rest of the data is measured in the METAPOLIS project (Baccini *et al* 1993).

The zinc enters the above defined region via the compartment of 'private household + industry and commodity' as a part of thousands of different products. One third flows through leaving the other two thirds to accumulate in the compartment. Of course each product has its own residence time in the household + industry and commodity resulting in individual time delays between their inputs and outputs.

From these waste treatment plants zinc is distributed to the environmental compartments and the landfill. This clearly shows one important aspect of these plants: they act as a centre of concentrating and collecting all material which was originally distributed in hundreds of different products.

The results are also presented in Fig. 1 and show the following:

i) The largest zinc fluxes are those from private household + industry and commodity to the incineration plants (31 %) and recycling (42 %).
ii) Another important amount of zinc flows to the atmosphere and to agriculture via fertilizer. We should point out that the data for the atmosphere are rough estimates. Measurements for specific times cannot be extrapolated to annual averages since the fluctuation in time is very large.
iii) A relative small part (≈2 %) flows to the sewage. However, this is important since it is a significant increase of concentration with regard to a geogenic reference value.
iv) There are three large storages in the system: the private household + industry and commodity, the landfill and the soil. In the households zinc is stored mainly in the construction material. The landfill is a very local reservoir and could in principle be recycled except for the part of the zinc that is washed out very slowly with a characteristic time scale of 10^4 - 10^5 years. In the soil on the other hand the zinc is distributed over the whole region. Measurements indicate that the zinc concentration in the soil is 20-100 mg/kg. Thus the increase from the anthroposphere is ≈ 1 % per year. This means that this increase is measurable only after many years. The same is true for attempts to control the anthropospheric outputs to the soil.

To conclude this section we show how the large zinc fluxes could be reduced.

A more detailed study (Baccini *et al.*, 1993) shows the following outputs from the private households to the incineration and sewage plants (Table 2).

Table 2. Relative zinc flows from household to incineration and sewage plant.

Incineration plant:	
household fittings	14 %
audio-visual electronics	28 %
kitchen equipment	10 %
batteries	25 %
various	23 %
Sewage plant:	
toilet flush water	31 %
car wash water	34 %
various	35 %

From this table it is easily seen that:

i) recycling of household fittings, audio-visual electronics, kitchen equipment and batteries would reduce the zinc output of the household to the incineration plant by 77 %,
ii) one third of the zinc in the sewage plant originates from washing cars.

This clearly demonstrates the powerful possibilities of such simple material flux analysis in controlling the material flows and resources. An important task for the future is to investigate also the confidence intervals. So far many cross checkings or comparisons with different measurements have been made.

Conclusions

We have discussed the material and energy flux analysis theoretically and illustrated it for the element zinc.

The analysis is based on the conservation laws for mass and energy, and it is simple enough to get a rough quantitative overview of the important energy and material fluxes in a system. This overview gives the possibility:

1) to detect large material fluxes in the anthroposphere which should be avoided in the long time before their effects can be measured in the environmental compartments.

2) to apply efficient (qualitative and quantitative) control to the material resources as demonstrated above, namely:

    i) by changing the material composition of the products. This reduces or eliminates undesired material fluxes in the anthroposphere.
    ii) by extracting and recycling the material from the waste and sewage in the waste treatment plants where most of the material passes through. In contrast to possibility i) this only reduces the input flow and the output flows to the environmental compartments but not the internal flows.

In this sense the material flux analysis is part of the so called environmental information system, namely the part that describes quantitatively the material flows from their extraction from the environment and production to the distribution to the environment by the waste treatment plants.

Acknowledgement

We thank Barbara Gamper for helpful discussions and Ruth Scheidegger for the layout of the script.

References

Baccini, P. & Brunner, P.H. 1991. *Metabolism of the anthroposphere*. Springer Verlag, New York.

Baccini, P. 1993. Stoffwechsel der Anthroposphäre. *Vorlesungsmanuskript*. Vertrieb: Polybuchhandlung ETHZ, Zürich

Baccini, P. & von Steiger, B. 1993. Die Stoffbilanzierung landwirtschaftlicher Böden. *Zeitschrift für Pflanzenernährung und Bodenkunde*. **156**, 45-54

Brunner, P., Daxbeck, H., Henseler, G., von Steiger, B., Beer, B., & Baccini, P. 1990. RESUB, Regionaler Stoffhaushalt des Unteren Bünztales. *EAWAG-Bericht*. EAWAG, Dübendorf, Switzerland

Baccini, P., Daxbeck, H., Glenck, E. & Henseler, G. 1993. Metapolis. Güterumsatz und Stoffwechselprozesse in den Privathaushalten einer Stadt. *Bericht 34A des NFP 'Stadt und Verkehr'*, Zürich, Switzerland

Brunner, P.H. & Baccini, P. 1992. Regional material management and environmental protection. *Waste Management and Research* **10**, 203-212.

H. P. Bader
Swiss Federal Institute for Water Resources and Water Pollution Control (EAWAG)
CH-8600 Dübendorf
Switzerland.

P. Baccini
Lehrstuhl für Stoffhaushalt und Entsorgungstechnik
ETH Hönggerberg
CH-8093 Zürich
Switzerland

Soil Monitoring, Monte Verità, © Birkhäuser Verlag Basel

AVAILABLE DATABASES FOR REGIONAL MASS BALANCES IN AGRICULTURAL LAND

B. von Steiger and J. Obrist

Agricultural activity always carries with it the risk of contaminating the soil, ground water and surface water, and we should attempt to estimate this risk. Although many data are collected with respect to soil pollution, little has been done to summarize the relative contributions of different pollution pathways.

A regional mass balance approach called 'Proterra' is applied here to estimate the contributions. Using existing databases, the method 'Proterra' can be applied in soil monitoring as an early warning system for detecting heavy metal inputs into soil by farming and atmospheric deposition. The model allows us to estimate on which farm type the heavy metal input poses intolerable risks given the atmospheric deposition.

Introduction / Objectives

Regional or national soil pollution surveys have become increasingly important in the environmental policy of most European countries. Combined with other monitoring operations and research programmes, survey results should help to improve environmental protection.

Pollution of the soil is usually surveyed by taking samples at selected locations, which are described as completely as possible, especially with respect to pedology, land use or natural vegetation, atmospheric deposition, climate and topography. The samples are analysed for pollutants as well as for characteristic soil properties. Subsamples are stored in an archive for future analysis. The results are interpreted so as to summarize the pollution status and change of the soil in the region or nation surveyed. By repetitive sampling at the same site it is possible to follow the development of soil pollution through time. In addition, the fluxes of contaminants into and out of the soil at the observation sites can be assessed between two sampling campaigns in order to determine mass balances on the plot scale as, for example in the Swiss National Soil Monitoring Network (Desaules, 1993).

If soil monitoring is restricted to permanent observation sites it is difficult to determine their spatial representativity beforehand. Although it is possible to optimize sampling schemes with time and to estimate interpolation errors, a preliminary choice has to be taken when the survey is designed. Monitoring sites may be selected with respect to a certain topographical situation or close to known sources of pollution, for example. The results of the survey will then be valid only for the sampled locations, and extrapolation to the entire region might not be feasible.

In this contribution a method for assessing regional mass balances in agricultural land is presented. The method, called 'Proterra', has been developed on the basis of the concept used in local studies described by Mayer & Ulrich (1974) for forest ecosystems. It is intended to supplement surveys described above to overcome the difficulties in extrapolation and to assess the relative importance of fluxes of pollutants into the soil from different sources. The objectives of mobilizing available databases for the assessment of regional mass balances in agricultural land are

a) early detection of soil pollution risks (risk assessment), and
b) knowledge of pollution pathways that can be used to improve environmental protection.

Methodological aspects

'Proterra' consists of procedures for acquiring data describing pollutant fluxes, a database with subroutines to manage and interpret the data on a personal computer, and documentation describing the method (von Steiger *et al.*, 1992). 'Proterra' is designed to determine or estimate fluxes of pollutants on to agricultural land and to assess the risk of contaminant accumulation in the soil using the mass balance approach. This approach is based on the idea that the system to be monitored is composed of compartments which are connected by the exchange of matter, consisting of chemical elements, between them. Baccini & Brunner (1991) call the interconnected compartments 'processes'. If the elementary composition of the materials passing through the processes and all fluxes are known then it is possible to calculate mass balances for the compartments. The mass balance for element i corresponds to the change in the amount of i in the observed compartment j. Following the notation given by Bader & Baccini (1993) in the preceeding contribution, we can write

$$\frac{dM_i^{(j)}}{dt} = \sum_k \Phi_i^{(k \to j)} - \sum_l \Phi_i^{(j \to l)} , \qquad (1)$$

where $M_i^{(j)}$ is the mass of the element i in the compartment j, $\Phi_i^{(k \to j)}$ is the element flow from the compartment k to the compartment j during dt and k and l are compartments distinct from the observed compartment j.

For assessing mass balances in agricultural land von Steiger & Baccini (1990) showed that 'net soil input' from the compartment called 'crop production' and 'deposition' from the atmosphere cause most pollution of agricultural land. Their concept of the system is given in Fig. 1.

Under the hypothesis that over a time dt no change of mass occurs in the compartment 'crop production' (stationary case), the input-fluxes are equal to the output-fluxes. Thus, the output-

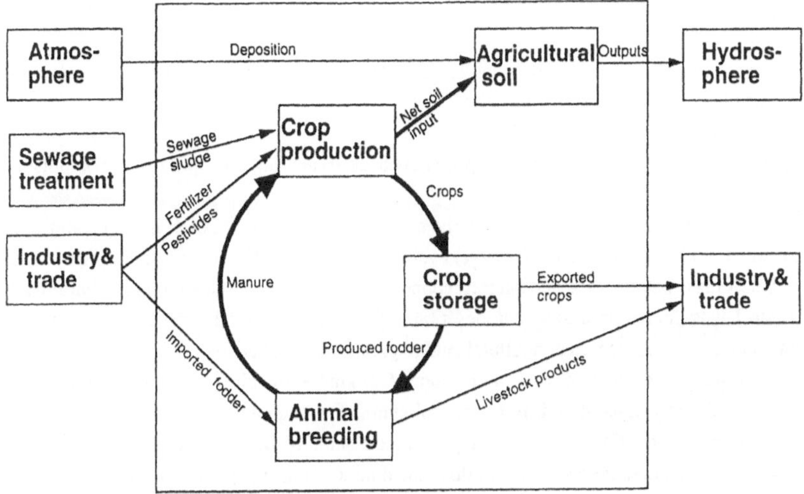

Figure 1. Exchange of goods between processes (sewage treatment, industry & trade, crop production, crop storage, animal breeding) and compartments (atmosphere, agricultural soil, hydrosphere).

flux $\Phi^{(\text{crop-production}\rightarrow\text{soil})}$, called 'net soil input', from the compartment 'crop production' into the soil can be calculated by

$$\Phi_i^{(\text{crop prod.}\rightarrow\text{soil})} = \sum_{1}^{k=3} \Phi_i^{(k\rightarrow\text{crop prod.})} - \Phi_i^{(\text{crop prod.}\rightarrow\text{crop storage})} \tag{2}$$

which follows from Equation (1) given that $dM/dt = 0$, $j\equiv$'crop production' and $l\equiv$'crop storage'. The index k in Equation (2) takes the values 1 for 'animal breeding', 2 for 'industry & trade' and 3 for 'sewage treatment', see Fig. 1.

The calculation of the mass balance for contaminants in the soil in principle requires that the outputs from the soil can be measured. The method 'Proterra' was developed for elements, that are fairly immobile in the soil. To assess the accumulation risk for these elements in the soil we can calculate the accumulation, which would be expected if no other outputs except by harvesting occurred and for which the term 'pollution risk ' r is used here:

$$r_i = \sum_k \Phi_i^{(k\rightarrow j)} = \Phi_i^{(\text{crop production}\rightarrow\text{soil})} + \Phi_i^{(\text{atmosphere}\rightarrow\text{soil})} \tag{3}$$

As Fig. 1 shows, this accumulation is given by the sum of the 'net soil input' and the 'atmospheric deposition'.

Available databases

There are two main reasons for using existing data bases in regional mass balances for soil pollutants instead of collecting new data:

a) it is cheap to transfer existing sets of data,
b) the data collected for other purposes such as trade, book-keeping and compulsory statistics, are reliable.

In a test study designed to develop the method 'Proterra' and to reach its applicability in practical work von Steiger *et al.* (1992) evaluated the availability of data sets for the calculation of mass balances for phosphorus, copper, zinc, cadmium and lead in agricultural land. Table 1 lists the required data both describing the region (e.g. crop area) and giving inputs and outputs with respect to agricultural land (e.g. fertilizer application).

The pilot study was in a region of about 550 km^2 embracing parts of the four Swiss Cantons: Aargau, Basel-Land, Bern and Solothurn. The availability of the data was not the same in each of the Cantons. For example, in the Canton of Bern detailed trade statistics on fertilizer application were at hand, while in the other Cantons the guide-lines for fertilization had to be used to estimate the fertilizer application. Table 1 shows this difference in the availability of the data required. To fill gaps, missing data were either estimated by extrapolation from other Cantons or had to be determined by specific investigations.

Moreover, some fluxes were quantified using values from farming guide-lines (e.g. phosphorus input with manure). Agricultural research should guarantee the up-dating of these values and they should be published regularly (e.g. FAP *et al.*, 1987; Menzi & Flückiger, 1991) in order to make these data available.

Table 1. Types and sources of data used for the assessment of regional mass balances in agricultural land by 'Proterra'.

variables		sources of information	
describing the region	unit	in *some* Cantons	in *all* Cantons
Crop area Livestock Cattle density Farm type	ha number ha^{-1} ha	Census by federal institute for statistics	Census by federal institute for statistics
input & output	unit		
Manure	kg $animal^{-1}$	Research results	Manure guide-lines
Sewage sludge	t $district^{-1}$	Average load (t/ha)	Compulsory statistics
Fertilizer	t $district^{-1}$	Trade statistics	Fertilization guide-lines
Pesticides	t $district^{-1}$	Compulsory statistics	Spray-plans
Crop production	kg ha^{-1}	Trade statistics	Book-keeping farms
Atmospheric deposition	kg ha^{-1} y^{-1}	Deposition survey	Interpolation of survey

Results

Average mass balances

Figure 2 shows the mass balances for phosphorus and the heavy metals copper, zinc, cadmium and lead in the agricultural soil of the district of Zofingen (56.7 km^2 agricultural land), calculated from Equations 2 and 3. Mass is indicated in Fig. 2 in dimensionless units obtained by dividing the actual masses by the units given on the base. This allows a combined representation for the five elements. The average mass balances show that there is a risk of pollution for each of these elements. The same holds also for every other district in the test region. The extent of the 'pollution risk' of copper, zinc, cadmium and lead exceeds 50 % of the total input in the district of Zofingen and in most of the other districts. The extent of the 'net soil input' of phosphorus (neglecting deposition of phosphorus) ranges between 0 % and 50 % of the total input in all the districts. The results for the other districts are shown in von Steiger *et al.* (1992).

Gaussian error propagation returns significance of the 'pollution risk' on the 95 %-level when the null hypothesis is tested against the hypothesis H_1 (confidence intervals are shown in Fig. 2), with

$$H_1\text{: 'pollution risk' } r > 0. \quad (4)$$

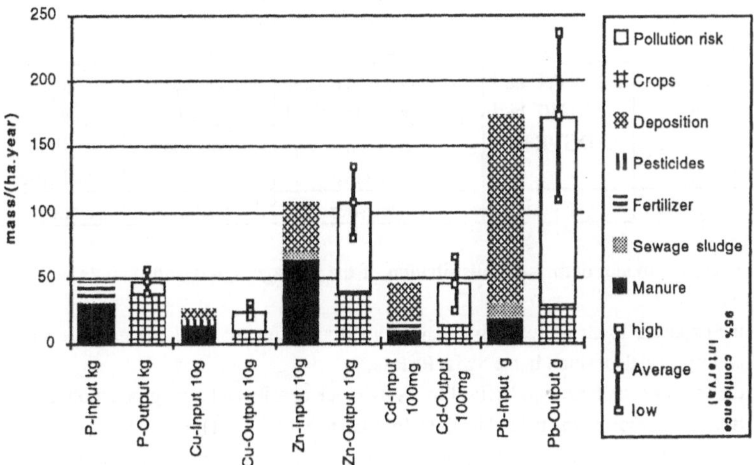

Figure 2. Average mass balances in the district Zofingen in mass ha^{-1} y^{-1} (see text)

The average mass balance for an entire district allows us to detect the sources of important inputs in agricultural land (e.g. lead input from deposition in Fig. 2) and therefore to know the main sources of pollution. Average mass balances showing the accumulation of a pollutant in the region signify that local pollution risks might be diminished by the redistribution of inputs (e.g. by transporting manure from one farm to another), but they cannot be completely avoided. If the average balance is equilibrated ($r=\pm 0$) then local pollution risks can be eliminated by a redistribution.

Local pollution risk

Since risk for soil and water pollution is not uniformly distributed over a region or nation, the method 'Proterra' includes a model that calculates the pollution risk r_q separately for different farm types and exposure to air pollution, defined by a set of categories listed in Table 2:

$$r_{i,q} = \sum_k \Phi_{i,q}^{(k \rightarrow \text{soil})} \quad , \qquad (5)$$

where k means 'crop production$_q$' and 'atmospheric deposition$_q$'. Here, q defines the subregion determined by the categories given in Table 2, and i indicates the element P, Cu, Zn, Cd or Pb.

Table 2. Categories of farm type and air pollution of agricultural land.

Farm type				Exposure to air pollution
Livestock animals	Livestock density[1]	Crop production	Sewage sludge quality	
no livestock cattle pig poultry	0-1 DGVE ha^{-1} 1-2 DGVE ha^{-1} >2 DGVE ha^{-1}	grass heavily fertilized grass lightly fertilized rotation special crops	no spreading good moderate bad	heavy average light background
4 categories	3 categories	4 categories	4 categories	4 categories

Thus, the model defines the distribution of the average inputs and outputs within the region:

- the exposure to air pollution is defined by extrapolating all available measurements of atmospheric deposition in the Swiss Plateau;
- the area of each farm type defined by its livestock density and crop production is known from the census by the federal institute for statistics (BfS, 1991);
- the distribution of manure on different crops is given choosing a weighting factor for every crop;

[1] 1 DGVE ha^{-1} corresponds to one cow of 600 kg weight per hectare

- for the distribution of sewage sludge and pesticides, an 'acceptance factor' (the part of the agricultural land, where the farmer accepts sewage sludge or pesticides) is considered for every crop;
- the distribution of fertilizer is modelled using the phosphorus uptake of the different crops and weighting factors describing how the farmers in the region plan their fertilizing (especially to which extent fertilizing by manure and sewage sludge is accounted for by the farmers).

The different factors in the distribution model of the method 'Proterra' are based on the knowledge of local agricultural consultants about the farming strategies. In order to calculate the pollution risk under different assumptions, the user of the model can adapt the factors in the distribution model (for details see von Steiger *et al.*, 1992).

The distribution model is applied to identify for which categories pollution risk is high and to determine the area of the affected agricultural land. In the following example we show the results for farms with pig fattening (>2 DGVE ha^{-1}), half of the area heavily fertilized grassland and the other half in crop rotation exposed to background air pollution, not accepting sewage sludge (Fig. 3).

This farm type covers 0.65% of the agricultural land in the district of Zofingen (37 ha). A high risk of copper and zinc pollution has been found for this farm type compared to the average pollution risk in the district.

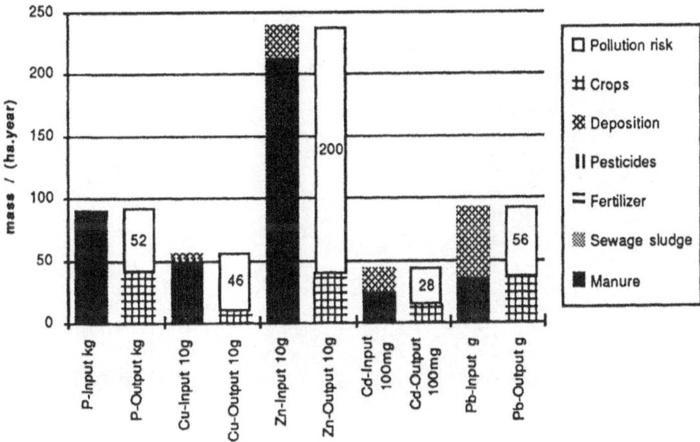

Figure 3 Mass balances in mass ha^{-1} y^{-1} of farm type 'pig fattening' (>2 DGVE ha^{-1}), half of the area are heavily fertilized grassland and the other half in crop rotation exposed to background air pollution, not accepting sewage sludge.

The different pollution risks in the district of Zofingen calculated using Equation (5) are classified into 'none', 'low', 'average', 'high' and 'very high' risk (see Fig. 4). The classification is based on the duration within which the guide levels according to the Swiss Ordinance relating to pollutants in soil (VSBo, 1986) are reached in the topsoil (0-20 cm, see Table 3), under the assumption that the accumulation is continuous at the given rates of contaminant fluxes.

High and very high pollution risk by copper occurs on 11.9 % of the total area of cultivated land. The reasons for very high pollution risks (occurring on 2.7 % of land) are farms with many pigs exposed to heavy or average air pollution.

Availability and reliability of data and of the distribution model

Data availability and reliability

We conclude from the experiences in the test study that, in most cases, available data allow us to assess regional mass balances in agricultural land. In Cantons where data on quantities of fertilizers or pesticides sold, distribution of fertilizers and sewage sludge, heavy metal content of manure or atmospheric deposition are not available, we have to deal with less reliable data listed at the right hand side in Table 1.

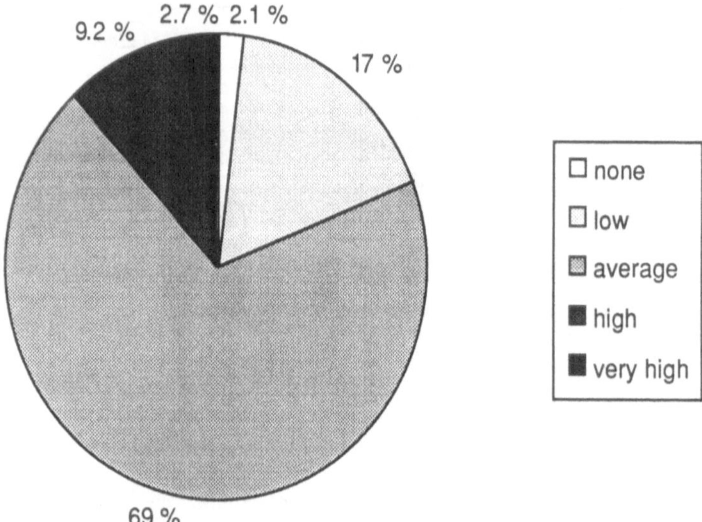

Figure 4. Classes of pollution risk for copper in district of Zofingen as percentages of total area

Table 3. Definition of classes for the pollution risk.

pollution risk	available time until Swiss guide values (VSBo, 1986) are reached[2]	range of calculated pollution risk in mass ha^{-1} y^{-1}			
		Cu 10g	Zn 10g	Cd 100mg	Pb g
none	>3162	<2.2	<11.8	<4.7	<22
low	1000-3162	2.2-6.9	11.8-37.4	4.7-15	22-68
moderate	316-1000	6.9-22	37.4-118	15-47	68-220
high	100-316	22-69	118-374	47-150	220-680
very high	<100	>69	>374	>150	>680

Reliability of data collected by someone other than the actual surveyor or monitor for purposes other than regional mass balances may be specified in different ways. Simple controls can detect implausible data, and knowledge about the exact procedure of data collection identifies possible reasons for systematic errors (e.g. overestimating contaminant uptake by plants using statistics on crop production).

Local estimation using the distribution model

First we give an example for the potential consequences of using regional input data for the estimation of local pollution risks. The quantity and the quality of sewage sludge distributed during a certain time from the sewage treatment plants in a district is fairly well known from compulsory statistics. It is much more difficult to determine the surface of agricultural land on which this mass of sewage sludge is distributed. The distribution model of 'Proterra' averages the phosphorus and heavy metal input over a specified fraction of the entire area where spreading sewage sludge is legal. This fraction is determined through an acceptance factor which is estimated from available data on farming in the specific region or by analogy to other regions. The area of agricultural land on which sewage sludge is applied might be either overestimated or underestimated.

We attempt to illustrate the reliability of both collected data and the distribution model of the method 'Proterra' by comparing the mass balance of the farm type specified in Fig. 3 with a mass balance determined by measuring the annual input and output fluxes of all the cultivated area on a farm of the same type (Fig. 5, data from von Steiger & Baccini, 1990). The results show that the local mass balances on the single farm compare well to the local estimation using the distribution model, except for phosphorus. Manure seemed to contain more heavy metal but less phosphorus on the single farm compared to the data from farming guide-lines available in the district of Zofingen.

---

2 the available time is based on an actual content of 23 mg Cu, 56 mg Zn, 0.23 mg Cd and 24 mg Pb per kg dry soil (Vogel *et al.*, 1989).

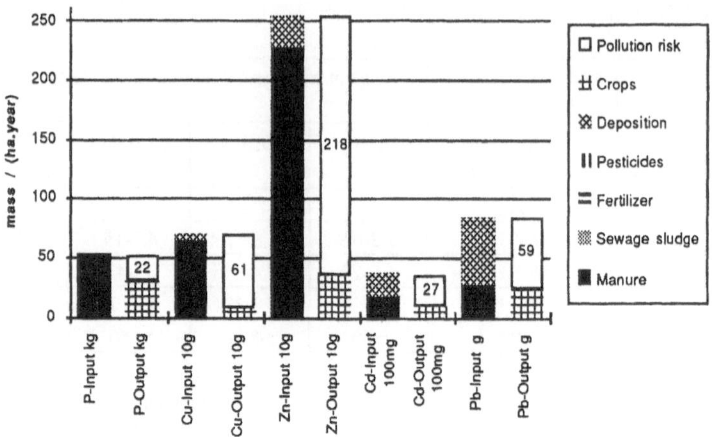

Figure 5. Mass balances of 'livestock' farm; data from von Steiger & Baccini (1990)

Contribution to long-term areal soil monitoring

Calculating contamination risk of agricultural land for defined categories, a regional mass balance approach may inform those responsible for long-term soil monitoring (see introduction) where and how often soil samples should be analysed for contaminants. Another application of the model would be to choose representative long-term monitoring locations using the categories associated with, for example, 'low' or 'high' pollution risk (cf. Fig. 4) after computing the size of the area where these pollution risks occur. Once the locations are chosen, exact and consistent data about the distribution of fertilizer and atmospheric deposition must be collected continuously in order to calculate the local pollution risk.

As Desaules (1993, Fig. 3) has shown at 14 observation sites of the Swiss national soil monitoring network (NABO), the fluctuation of the changes of topsoil contents after 5 years is much greater than fluctuations of the mean annual input ('pollution risk'). It seems that heavy metal inputs in agricultural soils are often masked by other processes and *a posteriori* soil analysis can not detect them.

Conclusions

Computing regional mass balances for agricultural land, as applied within 'Proterra', promises to help environmental policy. However, it cannot be over-emphasized that results are no better than available data. In many cases administrators can control the availability of data as part of their environmental policy.

Moreover, databases on inputs, outputs and associated farming type of agricultural land should be included in planning and running soil monitoring programmes. The method 'Proterra' is one possible way to get an integrated view of this information and thus help in soil monitoring both for extrapolation in time as for interpolation in space.

Acknowledgments

We thank Felix Schärer for preparing data on atmospheric deposition and Anja Pepels for reading a draft of our script, to all those people who made their data available in our pilot study, and to the Cantons Aargau, Basel-Land, Bern, and Solothurn for funding the project.

References

Baccini, P. & Brunner, P.H. 1991. *Metabolism of the anthroposphere*. Springer Verlag, Heidelberg.

Bader, H. P. & Baccini, P. 1993. Monitoring and Control of Regional Material Fluxes. In: *Soil Monitoring*. Schulin R., Desaules A., Webster R. & von Steiger B. (Eds). Birkhäuser Verlag, Basel, this volume.

BfS (Bundesamt für Statistik). 1991. Daten der Eidgenössischen Landwirtschaftszählung 1990. *Abteilung Agrarstatistik*, Bern.

Desaules, A. 1993. Soil Monitoring in Switzerland by the NABO-Network: Objectives, Experiences and Problems. In: *Soil Monitoring*. Schulin R., Desaules A., Webster R. & von Steiger B. (Eds). Birkhäuser Verlag, Basel, this volume.

FAP Reckenholz, RAC Changins, FAC Liebefeld. 1987. Düngungsrichtlinien für den Acker- und Futterbau.. *Landwirtschaftliche Beratungszentrale*, 8307 Lindau.

Mayer R. & Ulrich B. 1974. Conclusions on the filtering action of forests from ecosystem analysis. *Oecologia Plantarum*, 9, 157-168.

Menzi, H. & Flückiger E. 1991. Neue Richtwerte für Hofdünger: Aktueller Stand und Konsequenzen. *SVIAL-Weiterbildungskurs "Tierhaltung und Nährstoffkreislauf auf dem landw. Betrieb"*, Bern

Vogel, H., Desaules, A. & Häni, H. 1989. Schwermetallgehalte in den Böden der Schweiz. *Bericht Nr. 40. Nationales Forschungsprogramm "Boden"*, Liebefeld-Bern.

von Steiger, B. & Baccini, P. 1990. Regionale Stoffbilanzierung von landwirtschaftlichen Böden mit messbarem Ein- und Austrag. *Bericht Nr. 38. Nationales Forschungsprogramm "Boden"*, Liebefeld-Bern.

von Steiger, B., Obrist, J., Schärer, F. 1992. Die Stoffbuchhaltung als Früherkennungs-Instrument im Bodenschutz. Report prepared for the Cantons of Aargau, Bern & Solothurn. *ETH Zürich*, Schlieren.

VSBo 1986. Ordinance Relating to Pollutants in Soil. *Swiss Federal Office of Environment, Forests and Landscape (BUWAL)*, Bern.

B. von Steiger and J. Obrist
ETH Zürich, Institut für terrestrische Ökologie
Grabenstrasse 3
CH-8952 Schlieren
Switzerland

Discussion

Question by H. Menzi

Is the variation of the manure composition from one farm to another considered in the method 'Proterra'?

Answer by B. von Steiger

No, 'Proterra' calculates on the basis of average phosphorus production per livestock unit and year and applies average heavy metal:phosphorus ratios.

Question by R. Mayer

In Germany similar models are applied to estimate nitrogen load on agricultural land in several regions. But you seem to doubt the applicability of this kind of model in Switzerland. Why?

Answer by B. von Steiger

The model has to be validated by comparison with mass balances on a local scale. It is not widely applied at present, and some colleagues think, contrary to you, that we are forcing its application too soon without adequat validation.

OBJECTIVES OF LONG-TERM AREAL SOIL MONITORING

P. Lagas

The objective of long-term areal soil monitoring is to give a diagnosis and a prognosis of soil quality in order to protect the multifunctionality of soils. The basic soil functions were identified as follows

a. Production.
   The soil stores and provides water and nutrients, whether in agriculture, forestry or natural systems. Most of the nutrients become available when organic matter containing them decomposes in the soil.
b. Filtering.
   Groundwater is replenished and protected by soil
c. Ecology.
   The soil degrades organic material and recycles nutrient elements.
   It is the habitat of soil biota.
d. Foundation.
   The soil bears our houses and other buildings, roads, bridges, etc., and gives mechanical support to higher plants.

The environment is charged with increasing loads from industry, agriculture, traffic and other human activities. Chemical compounds are emitted into the environment, and many of them are toxic. They contaminate the soil and threaten its proper functioning which is important for assuring the quality of our food and drinking water.

From these considerations we may define the main objective of long-term areal soil monitoring to protect soil quality and thereby the multifuncitonality of the soil.

Diagnosis of soil quality and soil stress

For a diagnosis of soil quality concepts and methods have to be developed or improved to characterize, compare and monitor soil quality in relation to its multifunctionality as defined above. The concentrations of potential pollutants should be judged in relation to background values as well as ecotoxicological effects.

Prognosis of soil quality

Most soils have a large buffering capacity for contaminants, but the capacity is limited. The soil might function properly in all respects now. But how will it do so after 10 years, or after 100 years, when loads of toxic compounds have increased?

For a prognosis of soil quality in the future the trends in relation to guide levels should be studied. We need mechanistic knowledge of soil processes to build comprehensive models and statistically valid assessments to produce reliable statements.

In consequence, two kinds of approaches, the process- and the monitoring-oriented research, are needed, and soil monitoring should always be planned for the long term.

Mass balance

In several countries the air, rain, ground-water, and surface-water are being monitored. The soil is not, however, and this represents a missing link in environmental monitoring. The mass balance approach can provide information from one to another sector of environmental monitoring. For example the rates at which toxic materials are being deposited should be known for interpretation of soil monitoring data.

Information for policy-makers

Long-term monitoring is vital for successful management of our environment. Those who decide policy should know which contaminants are potentialy damaging and threaten the proper functioning of the soil and which of these are increasing in concentration. They should then be able to decide on preventive action.

Monitoring is needed for that. It is also necessary to ensure that regulations arising from political decisions are working. The information must be reliable, both to ensure that the decisions are sound and to convince those who must obey the regulations.

P. Lagas
National Institute of Public Health and Environmental Protection
NL-3720 BA Bilthoven
The Netherlands

Soil Monitoring, Monte Verità, © Birkhäuser Verlag Basel

GAPS OF CURRENT CONCEPTS IN SOIL MONITORING: POTENTIAL AND DRAWBACKS OF OTHER CONCEPTS

A. Desaules & B. von Steiger

The present state

The current concept of soil monitoring involves networks of sites at which soil pollution and physico-biological state of the soil are measured and monitored in the long term. Several networks have been installed since the mid 1980s in Europe.

Repetition of observations

Most measurements involve destructive soil sampling or result in severe disturbance of the sampling location. Thus, it is usually not possible to repeat sampling or to take supplementary measurements at exactly the same locations. Because of the heterogeneity of soil this poses a serious problem, a fact which is indicated already by the multitude of different sampling schemes that have been proposed to improve the representation and reliability of the measurements.

Temporal and spatial representativity

The number of long term monitoring sites is always limited by costs and is usually too small for geostatistical analysis. The spatial representativity of the individual sites with respect to their surroundings cannot be assessed rigourously and remain subject to personal judgement. Temporal changes of soil conditions are difficult to detect because in the short term they are usually much smaller than the variation from all the other sources in combination (spatial variation, sampling errors and analytical errors). Moreover, on the basis of soil inventories, the result of temporal change is obtained retrospectively.

Possible improvements of the measuring programmes

Current concepts of soil monitoring differ by their measuring programmes. Monitored properties are primarily the total contents of heavy metals and macro-nutrient elements. Less often organic compounds and biological properties are recorded. Other pedological factors, processes and properties which might be impaired by human activities (chemical and physical) could be considered in addition. It is not possible, however, to list comprehensively the relevant pedological factors because they depend on the specific human impact. The selection of properties for soil monitoring should be based on their indicator value soil functions such as buffer capacity, metabolic activity, and soil fertility.

Risk assessment

Soil monitoring is oriented mainly towards risks which are caused by human impacts including biological degradation. Assessing these risks often relies on guide values for specific properties. To minimize risks guide values should be set smaller than values at which undesirable impacts on the most sensitive soil function and soil organism occur.

The measurement errors in soil monitoring should be confined within a certain range in order to allow an acceptable quantification of risks. Similarly, monitoring sites as homogeneous as possible should be chosen.

Gaps of soil monitoring

We often do not know what the most sensitive pedological factors are, and a strategy of minimum risk would imply minimum human impact and management possibilities.

Two main gaps of current soil monitoring concepts are the lack of predictive value, and the lack of integration between various approaches. The 'partial approach' is often overstressed because it focuses on the distinction between *natural changes and man-made damage*.

Other ways of assessment

There are at least two other ways in which the accumulation and risk of pollution might be estimated. These are the methods of mass flux balances and biomonitoring, which are considered below.

Mass flux balances

The method is based on the rates at which materials, potential pollutants, enter and leave the soil. There is usually some underlying model on which the description of the physical processes is based. The data and the parameters of the model determine the rates of input and output of materials.

Such models serve two purposes. They force investigators to identify the sources of materials and the processes and to formulate clear hypotheses about the pathways of pollution and the ultimate fates of the pollutants. They lead directly to assessments of risk, at least qualitatively. They enable investigators to identify separate factors in complex systems, and by sensitivity analysis to assess their individual contributions. This should guide the search for new data, and be more economical than simple inventory based on survey data alone. The survey sites can be chosen in the knowledge of balances of mass fluxes because they relate characteristics of sites (e.g. land use) to an estimate of temporal change.

The models are well suited to forecasting. It is necessary only to think of a change in input, as it might be from a new factory or increased traffic, change the relevant quantities in the model and see what effect it predicts. The response is almost immediate.

The method has its disadvantages. All models are to some extent simplifications, and a model might oversimplify a complex set of processes. The biological effects of soil pollution are often ignored by the models actually in use. A model might also be inappropriate for other reasons:

- the processes may change with time,
- the model might not take account of interactions,
- the model might not be valid for extrapolating outside the range of conditions for which it was calibrated.

In addition, there is the risk that the method will be used because it is cheap rather than appropriate, and because it will always produce results. Finally, since the models are based on flux measurement, they might be hard and expensive to validate.

Biomonitoring

Instead of measuring the pollutants themselves one can monitor their presumed effect on the soil's flora and fauna. Thus, one might measure the microbial biomass, its activity or both, or the numbers of earthworms and other higher organisms. The merit of this is that the effects of the pollutants are measured directly.

Biomonitoring is the *most integrated* approach of soil monitoring in the sense that it monitors effects on organisms which are themselves part of the soil ecosystem. The selectivity of the pollutants for the different kinds of organisms or behaviour can be assessed. It might be cheaper to assay the effects of some organic chemicals in this way than to measure them directly.

The disadvantages are that one cannot be sure what the causes of any change are, and this in turn makes forecasting and abatement difficult.

Conclusions

Direct monitoring of soil pollutants is the most reliable, but it is not the only way of assessing risk, and in particular it does not forecast nor does it predict the effects of pollution on the ecosystem. Biomonitoring measures the effect of pollution, but it too fails to forecast future trends. Estimation of mass flux balances, despite uncertainty, has the potential to fill this gap, and the best way forward seems to be to combine these three approaches.

A. Desaules
Swiss Federal Research Station for Agricultural Chemistry and Hygiene of the Environment
Schwarzenburgstrasse 155
CH-3097 Liebefeld-Bern
Switzerland

B. von Steiger
Institut für terrestrische Ökologie
ETH Zürich
Grabenstrasse 3,
CH-8952 Schlieren
Switzerland

SECTION 2

Contaminant Fluxes and Mass Balances

CONTAMINANT MASS BALANCES IN SOIL MONITORING

Rainer Schulin

The assessment and balancing of mass fluxes is a promising method for predicting and detecting the accumulation of contaminants such as heavy metals in soils before it can be actually observed by direct soil surveys. Previous studies have demonstrated that this approach works well on a regional or larger scale, e.g. for monitoring farm land pollution by heavy metals such as lead, copper, zinc, and cadmium. However, the applicability of the mass balance approach to much smaller areas, such as the sites of a permanent soil monitoring network, is still an open question. With decreasing size of the mass balance compartment, the required degree of spatial and temporal resolution increases, rendering small scale variability increasingly important. This paper introduces the principles and problems of applying the mass balance approach in soil pollution monitoring and discusses in particular the conceptual difficulties of determining mass fluxes in the underground as well as their dependence on the choice of the system boundaries and the scale of spatial resolution. It is concluded that the mass balance approach is most problematic where fluxes across the highly structured soil-vegetation interface and fluxes between soil and groundwater have to be assessed.

Introduction

The assessment of mass balances on the basis of flux estimations has recently found to be a promising approach in routine soil monitoring for achieving a cheap, quick and early diagnosis of soil contamination by heavy metals and phosphorous on agricultural land (von Steiger & Baccini, 1990; von Steiger & Obrist, 1993). The spatial resolution of the method developped by von Steiger & Obrist (1993) is, however, limited to a regional scale. It allows us to make distinctive estimates between different strata of land management and site type to a limited degree, but it is not suited for estimating the accumulation of heavy metals and phosphorous by the soil of a specific site. Such a high resolution would, however, be necessary if the mass flux balance approach were to be applied to the sites of a permanent monitoring network.

In addition to the assessment of pollutant accumulation, the determination of mass fluxes and balances might have a high diagnostic value in its own right since rates of turnover and their deviation from steady state conditions may be viewed as primary indicators of ecosystem stress and change (Ulrich, 1987). The optimal design of monitoring and sampling strategies depends on the specific objectives. These will not be investigated here any further. Instead, the focus of the following discussion is on conceptual problems involved in environmental mass flux determination and the relevance of the scale of observation in setting up mass balance schemes for environmental pollution monitoring.

The mass balance principle

The idea to use mass balances for the purpose of monitoring processes is certainly not new, but it is a fundamental principle in the control of matter fluxes, e.g. in chemical engineering (Reklaitis, 1983), stock-keeping in ware-houses, nuclear inspection (Arenhaus *et al.*, 1983). Recently, it has also been adapted to the control and management of environmental quality (Hofmeister, 1989) and to the scientific analysis of matter fluxes in the 'anthroposphere' (Baccini & Brunner, 1991). It has its scientific basis in the concepts of thermodynamical systems. A thermodynamic system is simply a compartment of the environment that is completely delimited from its surroundings by real or imaginary boundaries.

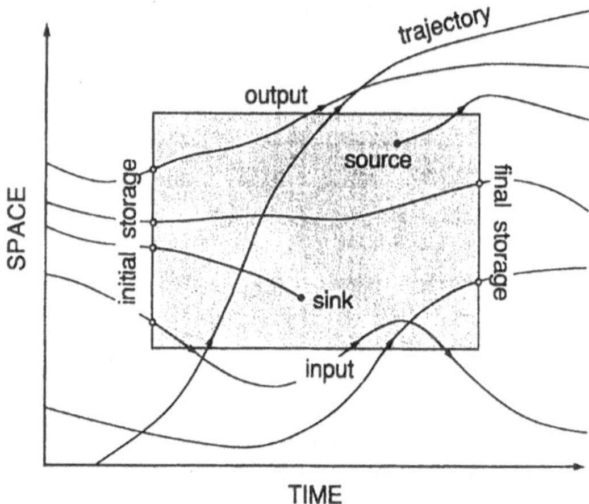

Figure 1. Space-time representation of a spatially invariant mass balance compartment. Lines represent trajectories of moving particles. The difference between entering and exiting trajectories is equivalent to the net balance of sources minus sinks.

The balance principle states that any change in the content of an extensive quality within the spatial and phase boundaries of the system is equivalent to the sum of net transport of the quality across the system boundaries and the net production or consumption of it within the system over a given time (Fig. 1). Steady state conditions produce zero-balances.

Any extensive quality of such a system can be subject of a balance. For certain physical entities such as mass, momentum, energy, mole numbers of chemical elements, and electrical charge, conservation laws require that no production or consumption occurs in the system, nor on the boundaries. Entropy is a special case in that it may be produced, but not be consumed. For conservative entities, balances can be used to detect 'leaks' or inconsistencies in the system boundaries, to check if all relevant processes have been identified, or to quantify parts of a balance indirectly by the complement of measured processes.

Mass transfer in environmental systems

Whereas the principles of material balances appear to be obvious, their application in environmental systems is in general not trivial at all. Problems start with the definition of the system, that is its compartment boundaries. By defining the system boundaries the transport processes to be taken into account are determined also. Transport across a system boundary can be assessed in various ways, directly as well as indirectly. Rarely, it is possible to monitor transport continuously in time and completely in space. In general, measurements will be restricted to samples representing discrete sections of a transport process both in space and in time.

One of the most favourable situations for the assessment of a flow is given when it is constricted to a few confined channels. Channelled flow exists, for example, in hydrological systems at the exit of a catchment where the outgoing flow of surface water can be measured through a water-gauge (Fig. 2). Similar situations of flow channelling are met in the case of heavy metal input onto farmland by way of commercial fertilizers. The production sites and distribution channels of these materials are highly concentrated in space and present themselves for control purposes (Fig. 3). As the distribution proceeds via the intermediary trade to the farmers, transport becomes more and more distributed and difficult to monitor. Since the distribution areas of traders are usually overlapping rather than disjoint, it is rather costly to determine on the basis of sale information the amount of commercial fertilizer arriving on a particular farm. With the help of some statistical information on fertilizer purchases by farmers, however, it is possible to estimate approximately the fertilizer import into a region by this approach (von Steiger & Obrist, 1993).

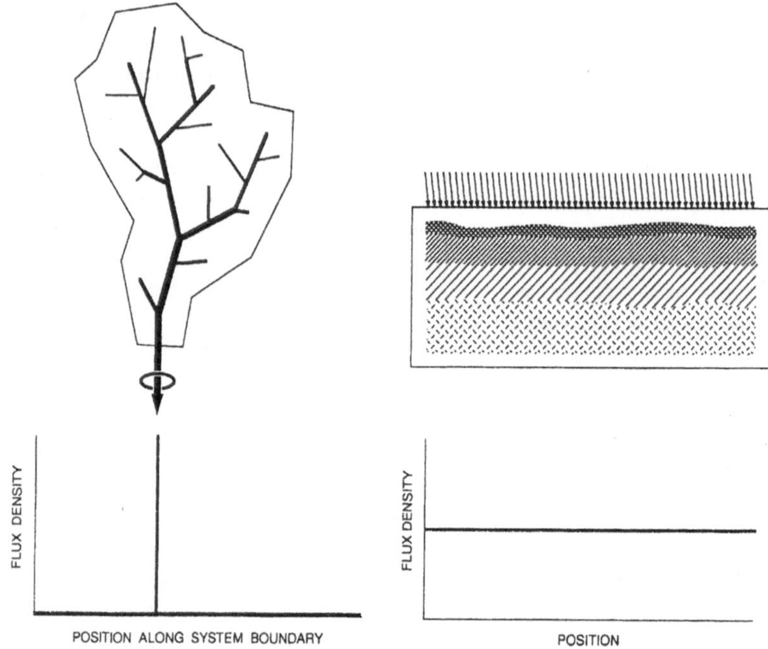

Figure 2. Schematic representation of the flux density distribution across control sections for channeled flow (left) and uniformly distributed flow (right).

Another favourable condition for the quantification of transport is the case where flow processes are not channelled, but evenly distributed across the system boundary so that the average flux is well represented by a fairly small sample in cross-section. An approximate example for such a situation may be the case of atmospheric deposition on surfaces which are homogeneous with respect to the scale of the observation area (Fig. 2).

Similar considerations hold with respect to the temporal dimension of transport processes. Channelling in space corresponds to a concentration of transport in a few events. The assessment of such pulses requires temporary interception of the translocated material as in the case of fertilizer application on farm land. Book-keeping of such events is in principle as practicable as the monitoring of transport processes which are close to a steady state in time. In the case of such quasi-stationarity, infrequent temporal sampling is sufficient to obtain representative results, although spatial heterogeneity may require dense sampling in space. Deep percolation in soil may serve as an approximate example for this situation.

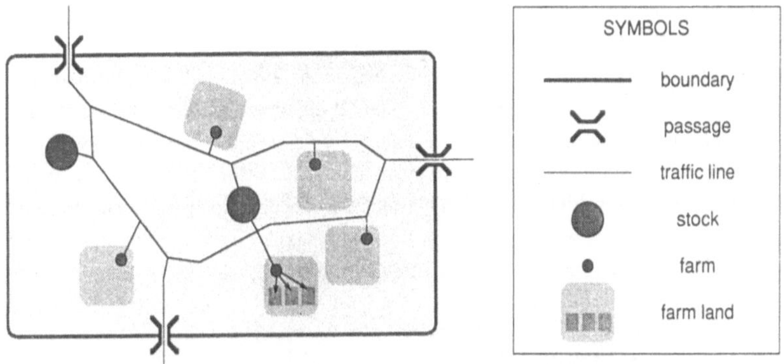

Figure 3. Basic distribution pattern of the commercial fertilizers or other agrochemicals. Intermediate storage by local dealers has been omitted for reasons of simplicity and clarity.

In reality, transport processes will rarely correspond closely to one of these idealized situations. The required sampling intensity depends on the spatial and temporal variability. Evaluation on the basis of available literature or pilot studies must then reveal if this approach is feasable in a routine soil monitoring programme not only in view of the cost-effectiveness, but also of the tolerability of disturbing impacts on the observations sites by the sampling activity.

Convection and dispersion

The availability of a method for direct and uninterrupted determination of a flow rate, as, for example, in the case of stream flow using a water-gauge is the exception rather than the rule. More often, methods are based on the interception of a flow over a defined period of time and analysis of the accumulated input into the receptor as in the case of a rain-gauge. Such a method can still be regarded as direct because the recorded signal corresponds to the integral of the intercepted material flux.

Most continuous as well as interceptive and interruptive methods represent indirect measurements of flow rates in the sense that the estimation of the flux of one substance is based on the measurement of the flux of another substance or of a mixture containing also other substances and on the measurement of the concentration ratio between the substances or the concentration of the substance in the mixture. The basic assumption behind this is that the transport is by passive convection, so that the transport rate can be calculated as the product of the concentration of the substance in the mixture and the flow rate of the convective mixture.

The convection assumption holds under two conditions, first if the mixture is homogeneous with no concentration gradients, and second if all streamlines have the same flow velocity. If both concentration and velocity gradients exist then they will inevitably result in dispersion effects since the gradient will move at different velocities in different stream lines. Because of the resulting correlation between the concentration profile and the velocity profile across the ensemble of stream lines or tubes, a dispersive flux component arises in addition to the mass flow or convection term (Fig. 4).

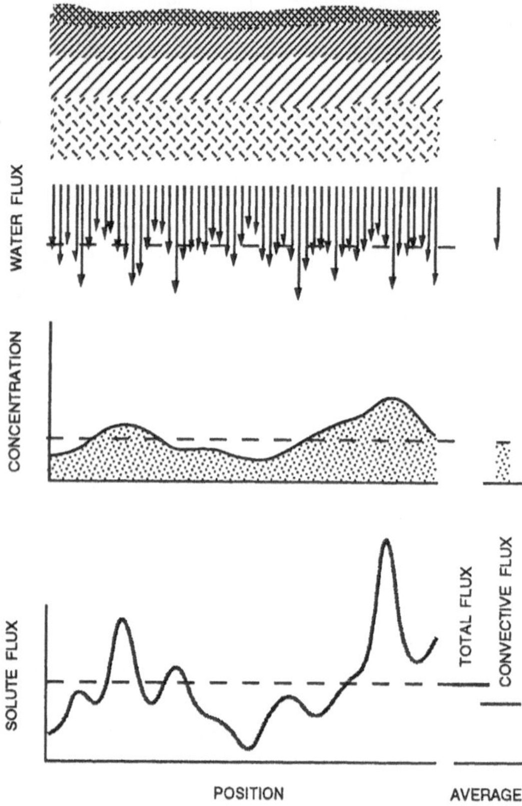

Figure 4. The dispersion effect in the assessment of mass flow under non-uniformly distributed flux conditions. On the right side convective mass flux density, calculated as the product of average flow with average concentration is compared with the average flux density of the total mass transfer.

The dispersion effect can be quite large with respect to the convective flux in soils (Schulin *et al.*, 1987). Its magnitude depends on the heterogeneity of the velocity field captured by a sample as well as on the steepness of the concentration gradients in the sampled flow field. In soils, pore water velocities usually vary over several magnitudes of order. Thus, the dispersion effect can be neglected only in the case of constant concentrations as e.g. under steady-state conditions. Under natural conditions, non-stationarity and heterogeneity prevail in the soil, and thus dispersive flow processes have to be dealt with.

The dispersion effect: a field example

Large dispersion effects are observed in soils with a 'dual-porosity' structure, e.g. a network of macropores bypassing and thus short-circuiting the bulk matrix in between. Transport characteristics typical for such a dual structure were observed in a tracer experiment done by B. von Steiger, W. Attinger and myself as part of the study of material balances on a regional scale by von Steiger & Baccini (1990) on the Swiss Plateau west of Zürich. The experiment was done on a loamy luvisol derived from glacial till above alluvial gravel. In early 1987 when the soil was still cold but not frozen a bromide tracer solution was applied evenly on two plots, each 50 m x 12 m. The flow of water through the root zone was then traced by monitoring the bromide.

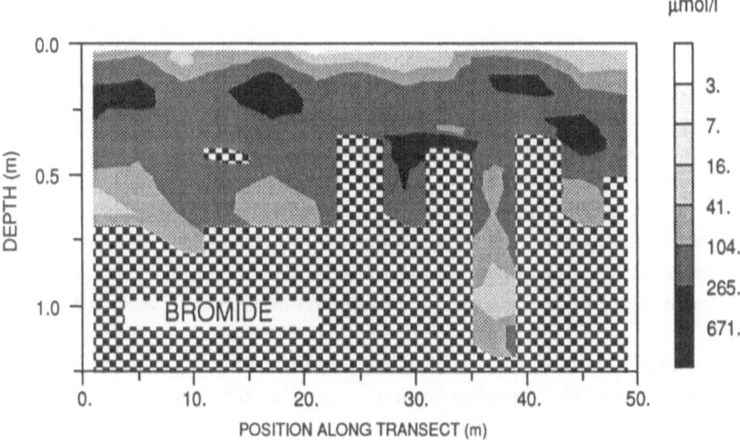

Figure 5. 'Contour' plot of the vertical distribution of bromide tracer along a transect through a loamy luvisol in arable land.

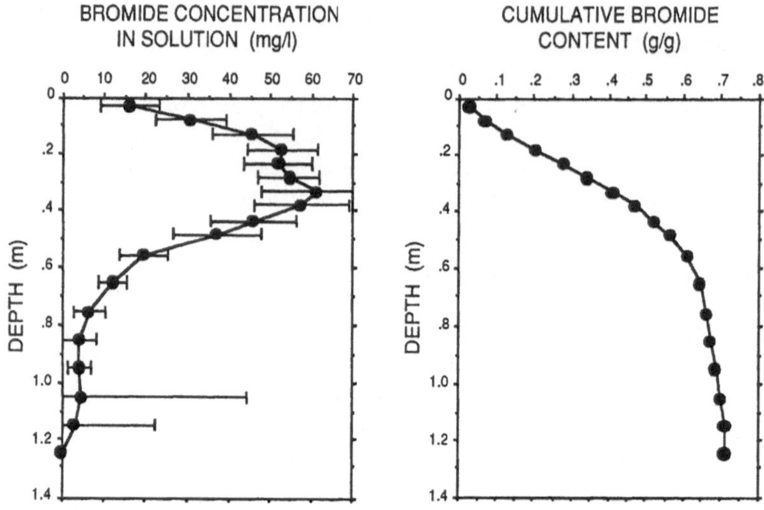

Figure 6. Averaged vertical distribution of bromide tracer on a 50 m x 12 m field plot: concentration of bromide in the soil solution (left), and cumulative tracer recovery relative to applied mass (right).

Figure 5 shows the distribution of the bromide tracer along a transect through one of the plots, located on fallow lying arable land, six weeks after the application of the tracer. The contour plot in this figure is based on soil core samples taken at intervals of 4 m and cut into 10 cm-sections. The spatial distribution of the tracer density displays a quite heterogenous pattern. A distinct front cannot be recognized.

Figure 6 shows the average profile of the vertical tracer distribution for the entire plot based on 53 core samples. It reveals that the recovered tracer had reached an average depth of approximately 25 cm below the soil surface. This was not surprising given that the top soil water content averaged approximately 35 % by volume and that a total of approximately 120 mm water precipitated onto the soil between tracer application and sampling.

It had not been expected, however, that only about 70% of the applied tracer mass could be recovered within a sampling depth of 1.3 m. The 30 % loss significantly exceeded the sampling error and could only to a very small degree be accounted for by volatilization of bromine or plant uptake. Thus, the only explanation was that a significant fraction had disappeared within the six weeks into the stony subsoil below 1.3 m where it could not be reached with the coring equipment at hand. Performing a similar tracer experiment with permanently installed suction cups, Roth *et al.* (1991) were able to provide also direct evidence that the postulated fast by-pass flow through preferential flow paths indeed occurs in this type of soil, although the sampling strategy of Roth *et al.* (1991) did not enable a mass balance to be computed.

The plausibility of the hypothesis that a small fraction of the porosity conducting the soil solution by an order of magnitude faster than the rest may dominate with respect to mass transport can be illustrated by a calculation based on the simplifying assumption that the water conducting porosity is composed of two regions, a fast and a slow region (Fig. 7). For simplicity, dispersion within these regions is assumed to be negligible. The values for the total water content θ, water flux q, and the pore water velocity of the slow region are chosen to match the case of the above tracer experiment. Table 1 shows that in this case a fraction of only about 3% of the water conducting porosity having a velocity ten times that of the other region is required in order to explain the observed effects. Estimating the total solute flux on the basis of the convection assumption, that is as the product of the average concentration with the average flux density, leads to an error by a factor of more than three. This example illustrates the problem of assessing mass fluxes in soils where no direct measurement by interception as in the situation of a lysimeter is possible.

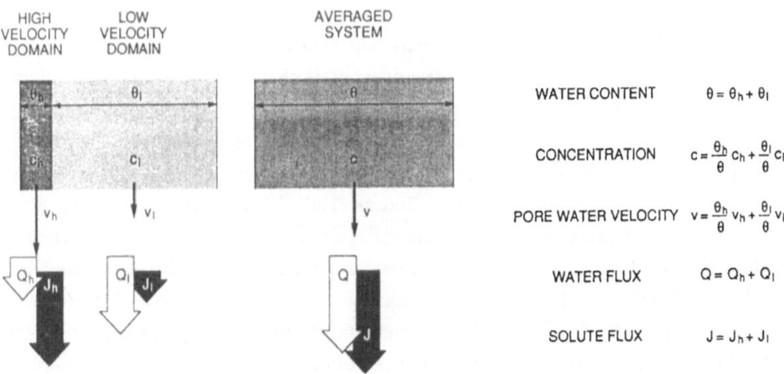

Figure 7. Flow and transport characteristics of a two-domain flow system (left), overall characteristics of an equivalent (averaged) one-domain system (middle), and their relationships (right).

Table 1. Hypothetical parameter values to illustrate the effect of a small domain with fast preferential flow according to the two-domain model shown in Fig. 7 on the error of solute flux estimates neglecting structural heterogeneity. The partitioning of water content, solute content and water flux has been chosen to fit the situation of the RESUB tracer experiment (Figs. 5 and 6). Other parameter values have been derived using the formula given in Fig. 7. For simplicity, solute transport within each of the two domains has been assumed to be purely convective.

Parameter	Unit	High Velocity Domain	Low Velocity Domain	Averaged System
Water Content	$cm^3\ cm^{-3}$	0.01	0.34	0.35
Concentration	$mg\ cm^{-3}$	0.400	0.025	0.036
Pore Water Velocity	$cm\ d^{-1}$	5.0	0.5	0.63
Water Flux	$cm\ d^{-1}$	0.05	0.17	0.22
Convective Solute Flux	$mg\ cm^{-2}\ d^{-1}$	0.02	0.00425	0.00792
Total Solute Flux	$mg\ cm^{-2}\ d^{-1}$	0.02	0.00425	0.02425

Modeling mass fluxes in soil

One approach to cope with the problem of spatial heterogeneity of flow velocities might be to increase the resolution of measurements to a degree that each sample represents a section of the flow field which is sufficiently homogeneous to allow dispersion within it to be neglected. Since in natural soils heterogeneity is characteristic even on the microscopic scale of individual pores this approach is of limited feasibility. A more promising approach is to calculate the fluxes on the basis of transport models that explicitly or implicitly take account of the statistical distribution of flow velocities and concentrations (van der Zee and van Riemsdijk, 1987; Butters *et al.*, 1989; Roth *et al.*, 1991). Such models need, of course, to be validated and calibrated for the particular system to be described. In soils, a popular mass transport model is the convection-dispersion equation which describes the dispersion effect by an analogy with Fickian diffusion. Other models taking better account of hierarchical structured porosities with restricted mixing between different pore regions which seems to be typical for many field soils have been described e.g. by Jury and Flühler (1992) as 'stochastic-convection models'.

One method for the calibration of transport models is to estimate model parameters by fitting model calculations to the results of tracer experiments. It is popular although it takes much time and labour, and contains many pitfalls, in particular with respect to field applications. As a general rule, it can not be recommended to fit several parameters simultaneously. Regardless of such problems, tracer experiments are very useful in providing information about residence time distributions and pathways. Such information is not bound to the validity of a specific model. Unfortunately, the number of tracer substances suited for field soil applications is very limited.

In addition to the more technical problems of model calibration, principal problems exist with the use of soil transport models for the assessment of transport processes on a soil monitoring site. One of the main problems is the dependence of dispersion parameter values on the spatial scales of the tracer experiments and of the observation sites (Gelhar, 1986; Schulin *et al.*, 1987).

Scale effects in mass flux balancing

Scale enters the problem of mass balancing in soil monitoring in various ways, many of which are closely related, however. With increasing length of the time period for which a balance is set up, the importance of storage changes often decreases relative to the turnover by transport and transformation processes, depending also on the storage capacity of the compartment. For sufficiently long periods, non-stationarities in the contaminant loading of the standing crop on farmland can be neglected as compared to the contaminant accumulation in soils. On the basis of this assumption it is possible to estimate heavy metal net transfer from agricultural plant production into soil from external inputs into the farm compartment. For the determination of short-term fluxes this method would yield large errors as the fluctuations of heavy metals stored in a farm may vary considerably over periods shorter than a year.

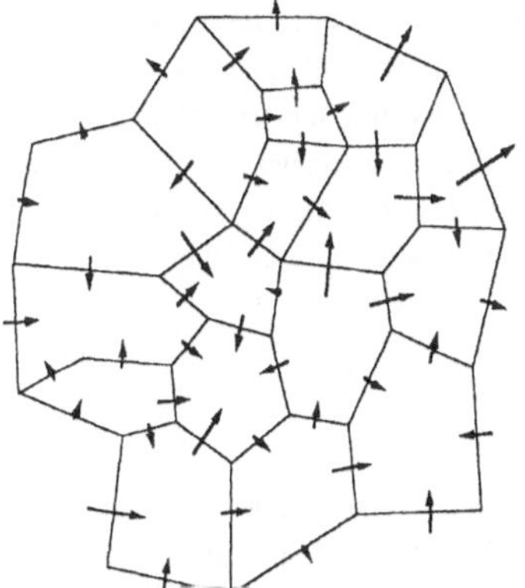

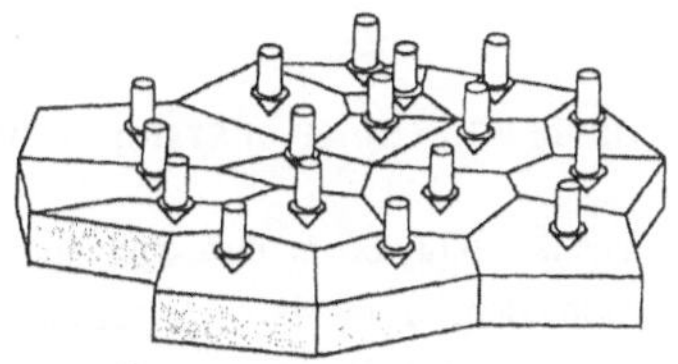

Figure 8. Lateral and vertical flows into a soil region and its sub-compartments. Lateral fluxes are less important for a mass balance over the entire region than for a sub-region, wheras vertical flows maintain their significance.

In soil monitoring the time scale of interest is of the order of years to decades and is largely independent of the spatial scale of observation. On the other hand, if balances have to be relevant on the scale of individual observation sites then the spatial resolution of the flow measurements entering into mass balances must be finer than the actual scale of interest. Thus, the temporal resolution required in soil monitoring will usually tend to be less demanding than the spatial resolution of the turnover processes on an observation site.

For the spatial scale, similar considerations hold as for the temporal scale, as in a larger compartment transfer processes across the system boundaries tend to decrease in importance relative to production and consumption processes in the same order as the surface to volume ratio. Soils are strongly anisotropic systems, displaying fundamentally different interactions in the lateral and vertical directions. The vertical dimension is characterized by a strong polarity as a result of gravity which does not exist horizontally.

Whereas the variability of the depth of a soil profile is very limited, the horizontal extension of an observation site in a monitoring programme may vary over many orders of magnitude. With increasing size the spatial variability of soil increases also, although at a decreasing rate (McBratney *et al.*, 1982; Burrough, 1983). In other words, scale effects in the spatial variability of soils are much more a matter of the horizontal dimensions than of depth.

The areal character of soil under a large-scale perspective also means that mass transfer processes in the vertical dimension such as deep seepage, capillary water rise, evapotranspiration, and precipitation will vary in proportion to the compartment size, whereas the importance of diffuse lateral exchange processes such as surface and interflow in the unsaturated zone tends to vary more or less in proportion to the length of the perimeter (Fig. 8). Such processes may thus be neglected for suffiently large perimeters. For other lateral transport processes, which proceed through a hierarchy of increasing channelling such as stream flow, this rather simplistic picture does not hold. However, as much as channelling enhances measurability and assessment of flows, the task of setting up a balance still tends to be easier for a larger system than for a smaller one. The relations between the characteristics of channelled flow processes, in particular their hierarchical structure, the size of the balance compartment, and monitoring efficiency merit more detailed evaluation.

Scale and choice of system boundaries

The feasibility of adequate flux measurements in the mass balance approach in soil monitoring depends to a large degree on the flexibility in the choice of the system boundaries. This flexibility is again related to the spatial and temporal scale of the system. If storage changes in neighbouring compartments of the soil, e. g. in the above-ground biomass and the adjacent atmosphere, as well as lateral transfer processes are insignificant relative to the accumulation of contaminants in soil as it may be assumed in many cases for regional soil contaminant balances (Fig. 9), then the problem of taking account of the input across the soil surface can be reduced to the determination of large-scale atmospheric deposition by shifting the system boundary to the interface between canopy and atmosphere without much loss of accuracy.

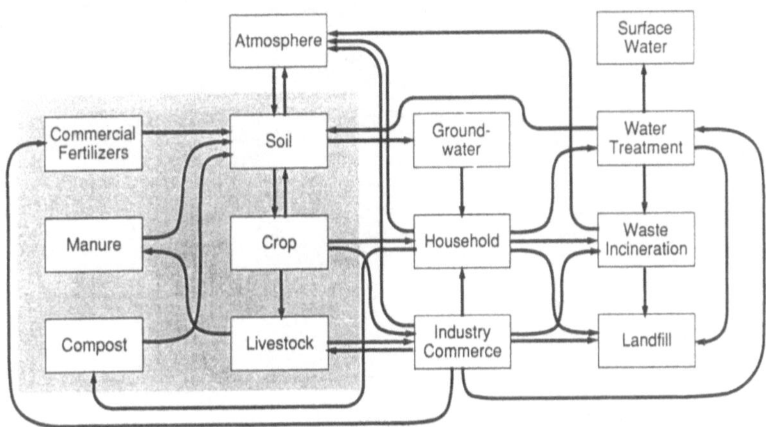

Figure 9. Basic flow pattern of elements circulating through farms (gray shading), their internal compartments, and their neighbour systems.

On the scale of a rather small monitoring site, transfer processes between soil and the above-ground biosphere and atmosphere as well as their spatial variability are much more significant relative to the total balance than for the regional balance. In particular, lateral transfer processes with a preferential direction such as sediment transport by erosion, surface flow, or interflow might not be negligible, independent of the chosen time span for the balance. Problems are expected to be greatest in forest ecosystems in which long successions will not average out storage changes over short times.

If the intersection for the measurement of fluxes is chosen along the soil surface the exchange of matter through the vegetation has to be taken into account (Fig. 10). This exchange is bidirectional as contaminants can be taken up by roots and transported into the above-ground parts and be returned to the soil through leaching, stemflow and litter fall together with atmospheric depositions which have been intercepted by the canopy. This cycling has the effect that fairly large fluxes may have to be determined, even if only their difference representing the net input into soil is of interest. As a result, a comparatively large relative error has to be expected.

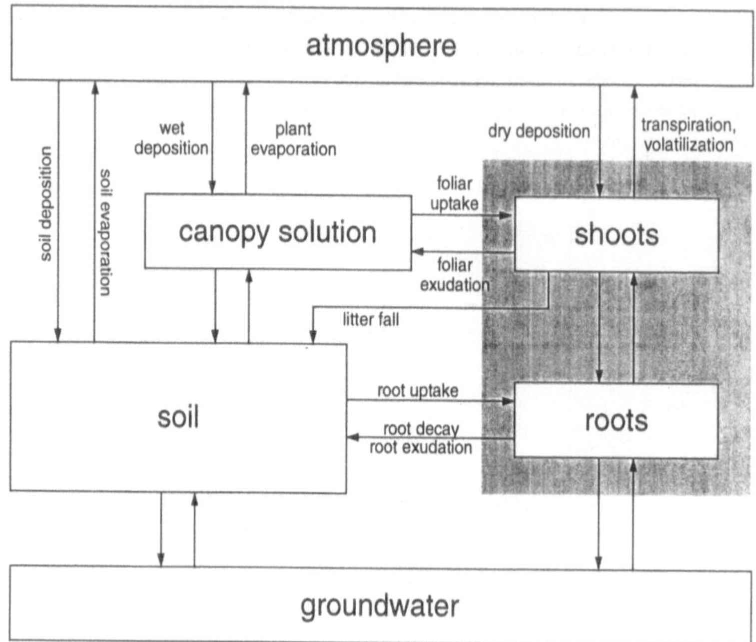

Figure 10. Mass transfer processes on a soil monitoring site. Mass balances for the soil compartment hinge on the assessment of fluxes into or out of the plant compartment (grey area). Processes such as harvest, soil erosion, or surface flow have been omitted for clarity.

As a general rule, the problem of backflow across system boundaries appears to be the more important, the closer the boundary is laid to the soil surface. Here, exchange appears most intense, and many of the processes are highly cyclic or fluctuating. This is particularly true of the turnover of organic carbon and mineral nutrients. Heavy metals are similarly subject to this biotic turnover, whether they function as micronutrients or as toxins.

Considerable fluctuations of material fluxes across the soil surface can also be due to abiotic processes. For example, changes in atmospheric pressure, soil water content and temperature can lead to fast gas exchange between topsoil and the above-ground atmosphere. These processes might be negligible for the balance of heavy metals in soils, but can be important for the mass balance of volatile contaminants. Similar fluctuations might also occur in convective transport by the soil water phase, in particular in the range of a fluctuating groundwater table.

A simple solution to the problem of flux assessment within the soil exists only where a flux can be neglected for soil chemical or hydrological reasons. Otherwise it is one of the most difficult problems in balancing soil contaminants. Because of the heterogeneity of the underground, an increase in scale cannot be expected to facilitate the task as it does aboveground, where the atmosphere functions as a comparatively homogeneous mixing cell.

Conclusions

A soil monitoring strategy aiming at the early detection of soil pollution by means of the mass balance approach has to deal with combinations of different types of mass fluxes with respect to the spatial and temporal pattern of intensity (channelled or diffuse in space, discrete or continuous in time). Except for a few ideal cases, the determination of the relevant fluxes is very demanding. Direct methods by means of gauges are in general not available. With respect to the soil system, the assessment of mass fluxes is usually based on sampling by interceptive devices (lysimeters) or on indirect estimates on the basis of the convection assumption, that is by multiplying the results of resident concentration and phase flow (e.g. soil water flow, sediment transport) determinations. The latter approach is, however, valid only if at least one of the two factors has a homogeneous distribution over the flow cross section under consideration. For heterogeneous distributions of mass fluxes as found in structured soils, the convection assumption may lead to substantial errors.

The feasibility of the flux balance approach to soil monitoring depends very much on the scale of the problem. Because boundary effects are less important, it is in general easier to monitor large scale than small scale compartments. A major unresolved problem, besides the assessment of mass fluxes in the soil, is the accounting of mass fluxes between soil and atmosphere through a vegetation canopy because of the complex structure of the interfaces and a complex mixture of bi-directional exchange processes. In order to provide a systematic framework for the analysis of errors and uncertainties and the evaluation of scale effects in this context, a general theory of environmental flux assessment and material balancing is desirable. Such a theory is still lacking.

References

Arenhaus, R., Beedgen, R. & Goeres, H.-J. 1983. *Sampling for the verification of materials balances*. Report KFK-3570, Kernforschungszentrum Karlsruhe, 80 p.

Baccini, P. & Brunner, P. H. 1991. *Metabolism of the anthroposphere*. Springer-Verlag, Berlin, 157 p.

Burrough, P. A. 1983. Multiscale sources of spatial variation in soil. I. The application of fractal concepts to nested levels of soil variation. *Journal of Soil Science* **34**, 577 - 597.

Butters, G. L., Jury, W. A. & Ernst, F. F. 1989. Field scale transport of bromide in an unsaturated soil. 1. Experimental methodology and results. *Water Resources Research* **25**, 1575 - 1581.

Gelhar, L. W. 1986. Stochastic subsurface hydrology from theory to applications. *Water Resources Research* **22**, 135S - 145S.

Hofmeister, S. 1989. Stoff- und Energiebilanzen: Zur Eignung des physischen Bilanz-Prinzips als Konzeption der Umweltplanung. *Landschaftsentwicklung und Umweltforschung* **58**, Technische Universität Berlin.

Jury, W. A. & Flühler, H. 1992. Transport of chemicals through soil: mechanisms, models, and field applications. *Advances in Agronomy* **47**, 141 - 201.

McBratney, A. B., Webster, R., McLaren, R. G. & Spiers, R. B. 1982. Regional variation of extractable copper and cobalt in the topsoil of south-east Scotland. *Agronomie* **2**, 969 - 982.

Reklaitis, G. V. 1983. *Introduction to material and energy balances*. John Wiley & Sons, New York, 683 p.

Roth, K., Jury, W. A., Flühler, H. & Attinger, W. 1991. Transport of chloride through an unsaturated field soil. *Water Resources Research* **27**, 2533 - 2541.

Schulin, R., van Genuchten, M. Th., Flühler, H. & Ferlin, P. 1987. An experimental study of solute transport in a stony field soil. *Water Resources Research* **23**, 1785 - 1794.

Ulrich, B. 1987. Stability, elasticity, and resilience of terrestrial ecosystems with respect to matter balance. *Ecological Studies (Springer)* **61**: 11 - 49.

van der Zee, S. E. A. T. M. & van Riemsdijk, W. H. 1987. Transport of reactive solute in spatially variable soil systems. *Water Resources Research* **23**, 2059 - 2069.

von Steiger, B. & Baccini, P. 1990. Regionale Stoffbilanzierung von landwirtschaftlichen Böden mit messbarem Ein- und Austrag. *Berichte des nationalen Forschungsprogramm 22: "Nutzung des Bodens in der Schweiz"*, Nr. **38**, Eidgenössische Forschungsanstalt für Agrikulturchemie und Umwelhygiene, Liebefeld-Bern.

von Steiger, B. & Obrist J. 1993. Available databases for the assessment of regional mass balances in agricultural land. In: R. Schulin, A. Desaules, R. Webster, and B. von Steiger (eds), *Soil Monitoring - Early Detection and Surveying of Soil Contamination and Degradation*. Birkhäuser Verlag, Basel, this volume.

Rainer Schulin
ETHZ-Institut für terrestrische Ökologie (ITÖ)
Grabenstrasse 3/11a
CH-8952 Schlieren
Switzerland

Discussion

Question by W. Jury
Your field experiment demonstrated by mass balance that preferential flow can be a dominant form of transport even though it cannot be predicted or measured. How can we help policy makers to determine the importance of this fast flux for groundwater contamination?

Answer by R. Schulin
We need a reliable estimate of the flux in a realisic setting to use as a reference. One possible approach is a large tunnel underneath a field, instrumented to capture all of the water and chemical flux. Such a device has been proposed for our future studies in Switzerland.

Question by P. Baccini
Instead of building large lysimeters we could choose a soil compartment above an isolated groundwater pool. According to your conclusion it would be easier to balance such a large system than a 'small' lysimeter.

Answer by R. Schulin
With 'natural systems' it is hardly possible to exclude leakage. Furthermore, large systems show time scales of decades, and therefore do not permit us to study variations within reasonable times.

Question by E. Matzner
Is the importance of preferential flow related to the input concentration of a pollutant? Will it be of less importance to pollutant input from the atmosphere as compared to inputs by pesticide and fertilizer applications?

Answer by R. Schulin
It might be important under both conditions. In cases where pollutants are washed into the soil by heavy rain flux estimates based on samples of slow flowing water will tend to be biased towards underestimation of the seepage flux. On the other hand, leaching of solutes from the soil matrix may more probably be overestimated.

Soil Monitoring, Monte Verità, © Birkhäuser Verlag Basel

ATMOSPHERIC DEPOSITION PROCESSES

J. L. Collett, Jr.

Dry, wet, and occult (cloud drop) deposition all contribute to the removal of compounds from the atmosphere. Many atmospheric species undergo chemical and physical transformations which affect their removal rates. Deposition rates depend on the nature of both the compounds and the surfaces at which they deposit. Both dry and occult deposition depend strongly on the nature of the surface, leading to variations in deposition fluxes over small distances. Wet deposition depends less on fine surface structure. Predictions of wet fluxes are complicated, however, by variability in cloud microphysical conditions which govern precipitation growth and, ultimately, the efficiency of pollutant scavenging by precipitation in clouds. While knowledge of the relative importance of the different deposition pathways has improved significantly for many inorganic compounds, much uncertainty remains in describing surface sinks for most organic species.

Sources, Transformations, and Sinks of Trace Atmospheric Compounds

The atmosphere contains a complex mixture of chemical compounds. In gaseous form, the mixture ranges from major compounds such as O_2 and N_2 to trace species such as NO, NO_2, O_3, H_2O_2, HNO_3, SO_2, CH_3SCH_3, CH_4, HCOOH, chlorofluorocarbons (CFCs), and a wide variety of other hydrocarbons. Many particulate compounds are also found in the atmosphere (Prospero et al., 1983). These include soil dust, pollen, sea spray, soot and fly ash, and products of gas-to-particle conversion such as NH_4NO_3, H_2SO_4, NH_4HSO_4, and a variety of hydrocarbons. Particle sizes range from sub-micrometre for particles produced by gas-to-particle conversion up to several micrometres for some mechanically generated particles including dust and sea spray.

The compounds in the atmosphere come from a variety of sources. Some are of natural origin while others are manmade. Many compounds derive from both. Pollutants emitted to the atmosphere (e.g. NO and NH_3) are termed primary pollutants; others that are produced via atmospheric reactions (e.g. O_3 and H_2SO_4) are termed secondary pollutants.

Many species undergo chemical or physical transformations or both in the atmosphere. Sulphur dioxide, for example, can be oxidized in the gas phase by hydroxyl radical to form H_2SO_4. This chemical transformation is accompanied by a phase transformation; while SO_2 resides in the gas phase, H_2SO_4 forms liquid droplets or condenses onto the surface of existing particles. Other gas phase species also can react to form particulate products. The reaction between HNO_3 and NH_3

$$HNO_3 + NH_3 \Leftrightarrow NH_4NO_3$$

is a common example. Atmospheric particles grow through a variety of mechanisms including vapour condensation, aggregation, and processing by clouds. In the latter case individual cloud drops combine individually scavenged particles and soluble gases into one larger particle released when the cloud drop evaporates. As a result of the several possible transformations, many species emitted to the atmosphere in one form eventually redeposit at the surface in an altered state. Since rates of deposition can differ greatly between gases and particles, as well as among particles of different sizes and among different gas types, such transformations greatly influence atmospheric residence times of many compounds.

There are three principal pathways for the deposition of atmospheric compounds: dry deposition, wet deposition, and occult (cloud drop) deposition. Dry deposition concerns the loss of particles and gases at the surface. Wet deposition is the removal of compounds in precipitation, either snow or rain. Occult deposition is the term frequently applied to the process of direct capture of cloud or fog drops by the surface. This often ignored pathway can contribute significantly to total pollutant deposition in mountainous or coastal regions where interception of clouds by plant canopies occurs frequently. A more detailed discussion of these three deposition pathways will comprise much of this paper.

The wide variety of chemical compounds in the atmosphere is reflected in the types of compounds which are deposited at Earth's surface. Over the past 15 to 20 years a great deal of emphasis has been placed on evaluating the deposition of acidic species, including nitric and sulfuric acids (see e.g. NAPAP, 1991). These species were targeted because of their potential adverse effects on materials, plants, and surface waters. Many studies of acid deposition also focused on deposition of NH_3 and soil dust which represent substantial reservoirs of acid neutralizing capacity in many atmospheric environments. Deposition of nitrogen-containing compounds is also of interest because of potential fertilization effects for many plant communities (Kauppi et al., 1992). Some of the studies also examined the deposition of trace metals. More recently increased efforts have been directed toward determining rates of deposition of carbon-containing species including PCBs and pesticides (Meyers and Hites, 1982; Murray and Andren, 1992).

Deposition rates depend strongly on the nature of the species being deposited. As we shall see below, meteorological influences are also important. These range from the presence or absence of precipitation to the stability of the lower atmosphere. In the case of both dry and occult deposition, the nature of the surface is quite important. Heterogeneity in surface structure leads to large differences in deposition rates, often over spatial scales of metres or less. Wet deposition, by contrast, tends to be more uniform over such small distances. While monitoring networks might be reasonably successful in depicting regional patterns of wet deposition, construction of a monitoring network to determine regional patterns of dry or occult deposition is a far more formidable task.

Deposition Pathways

Although there are common components of the three deposition pathways, a thorough understanding of the important properties affecting the relative importance of dry, occult, and wet deposition for any compound requires that we examine the individual features of each mechanism. That is done below, using deposition of HNO_3 and NO_3^- to illustrate some of the most important features in a real system. The emphasis here is on gaining an understanding of the deposition processes. By understanding the key steps influencing each deposition pathway, differences between the behaviour of the example nitrate compounds and other types of compounds exhibiting quite different physico-chemical behaviour can be addressed fairly straightforwardly. A brief overview of some of the monitoring techniques for each deposition pathway accompanies the discussion.

Dry Deposition

Dry deposition of atmospheric material involves the transport of gases and particles (including liquid aerosol droplets) from the free atmosphere to the surface. Rates of dry deposition depend upon the characteristics of the atmosphere, the properties of the depositing substance, and the nature of the surface where deposition occurs. Surfaces of interest include materials, lakes, soils, and plant canopies.

Mathematically the dry deposition flux is often represented as the product of the concentration, $c(x,y,z_1,t)$, of the depositing material at some reference height, z_1, and a term known as the deposition velocity, v_d.

$$\text{Flux} = c(x,y,z_1,t) \cdot v_d$$

The concept of a deposition velocity (length per unit time) is readily apparent when dividing the units of flux (mass per unit area per unit time) by the units of concentration (mass per unit volume). An analogy is often drawn between pollutant transport to the surface by dry deposition and the flow of electricity in a circuit. The concentration can be compared to a voltage and the flux to an electrical current. Application of Ohm's law determines a resistance to transport, analogous to an electrical resistance. This resistance is the reciprocal of the deposition velocity. Individual resistances to transport, occurring in parallel or in series, can then be modelled in a fashion completely analogous to an electrical circuit. Figure 1 illustrates the resistance analogy for dry deposition of a gas to a vegetative surface.

Three key steps are often identified in the transport of a gas or particle to the surface. First, material must be transported through the lower atmosphere to a quasi-laminar sublayer which exists very close to the surface. This step, often referred to as the aerodynamic transport step, occurs primarily through turbulent diffusion and sedimentation. Sedimentation is generally important only for supermicrometre particles since settling velocities for smaller particles are too small to be important. The rate of turbulent diffusion depends on surface roughness, wind speed, and atmospheric stability. The second step involves transporting the substance through the quasi-laminar sublayer that exists immediately adjacent to the surface and is often termed the surface component of the transport. Several mechanisms contribute to this component of the overall transport including diffusion, inertial motion, interception, sedimentation, electrophoresis, thermophoresis, and diffusiophoresis. The final transfer step occurs at the surface itself. Once a particle reaches the surface it can physically adhere to the surface, chemically bind to the surface, or bounce off. The final step in dry deposition of a gas is the chemical reaction or physical adsorption of the gas at the surface. The physico-chemical natures of both the surface and the gas play important roles in determining the efficiency of this step. In the case of interactions with plant surfaces, opening and closing of the stomata, size of the stomatal openings, characteristics of the cuticle and mesophyll, and the presence or absence of surface moisture can significantly influence rates of dry deposition.

A substantial portion of NO_x emitted to the atmosphere is oxidized to HNO_3, primarily by hydroxyl radicals in the daytime and nitrate radicals at night. While NO and NO_2 have only modest dry deposition velocities, HNO_3 is much more chemically reactive toward many surfaces and therefore is removed rapidly from the atmosphere via dry deposition. Incorporation of HNO_3

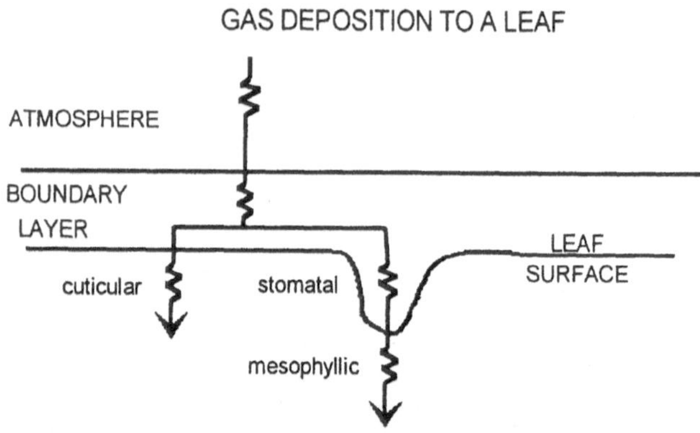

Figure 1. Conceptual illustration of several resistances to transport in the dry deposition of a gas to a leaf. Illustrated are the resistances to transport through the lower atmosphere and through the quasi-laminar boundary layer adjoining the leaf surface as well as the resistances to uptake of the gas at the surface. Note how resistances to surface uptake can occur in parallel or in series or both.

into particulate form alters its deposition velocity again. Reaction of HNO_3 at the surface of large, basic particles such as soil dust results in relatively rapid removal of the nitrate because these particles have substantial sedimentation rates. Reaction of HNO_3 with NH_3 reduces the rate of dry deposition because the resulting submicrometre particles formed by rapid growth of the initial gas-to-particle conversion products are too small to be transported rapidly through the lower atmosphere by sedimentation yet too large to experience significant transport rates from brownian diffusion. The ultimate residence times of these nitrogen-containing species clearly depend on both the oxidative capacity of the atmospheric environment to which they are emitted and the extent of HNO_3 incorporation into small and large particles.

Dry deposition rates of other compounds reflect similar considerations. Reactive, water-soluble gases tend to be removed rapidly, while nonreactive gases, such as CFCs, do not interact significantly with surfaces and therefore deposit and reside in the atmosphere for long times. Dry deposition velocities of species found primarily in particulate form, including trace metals, many organics, and sulphates, depend on the size of the particles as well as the nature of the surface and meteorological conditions. The slowest dry deposition is associated with particles in the size range from approximately 0.1 to 1 μm diameter, since both sedimentation and brownian diffusion are inefficient in this size range. The slow removal of particles in this size range has led them to be known as accumulation mode particles. They are formed largely from coagulation of smaller particles produced by condensation of hot vapours during combustion and others produced by gas-to-particle conversion in the atmosphere. Condensation of vapours on to these smaller particles in the atmosphere can also result in their growth into the accumulation mode. Trace elements emitted from combustion, such as As, Se, Cd, Pb, Cu, Ni, and V, tend to reside in these smaller particles and therefore are removed inefficiently by dry deposition. Erosion of crustal material generally produces supermicrometre particles that deposit quickly due to rapid sedimentation. Trace elements found in crustal material, such as Fe, Al, and Ca, therefore tend to be removed efficiently by dry deposition. Müller (1984) has estimated atmospheric residence times of several particulate substances under typical Middle European conditions. His findings reflect the general pattern discussed here, with residence times ranging from 0.9 days for Ca to3.0 days for Pb to 4.2 days for SO_4^{2-}. The fraction of each compound wet deposited was found to increase along with it's residence time, from only 13% for Ca to 49 % and 59% for Pb and SO_4^{2-}, respectively.

Dry deposition rates for organic material depend to a large extent on the nature of vapour-particle partitioning for individual species. Duce et al. (1983) and Bidleman (1988) have reviewed much of the information on dry deposition of these species. Substances that are of interest from a contaminant standpoint include semivolatile organic compounds such as pesticides, PCBs, and PAHs. Vapour-particle partitioning for individual species depends on the compound's vapour pressure, the total suspended particulate concentration, and ambient conditions. Small organic particles formed by vapour condensation in the atmosphere are expected to deposit slowly, by analogy with fine-particle inorganic species. Deposition rates for organic vapours depend on their affinity for surfaces. Those that are soluble or react with the surface are removed most efficiently. Low molecular weight organic acids, for example, are readily deposited onto wet surfaces and alkaline soil. Some organic species that deposit slowly, such as alkanes and alkenes, are eventually oxidized in the atmosphere to inorganic forms including CO and CO_2.

Several approaches have been used for determining rates of dry deposition. Surface analysis techniques include measurement of accumulation rates on natural or artificial surfaces. Atmospheric flux techniques include the gradient method, which utilizes vertically resolved measurements of contaminant concentrations, and eddy correlation, which involves high resolution measurements of turbulent fluctuations in wind velocity and contaminant concentrations. Each technique has its own strengths and weaknesses as discussed by Davidson and Wu (1990); however, the greatest problem associated with estimating rates of dry deposition for areas of interest is probably the tremendous influence of surface heterogeneity. Variations in plant canopy structure, for example, can affect resistances to transfer through changes in surface roughness which affect rates of turbulent transport through the lower atmosphere, changes in the nature of the quasi-laminar sublayer which affects the surface resistance, and changes in physico-chemical interactions between the contaminant and the plant surface itself. Changing meteorological conditions add yet another level of complexity as do interactions between co-depositing contaminants. Experimental measurements of dry deposition of several compounds to European ecosystems are described by several authors (Barth and Klockow, 1989; Herrmann and Jaeschke, 1989; Michaelis et al., 1989; Vermetten and Hofschreuder, 1989; Waraghai and Gravenhorst, 1989).

Occult Deposition

Cloud drops, which typically have diameters of 3 to 50 μm, are efficient scavengers of many atmospheric contaminants. When water vapour becomes supersaturated in the atmosphere, certain aerosol particles termed cloud condensation nuclei (CCN) serve as centres of heterogeneous nucleation. Water vapour deposits on the surfaces of these particles and causes them to grow. If the maximum supersaturation experienced by these particles is large enough they grow unstably into cloud drops, a process known as activation (Pruppacher and Klett, 1980). The critical supersaturation required for activation is a function of both particle size and composition. Larger, more soluble particles activate more easily. Soluble species within the CCN dissolve in the cloud drop, thus determining its initial chemical composition. This pathway for particle scavenging by cloud drops is known as nucleation scavenging. Soluble gases and non-activated aerosol particles are subsequently scavenged by the cloud drops, contributing further to their contaminant loading. Cloud drops have been observed to scavenge a significant fraction of the inorganic contaminant loading in the air in which the cloud forms (Schumann, 1991). Fewer data are available for organic species, although substantial cloudwater concentrations of several organic compounds, including pesticides, have been detected in several studies (Glotfelty et al., 1987; Munger et al., 1989a; Capel et al., 1991).

The process of cloud drop deposition is in many ways analogous to the dry deposition of supermicrometre particles (see e.g. Waldman and Hoffmann, 1987). Sedimentation contributes significantly to deposition rates for ground-based fogs. Where clouds frequently intercept plant canopies along the coast or in the mountains, wind-driven impaction and interception can lead to substantial deposition rates of both water and the contaminants contained within the cloud drops. Recent studies suggest that in such regions occult deposition contributes significantly to total pollutant deposition (Schmitt, 1989; Collett et al., 1990; Saxena and Lin, 1990). The rate of cloudwater deposition depends strongly on wind speed, the cloud drop size distribution, and the structure of the plant canopy. Highly exposed conifers on ridges directly exposed to intercepting clouds driven by strong winds probably experience the highest rates of occult deposition of contaminants (Collett et al., 1991a).

Few direct measurements of cloudwater deposition have been made. A more common approach has been to measure cloudwater contaminant concentrations, drop size distributions, canopy structure, and meteorological conditions, and to predict deposition rates based on a transfer resistance model (see e.g. Lovett, 1984). Most of the research has centred on species contributing to acid deposition. Kroll and Winkler (1989) have combined cloud chemistry measurements with model estimates of cloudwater deposition rates to estimate occult deposition for several of these species, as well as some trace metals, to coniferous forests in Germany. Some spatially resolved measurements of deposition rates to artificial collectors have been made (Collett et al., 1991a), but it is difficult to relate these directly to deposition rates to natural surfaces. As in the case of dry deposition, extrapolation of measured occult deposition rates to wider areas is plagued by surface heterogeneity. Lovett and Reiners (1986) have illustrated the importance of canopy gaps and edges in influencing rates of occult deposition.

Our example compounds, HNO_3 and NO_3^-, both participate actively in occult deposition. Soluble particles containing NO_3^- are usually scavenged efficiently by cloud drops. HNO_3 is very soluble in water and therefore is nearly completely scavenged by the drops. Recent evidence suggests that the distribution of solutes, including nitrate, in the cloud drops might not be uniform over the cloud drop size distribution (Noone et al., 1988; Munger et al., 1989b; Ogren et al., 1989; Collett et al., 1993). Since droplet deposition also depends on drop size, this evidence suggests that a simple model that multiplies total droplet flux to the surface by the average concentration of cloudwater contaminant might be inappropriate.

The participation of any species in occult deposition depends on its propensity to be scavenged by the cloud drops. Solubility is the key for gases. Participation of dissolved species in subsequent chemical reactions, such as the deprotonation of dissolved HNO_3, can enhance the pure physical solubility substantially and result in much greater partitioning to the liquid phase. The presence of surface-active organic material in cloud and fog drops has been hypothesized to result in formation of an organic surface film which can enhance partitioning of many volatile and semi-volatile organic compounds to the liquid phase. Pure organic particles are not expected to undergo nucleation scavenging, although if these particles become coated with or mixed with a soluble material in the atmosphere scavenging might occur. The presence of substantial suspended organic material in samples of fogwater suggest that this process is important. Substantial concentrations of many trace metals in cloud and fogwater are consistent with expectations for scavenging of atmospheric particles containing these compounds.

Wet Deposition

Wet deposition, the deposition of materials via snow and rain, is the most studied deposition pathway. From a monitoring perspective, it is fairly simple to measure the amount of precipitation falling on a given surface (although making accurate measurements can be difficult when it is windy, particularly for snow) and multiply that by the concentration of a contaminant in the precipitation to determine the wet deposition flux of that contaminant. Because snow crystals and raindrops fall quickly, the nature of the surface at which they deposit is less important than for occult or dry deposition. Consequently, contaminant fluxes from wet deposition tend to vary much less over short distances than fluxes from dry or occult deposition.

While monitoring wet deposition fluxes might be simple, understanding the detailed mechanisms that determine those fluxes in a given event certainly is not (Scott, 1981a). Figure 2 depicts several of the important processes influencing the composition of wet deposition. We can approach the problem most conveniently by dividing the scavenging of pollutants by precipitation into in-cloud and below-cloud regimes. Within the cloud contaminant scavenging by cloud drops, described above, is still pertinent. More complexity is added, however, when we consider the relation between the composition of the cloud drops and the precipitation. Within warm clouds precipitation is formed primarily through the collision and coalescence of cloud drops. In this situation we expect the precipitation chemistry to reflect the composition of the cloud drops involved. In most temperate latitudes, however, precipitation is generally produced in mixed-phase (ice-water) clouds where supercooled liquid cloud drops coexist with ice crystals. The ice crystals grow by three mechanisms: diffusion of water vapour to their surface (vapour deposition), inertial capture of cloud drops (accretional growth), and aggregation. The supercooled cloud drops provide a source of water vapour for vapour depositional growth. This process occurs because the saturation vapour pressure of water is greater above liquid water than above ice at the sub-freezing temperatures found in these clouds. The transfer of water vapour from cloud drops to ice crystals acts as a distillation process, concentrating contaminants in the cloud drops while producing clean snow crystals. The result is the establishment of a large gradient in contaminant concentrations between the liquid and ice in the cloud. Where conditions

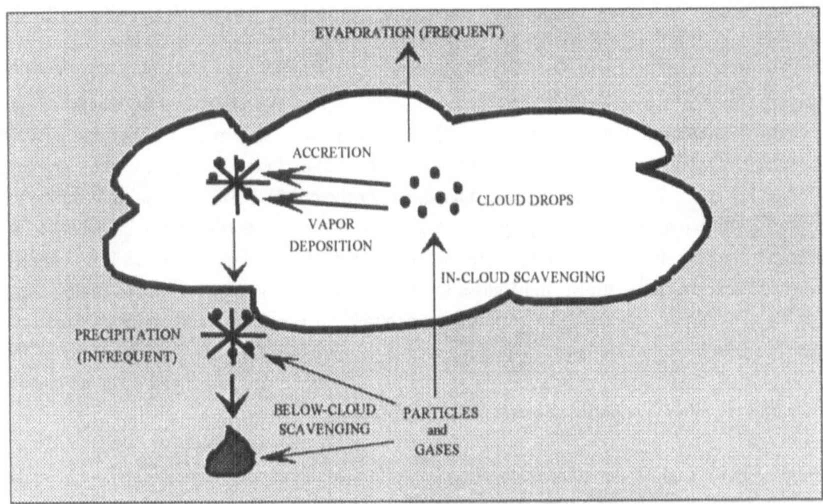

Figure 2. Some of the main processes influencing the composition of wet deposition at temperate latitudes. Most clouds evaporate without ever producing precipitation, acting primarily as processors of trace atmospheric species. See text for a discussion of the various scavenging processes. Chemical oxidation processes and direct scavenging of particles and gases by precipitation within the cloud are not depicted.

are appropriate (large cloud drops and ice crystals) accretional growth of the ice crystals can occur. The extreme example of accretional growth is formation of graupel or hail. Because many of the contaminants in the air are scavenged by the cloud drops, accretional growth efficiently transfers these contaminants to the precipitation and greatly increases contaminant fluxes in wet deposition (Scott, 1981b; Borys et al., 1988; Collett et al., 1991b). Aggregation of ice crystals alone is not very interesting chemically, although the greater fall speeds of the aggregates can increase the probability for accretional growth.

Additional contaminant scavenging by the precipitation can occur within the cloud, although the extent of accretional growth is the primary factor determining in-cloud pollutant scavenging. Significant scavenging of aerosol particles and gases can occur, however, below the cloud as the precipitation falls toward the surface. While below-cloud scavenging is often thought to be less important than the in-cloud processes, this conclusion depends greatly on the extent of accretional growth occurring within the cloud and the vertical distribution of pollutant concentrations in the atmosphere. Where accretional growth is limited or pollutants are very concentrated in layers confined below the clouds, below-cloud scavenging can make a substantial contribution to the total wet deposition flux (Zinder et al., 1988).

Precipitation can scavenge both particles and gases below the cloud. The extent of gas scavenging once again depends on the solubility of the gases. Slow mass transfer can prevent attainment of gas-liquid equilibrium when raindrops are falling through a concentration gradient (Oberholzer et al., 1992). In these cases knowledge of the raindrop size distribution is necessary to predict the extent of gas scavenging accurately. The extent of gas uptake can be enhanced by chemical sinks for the dissolved gas, as for cloud drop scavenging of gases. Much less is known about gas scavenging by snow; most gases are thought to be scavenged less efficiently by snow than by rain. Snow scavenging of particles, by contrast, is considered much more efficient than particle scavenging by raindrops. The difference is attributable to the large surface area of the snow crystals. As for dry deposition, large and small particles are scavenged most efficiently, while scavenging of accumulation mode particles is slow.

Together HNO_3 and NO_3^- participate in all aspects of precipitation pollutant scavenging. As in the preceding section, both HNO_3 and NO_3^- can be scavenged by the cloud drops. If the cloud drops are transferred to the growing hydrometeors, any species contained in the cloud drops are transferred as well. Some direct scavenging of HNO_3 and particle NO_3^- by precipitation can also occur within the cloud. Below the cloud raindrops are very efficient scavengers of the highly soluble HNO_3. Snow has also been observed to scavenge HNO_3 (Mitra et al., 1992). Particles containing NO_3^- tend to fall predominantly in the accumulation mode and therefore are not scavenged efficiently; however, some NO_3^- is often found in larger particles which are removed efficiently. In the humid environment of a rainstorm, some of the hygroscopic accumulation mode particles containing NO_3^- may grow large enough to be scavenged with greater efficiency.

Precipitation scavenging of other atmospheric contaminants depends on the phase in which they reside. Gaseous species are scavenged according to their effective solubilities. Henry's Law constants, which predict the vapour-liquid distribution of a gas at equilibrium, are strongly dependent on temperature. Variations of the constants with temperature are not well established for many organic vapours; consequently much uncertainty is present in estimating the extent to which they will be scavenged in real systems. Bidleman (1988) has reviewed what is known about the importance of gas scavenging for several organic vapours. Most of the organic particulate material in the atmosphere is contained in fine particles; a similar situation is found for many trace elements emitted from combustion processes. These particles, which often fall in the accumulation mode, are removed inefficiently by scavenging below the clouds.

Whereas the physico-chemical properties of the gaseous compounds influence stongly the extent of below-cloud scavenging, particle scavenging below the clouds depends solely on the physics of particle-particle interactions and not on particle composition. Since many organic particles are contained in the inefficiently scavenged accumulation mode, however, their water solubility still plays an important role. Hygroscopic particles grow in the humid environment below a precipitating cloud, increasing the efficiency of scavenging through inertial impaction and interception. Furthermore, hygroscopic particles are more likely to serve as CCN, providing an alternate pathway for incorporation into precipitation. Czuczwa et al. (1988) and Levsen et al. (1991) discuss organic compounds measured in precipitation collected at sites in Switzerland and Germany.

Summary and Assessment of Needs

The removal of most atmospheric contaminants is a complex function of the properties of the contaminants, the nature of surface sinks, and the meteorology of the atmosphere. Many compounds are removed through a combination of dry and wet deposition. Where clouds or fog frequently intercept the surface, occult deposition can contribute significantly to contaminant removal. Tracking the fates of contaminants emitted to the atmosphere is complicated by the numerous chemical and physical transformations that these compounds can undergo. Many compounds are oxidized in the atmosphere; several gases react to form particulate matter. The fates of even those species that occur only in particulate form can be complicated by particle coagulation, particle growth by condensation, and particle processing by clouds.

Both dry and occult deposition depend strongly on the physical structure of the surface. Dry deposition of gases also depends on the chemical nature of the surface. Spatial variability in the nature of deposition surfaces creates a tremendous problem in extrapolating dry and occult deposition measurements over even modestly sized areas. Techniques for characterizing the extent of dry or occult deposition at a given site are also unsatisfactory in many ways. The use of artificial collection surfaces, for example, can be unsatisfactory if they do not accurately represent deposition rates to natural surfaces. Alternatively, it is difficult to compare results from different areas for studies which utilize natural surfaces. Micrometeorological techniques are promising for some compounds such as O_3 and CO_2, but these techniques are expensive and suitable instrumentation is available for only a few compounds.

Wet deposition depends less on fine scale surface structure than the other deposition pathways but involves a complex series of steps which determine the efficiency of overall contaminant removal. Few studies have been able to provide a comprehensive view of the many processes that act simultaneously to determine the chemical composition of precipitation. The incorporation of cloud drops during precipitation growth, gas scavenging properties of snow crystals, vertical distributions of pollutants in the atmosphere, the temperature dependence of the solubility of many organic vapors, and the nature of partitioning between the particulate and vapour phases for many organic compounds all need further investigation.

References

Barth, S. & Klockow, D. 1989. A contribution to the experimental quantification of dry deposition to the canopy of coniferous trees, in *Mechanisms and Effects of Pollutant-Transfer into Forests.* H.-W. Georgii (ed), Kluwer Academic Publishers, Dordrecht, The Netherlands, 87-95.

Bidleman, T.F. 1988. Atmospheric processes. *Environmental Science & Technology* **22**, 361-367.

Borys, R.D., Hindman, E.E. & Demott, P.J. 1988. The chemical fractionation of atmospheric aerosol as a result of snow crystal formation and growth. *Journal of Atmospheric Chemistry* 7, 213-239.

Capel, P.D., Leuenberger, C. & Giger, W. 1991. Hydrophobic organic chemicals in urban fog. *Atmospheric Environment* **25A**, 1335-1346.

Collett, J.L., Jr., Daube, B.C., Jr. & Hoffmann, M.R. 1990. The chemical composition of intercepted cloudwater in the Sierra Nevada. *Atmospheric Environment* **24A**, 959-972.

Collett, J.L., Jr., Daube, B.C., Jr. & Hoffmann, M.R. .1991a. Spatial and temporal variations in precipitation and cloud interception in the Sierra Nevada of central California. *Tellus*, **43B**, 390-400.

Collett, J.L., Jr., Prevot, A.S.H., Staehelin, J. & Waldvogel A. 1991b. Physical factors influencing winter precipitation chemistry. *Environmental Science & Technology* **25**, 782-788.

Collett, J., Jr., Oberholzer, B. & Staehelin, J. 1993. Cloud chemistry at Mt. Rigi Switzerland: dependence on drop size and relationship to precipitation chemistry. *Atmospheric Environment* **27A**, 33-42.

Czuczwa, J., Leuenberger, C. & Giger, W. 1988. Seasonal and temporal changes of organic compounds in rain and snow. *Atmospheric Environment* **22**, 907-916.

Davidson, C.I. & Wu, Y.L. 1990. Dry deposition of particles and vapors, in *Acidic Precipitation, Volume 3: Sources, Deposition, and Canopy Interactions.* S.E. Lindberg, A.L. Page, and S.A. Norton (eds), Springer-Verlag, New York, 103-216.

Duce, R.A., Mohnen, V.A., Zimmerman, P.R., Grosjean, D., Cautreels, W., Chatfield, R., Jaenicke, R., Ogren, J.A., Pellizzari, E.D. & Wallace, G.T. 1983. Organic material in the global troposphere. *Reviews of Geophysics and Space Physics* **21**, 921-952.

Glotfelty, D.E., Seiber, J.N. & Liljedahl, L.A. 1987. Pesticides in fog. *Nature* **325**, 602-605.

Herrmann, U. & Jaeschke, W. 1989. Determination of dry deposition of gases over tree tops by measured data and a numerical model, in *Mechanisms and Effects of Pollutant-Transfer into Forests.* H.-W. Georgii (ed), Kluwer Academic Publishers, Dordrecht, The Netherlands, 13-20.

Kauppi, P.E., Mielikäinen, K. & Kuusela, K. 1992. Biomass and carbon budget of European forests, 1971 to 1990. *Science* **256**, 70-74.

Kroll, G. & Winkler, P. 1989. Trace substance input to coniferous forests via cloud interception, in *Mechanisms and Effects of Pollutant-Transfer into Forests.* H.-W. Georgii (ed), Kluwer Academic Publishers, Dordrecht, The Netherlands, 205-211.

Levsen, K., Behnert, S. & Winkeler, H.D. 1991. Organic compounds in precipitation. *Fresenius' Journal of Analytical Chemistry* **340**, 665-671.

Lovett, G.M. 1984. Rates and mechanisms of cloud water deposition to a subalpine balsam fir forest. *Atmospheric Environment* **18**, 361-371.

Lovett, G.M. & Reiners, W.A. 1986. Canopy structure and cloud water deposition in subalpine coniferous forests. *Tellus* **38B**, 319-327.

Meyers, P.A. & Hites, R.A. 1982. Extractable organic compounds in midwest rain and snow. *Atmospheric Environment* **16**, 2169-2175.

Michaelis, W., Schönburg, M. & Stössel, R.-P. 1989. Deposition of atmospheric pollutants into a north German forest ecosystem, in *Mechanisms and Effects of Pollutant-Transfer into Forests.* H.-W. Georgii (ed), Kluwer Academic Publishers, Dordrecht, The Netherlands, 3-12.

Mitra, S.K., Diehl, C. & Pruppacher, H.R. 1992. A laboratory study of the uptake of SO_2 and HNO_3 by snow crystals, in *Proceedings of the 11th International Conference on Clouds and Precipitation.* Montreal, Quebec, 851-854.

Müller, J. 1984. Atmospheric residence times of carbonaceous particles and particulate PAH-compounds. *Science of the Total Environment* **36**, 339-346.

Munger, J.W., Collett, J., Jr., Daube, B.C., Jr. & Hoffmann, M.R. 1989a. Carboxylic acids and carbonyl compounds in southern California clouds and fogs. *Tellus* **41B**, 230-242.

Munger, J.W., Collett, J., Jr., Daube, B., Jr. & Hoffmann, M.R. 1989b. Chemical composition of coastal stratus clouds: dependence on droplet size and distance from the coast. *Atmospheric Environment* **23**, 2305-2320.

Murray, M.W. & Andren, A.W. 1992. Precipitation scavenging of polychlorinated biphenyl congeners in the Great Lakes Region. *Atmospheric Environment* **26A**, 883-897.

NAPAP 1991. *National Acid Precipitation Assessment Program - 1990 Integrated Assessment Report*, Washington, DC.

Noone, K.J., Charlson, R.J., Covert, D.S., Ogren, J.A. & Heintzenberg, J. 1988. Cloud droplets: solute concentration is size dependent. *Journal of Geophysical Research* **93**, 9477-9482.

Oberholzer, B., Collett, J., Jr., Steiner, M., Staehelin, J. & Waldvogel, A. 1992. Precipitation scavenging of ammonia on a mountain slope: experimental and model comparison, in *Precipitation Scavenging and Atmosphere-Surface Exchange*, Volume 1, S.E. Schwartz and W.G.N. Slinn (eds) Hemisphere Publishing Corporation, Washington, 295-306.

Ogren, J.A., Heintzenberg, J., Zuber, A., Noone, K.J., & Charlson, R.J. 1989. Measurements of the size-dependence of solute concentrations in cloud droplets. *Tellus* **41B**, 24-31.

Prospero, J.M., Charlson, R.J., Mohnen, V., Jaenicke, R., Delany, A.C., Moyers, J., Zoller, W. & Rahn, K. 1983. The atmospheric aerosol system: an overview. *Reviews of Geophysics and Space Physics* **21**, 1607-1629.

Saxena, V.K. & Lin, N.-H. 1990. Cloud chemistry measurements and estimates of acidic deposition on an above cloudbase coniferous forest. *Atmospheric Environment* **24A**, 329-352.

Schmitt, G. 1989. Comparison of wet deposition inputs via rain and fog interception at a high elevated forest site, in *Mechanisms and Effects of Pollutant-Transfer into Forests*. H.-W. Georgii (ed), Kluwer Academic Publishers, Dordrecht, The Netherlands, 213-220.

Schumann, T. 1991. Aerosol and hydrometeor concentrations and their chemical composition during winter precipitation along a mountain slope - III. Size-differentiated in-cloud scavenging efficiencies. *Atmospheric Environment* **25A**, 809-824.

Scott, B.C. 1981a. Modeling of atmospheric wet deposition, in *Atmospheric Pollutants in Natural Waters,* Steven J. Eisenreich (ed) Ann Arbor Science Publisher, 3-21.

Scott, B.C. 1981b. Sulfate washout ratios in winter storms. *Journal of Applied Meteorology* **20**, 619-625.

Vermetten, A.W.M. & Hofschreuder, P. 1989. Deposition of gaseous pollutants in a douglas fir forest: First results of the ACIFORN project, in *Mechanisms and Effects of Pollutant-Transfer into Forests.* H.-W. Georgii (ed), Kluwer Academic Publishers, Dordrecht, The Netherlands, 61-68.

Waldman, J.M. & Hoffmann, M.R. 1987. Depositional aspects of pollutant behavior in fog and intercepted clouds. *Advances in Chemistry Series: Sources and Fates of Aquatic Pollutants* **216**, 79-129.

Waraghai, A. & Gravenhorst, G. 1989. Dry deposition of atmospheric particles to an old spruce stand, in *Mechanisms and Effects of Pollutant-Transfer into Forests.* H.-W. Georgii (ed), Kluwer Academic Publishers, Dordrecht, The Netherlands, 77-86.

Zinder, B., Schumann, T. & Waldvogel, A. 1988. Aerosol and hydrometeor concentrations and their chemical composition during winter precipitation along a mountain slope - II. Enhancement of below-cloud scavenging in a stably stratified atmosphere. *Atmospheric Environment* **22**, 2741-2750.

J. L. Collett, Jr., Institute for Environmental Studies, University of Illinois, 1101 West Peabody Drive, Urbana, IL 61801, USA

Discussion

Question by P. Baccini

1) What "screening models" do you use to assess the order of magnitude of deposition of a specific compound, and

2) is it possible to quantify, with the field methods available, total deposition in a certain region?

Answer by J. Collett

1) I would begin by determining the phase (particle or gas) that the compound exists in. If it is gaseous, I would examine its reactivity toward various surfaces. If it is a particle, I would determine the most probable particle size, since that will govern to a large extent the lifetime of the compound in the atmosphere.
2) I believe it is possible to make at least rough estimates of deposition fluxes for many compounds of interest. The method chosen for determining the fluxes should depend on the species you are interested in, the size of the target region and the degree of surface heterogeneity within the region, and the time period over which you want to calculate the flux.

Question by G. Furrer

Is it possible to estimate deposition rates originating from a single point source?

Answer by J. Collett

In theory this is possible. Making such an estimate will require detailed representations of transport, chemical and physical transformations of the compound in the atmosphere, and the deposition processes. The situation will be complicated if multiple deposition pathways act simultaneously. In practice it is difficult, and expensive, to verify these types of estimates.

Soil Monitoring, Monte Verità, © Birkhäuser Verlag Basel

FILTERING OF AIR-BORNE CONTAMINANTS BY VEGETATION CANOPIES

R. Mayer

Different approaches and methods for the assessment of deposition fluxes are discussed. At present the micrometeorological approach is too laborious for routine measurements of dry deposition in a monitoring network, and deposition models used in connection with micrometeorological measurements have been developed only for a few compounds. This approach is useful for the control and calibration of other methods in selected climatic situations. The canopy flux balance is suggested for obtaining reasonable *estimates* of dry and total deposition rates. This method applies best to tall canopies such as forests and it yields estimates for average deposition rates for longer times (minimum 1 year).

Introduction

The atmosphere has a certain capacity to store substances, and when this capacity is filled substances return to the ground. In the case of water we call this flux precipitation (snow included), but for other substances we call it deposition. Vegetation, extending its above-ground parts into the atmospheric boundary layer, changes the wind field in this layer, thus influencing deposition. We shall call this part of the vegetation, irrespective of the type of vegetation, the "canopy layer". The capability of vegetation canopies to remove atmospheric substances, in excess of the flux to a smooth, inert surface, will be called "filtering". This paper shows the possibilities and restrictions in quantifying the filtering (excess deposition) of air-borne substances by vegetation canopies.

Substances entering the canopy by deposition will either be stored temporarily or permanently in the biotic part (leaves, bark, wood, etc.), or be carried to the soil surface together with precipitation and litter fall. Assessment of the net filtering flux, as part of total atmospheric deposition, is of the greatest importance in the determination of contaminant of input to soil covered by vegetation.

Processes leading to filtering of substances

When we deal with canopy filtering we have to tackle meteorological as well as biological and ecological problems. Reviews in this field of research are given by Mayer & Ulrich (1980), Garlang *et al.* (1988), and by Fuhrer & Slanina (1988). Meteorologists distinguish two types of deposition by which substances are removed from the atmosphere and transported to land and water surfaces:

(1) wet deposition — including all material (including snow) reaching the surface with precipitation,

(2) dry deposition — including particles (dust, aerosols) and gases

Wet and dry deposition cannot always be clearly distinguished. Some authors (e.g. Unsworth & Fowler, 1988) call the capture of cloudwater by surface structures occult or hidden deposition (because it does not appear in the rain gauges used for collecting wet deposition). This is important at higher altitudes, in mountains and in areas with frequent fog. Other authors include this type of atmospheric input in either wet or dry deposition.
Dry deposition of large particles is also called sedimentation.
When assessing deposition fluxes, we have to look closely at single processes governing the removal of materials from the atmosphere.

Wet deposition processes
Before being carried down to the surface in rain, pollutants have been collected from the atmosphere either within the clouds (rainout), or below the clouds during the descent to the ground (washout). The distinction between these is arbitrary because the underlying processes are essentially the same: condensation, impaction and interception, thermo- and diffusiophoresis, Brownian diffusion and dissolution.
These processes must be considered if we discuss the residence time of materials in the atmosphere, but, for our purpose we need not to go into further details here. Wet deposition is governed by gravitational forces and turbulent transport (air movement) through the boundary layer.

Dry deposition processes
Atmospheric gases and aerosols can be transported to the surface by diffusion, gravity and air movement, but a significant net material flux to the surface develops only when the substances are filtered out and remain attached to the surface by adsorption, assimilation, and chemical reaction (Hosker & Lindberg, 1982). So the term "filtering" is linked mainly to dry deposition rather than to wet deposition.
The physical processes governing the total flux of gases to a vegetation surface have been described by Fowler (1980). The rate of dry deposition is controlled by resistances, acting in series or in parallel. It can be expressed in a form analogous to Ohm's law: flux equals

potential difference (concentration gradient) divided by resistance. Following Hough (1988), we may distinguish types of resistance:

r_a = atmospheric (or aerodynamic) resistance, depending on air turbulence within the tropospheric boundary layer,
r_b = resistance of the viscous sub-layer adjacent to the surface (with distances of the order of 1 mm), depending on surface roughness and molecular diffusivity,
r_c = canopy or surface resistances (including cuticular and stomatal resistances), depending on the affinity of the surface for the gas or aerosol concerned; when the surface is covered by water, snow or mineral particles, the surface resistance is determined by these materials.

The reciprocal of the total resistance is the deposition velocity, V_g, defined as constant vertical flux to a surface. It is obvious that r_b and r_c depend on the quality of the surface, which has to be taken into account when measuring deposition fluxes. For example, in the case of very reactive gases r_c will become very small. The stomatal component of r_c will depend upon stomatal movements and will thus show diurnal variability. Changing foliage in deciduous forests will cause seasonal variation of r_c.

The mechanisms by which particles are removed from the atmosphere depend upon particle size: large particles (> 5 μm diameter) obey more or less the same mechanisms as wet deposition ("sedimentation"), being transported through the viscous boundary sub-layer by gravity and impaction, while small particles (aerosols: < 1 μm diameter) cross it by Brownian diffusion.

Particles that have reached the surface can return to the free atmosphere by being blown or bounced off. The rate of re-suspension depends on surface structure and reactivity.

Summarizing, there are processes through which net deposition rates of atmospheric substances are influenced by the accepting surface. This influence is large for dry deposition of aerosols and gases, including the capture of cloud droplets by surface structures, described above as occult or hidden deposition. Any attempt to measure net deposition rates must therefore take into account the specific conditions at the accepting surface.

Methods for the assessment of deposition fluxes

We want methods to quantify the filtering of airborne substances by vegetation canopies. This means that we want to know the excess deposition to a specific, vegetation-covered surface compared to a smooth, inert surface at the same spot. We can express our objective in the following equation:

$$F_v = (WD_v + DD_v) - (WD_o + DD_o) \qquad (1)$$

where F_v is the filtering (excess deposition), *WD* and *DD* are wet and dry deposition onto an inert (o) and to a vegetation-covered (v) surface, respectively. All terms are expressed in mass per unit area per unit time.
Since we need this excess input from the atmosphere to survey soil contamination in a mass balance approach, we are looking for long-term averages of mass fluxes. The flux F_v must therefore represent the integral in space and time over various boundary conditions such as canopy wetness, leaf index, and stomata opening.
As can be seen from the theoretical considerations above, wet deposition to the canopy does not depend very much on acceptor properties because it is governed by gravity and turbulent transfer. Therefore, as long as the rain collector does not severely disturb the wind field, we can take

$$WD_v \approx WD_o$$

as a good approximation. Equation (1) then reduces to

$$F_v = DD_v - DD_o \tag{2}$$

For the reasons mentioned, it is fairly easy to measure the wet deposition flux to a uniform surface. Collectors are used which are open only during rain (wet-only collectors). There are several pitfalls which might influence the reproducability of wet-only measurements, but for the purpose of monitoring atmospheric input, the task can be done with sufficient accuracy. The problem becomes more complicated if vegetation forms a very rough canopy (e.g. forests with different tree species and age classes), if the surface topography is not sufficiently uniform, or if precipitation comes as fog or snow. for a review of literature and methods see Sevruk (1986). Snow collection is discussed by Glatzel *et al.* (1986), Unsworth & Fowler (1988), and by Grosch & Georgii (1989). A method of measuring atmospheric deposition via fog using a fog interceptor is given by Kroll & Winkler (1988).
Wet-only sampling requires elaborate equipment which is often not available for routine monitoring on a large scale. Very often we have only collectors which are continuously open, thus collecting not only wet deposition, but also dry deposition of large particles (sedimentation). Small particles and gases are collected by these devices according to their specific surface properties. Since these collectors are constructed in such a way that they behave as inert surfaces (in the sense expressed above) we are measuring the sum of wet and dry deposition

$$WD_o + DD_o$$

according to equation (1).

Precipitation collected by this method is usually called "bulk precipitation". As we have seen, bulk collectors are useful for comparing, for example, pollutant fluxes at different sites with a standardized collector surface, thus giving an idea of the relative input in an area.

Where we have measured WD_0 using a wet-only collector we can use the bulk collector for calculating DD_0. But this is not enough to solve equation (2) in which, for most atmospheric substances and vegetation types, the term DD_0 (dry deposition to an inert, smooth surface) is almost negligible compared to DD_v (dry deposition to canopy).

From this it follows that we must determine dry deposition to the vegetation canopy to estimate the filtering. Unfortunately, there is no substitute collector with the same acceptor properties as the vegetation canopy that could be used for direct collection of dry deposition fluxes. There have been attempts to develop such devices (for review see Galloway & Parker, 1980; also Dröscher, *et al.*, 1989), but in view of the complex structure of the canopy with rapidly changing conditions at the plant surface this is obviously impossible to achieve. Until now two general approaches have been used to determine dry deposition to vegetation surfaces, DD_v, indirectly. They are as follows.

1. Micrometeorological methods. These include physico-chemical measurements in the boundary layer above the vegetation canopy and the use of mathematical models describing the deposition processes.

2. Estimation of total deposition by using the canopy itself as the collecting surface and measuring the flux of substances below the canopy. Dry Depositon DD_v is then calculated as the difference between Total Deposition ($DD_v + WD_v$) and Bulk Deposition above the canopy ($DD_0 + WD_0$).

Micrometeorological methods for assessment of filtering

Depending upon the mathematics used to describe the transfer of the pollutant to the surface, there are several methods in use with different properties to be measured in the field. The most important are as follows.

1. The gradient method: the concentration gradient of the pollutant is measured in a vertical profile above the acceptor surface. The diffusivity of the pollutant is estimated from wind speed and temperature measurements. The deposition flux is calculated as the product of concentration gradient and diffusivity. This approach is used most frequently but it has been applied only to extensive, uniform surfaces and it requires empirical assessment of the flux-gradient relationship. This method has been used by Fowler (1980) to describe dry deposition of gaseous sulphur dioxide. The mathematical model takes into account atmospheric and surface resistances, as described above (page 90). Other examples are presented by Vermetten *et al.* (1986) for SO_2, NO_x, NH_3, and O_3, by Enders & Teichmann

(1986), by Michaelis *et al.* (1989), by Herrmann & Jaeschke (1989), and by Enders & Teichmann (1989) for the same species.

2. Eddy-correlation method: the flux of the pollutant is calculated from gas concentration measurement simultaneously with wind speed measurement, at a certain distance above the acceptor surface. The time resolution of the measurement has to be very fine (time scale 1 s or less).

The method is described by Dyer (1975) and Kaimal (1975). It is used, for example, by Enders *et al.* (1989) for determining ozone deposition.

3. Tracer methods: the deposition rate is measured as uptake of a radioactive tracer or a stable isotope by an acceptor surface. It is not used for large natural surfaces.

Hough (1988) reviews dry depositon velocities for various gaseous species and surface conditions derived using these methods. When measurements of the windfield and concentration gradients above the canopy are available the velocities can be used for calculating dry deposition flux. Since conditions in the canopy change rapidly in space and time a very dense set of data is required. This is especially true for canopies with complex architecture and a rough surface. Therefore this approach is too laborious for routine measurements in a monitoring network and it has never been used for this purpose. Also the deposition models have been developed for only few compounds. The situation is even worse for the deposition of particulate matter (aerosols).
We must therefore conclude that the micrometeorological approach does not, at present, lend itself to a monitoring routine. However, it is useful for the control and calibration of other methods in selected climatic situations. The investigations cited in the context of the gradient method are based on elaborate field installations with meteorological measuring towers, which are usually part of a large scale monitoring network with less elaborate equipment.

Assessment of filtering by use of the canopy flux balance
This method is widely used in the study of element cycling in forest ecosystems as described by Mayer & Ulrich (1974). It includes the collection of precipitation above and below the canopy (canopy drip, throughfall and stemflow) as well as the collection of the litter fall. Methods and equipment used in many European forest investigations are described by Meiwes *et al.* (1984).
Precipitation collected below the canopy contains not only substances removed from the atmosphere, but also materials leached from the interiors of the plants. These materials cannot be attributed to atmospheric input because they have been taken up by plant roots from soil. To use the canopy as a collecting surface for atmospheric substances, and to take

soil input with precipitation and litter fall as the medium in which total deposition is analyzed, it is necessary to separate external input of substances from internal leaching and root uptake. This partitioning is not easily done, as the rate of internal leaching depends upon the ionic composition of precipitation itself as well as on the nutritional status of the plants.

Many investigations show the importance of internal leaching and explain the mechanisms involved. Matzner (1986), Seufert (1986), Laitat (1987), Kreutzer (1988), Leonardi & Flückiger (1988), and Seufert (1989) have shown that acid precipitation and fog cause the leaching of cations Ca, Mg, K, and Mn. Protons in precipitation are buffered by this mechanism. On the other hand, in most regions the contribution of dry deposition of sulphate to soil input with precipitation is much more important than internal leaching.

Another complication is that some atmospheric substances are permanently retained within the plant, either being adsorbed or precipitated, or being assimilated by entering the plant's interior spaces. This is especially the case with many heavy metals, which are strongly adsorbed on organic surfaces (Godt *et al.*, 1986; Queirolo *et al.*, 1989). But the canopy may also retain protons and nutrients (Stewart *et al.*, 1987; Raven, 1987; Roelofs *et al.*, 1987; van Hove *et al.*, 1988; Kreutzer, 1988; Baumann & Baumbach, 1989). In forests most of these substances reach the ground with litter where they can be detected together (Mayer, 1985). As a consequence the assessment of dry deposition by use of the flux balance for the canopy layer is possible only for longer time (minimum 1 year).

Conclusions and recommendations

There is no method satisfactory for all (regional) conditions and chemicals. But until now the canopy flux balance approach seems the only one which yields reasonable estimates for dry and total deposition rates. This is at least true for forest vegetation, in which filtering of pollutants is most important. Bredemeier *et al.* (1988) have shown how the method can be used for assessing total deposition rates to forest ecosystems. For atmospheric nitrogen input a combination of canopy flux balance and micrometeorological methods seems to be at present the only way to arrive at reliable estimates (Grennfelt & Thörnelöf, 1992).

The interpretation of the canopy flux balance may be supported by plant physiological arguments, giving information about internal leaching and assimilation of atmospheric substances, as well as by meteorological measurements in connection with modelling (e.g. Hoefken & Gravenhorst, 1982; Waraghai & Gravenhorst, 1989; Barth & Klockow, 1989).

With regard to establishing a monitoring network, the following fluxes have to be measured continuously and be representative of the area:

1. Bulk precipitation
2. Precipitation below vegetation canopy (throughfall, canopy drip, stem flow)
3. Litter fall

There are several investigations which are very similar, and a selection of these studies is listed below.

Review of some field studies

Below is a selection of field studies in which dry deposition to vegetation canopies was investigated. Studies listed are only those for which the experimental setup has been published. The list is not complete, but it gives an idea of current research and monitoring efforts in different countries. Emphasis is on investigations on forest canopies, which are the most common. Further reviews are given in Adriano & Havas (1989) and in Bresser & Salomons (1990). For each study a short characterization of experimental design will be given.

Belgium

Location: Vielsalm, Wallonia.

Reference: Laitat & Fagot (1988).

Spruce forest, one of four stations established to study acid rain in Wallonia. Research objectives: to determine the amount of acid deposition to forests. Equipment: bulk collectors below the canopy and in the open. Methods for deposition assessment: comparison above and below canopy deposition; no further information.

France

Location: Vosges (Station laboratoire du Donon).

Reference: Biren *et al.* (1988).

Part of the French Research Programme DEFORPA. Monitoring and experimental air pollution station in the Vosges mountains, altitude 750 m. Vegetation: spruce and fir, height 29 to 34 m, in various stages of forest decline. Research objectives: to determine the amount of acid deposition with regard to height of vegetation. Equipment: permanent equipment for measurement of precipitation (wet/dry-collectors) and gaseous pollutants; meteorologic tower with 4 platforms at heights 7, 16, 30 and 44 m. Methods for deposition assessment: no information.

Location: Mont Lozère (département de Lozère, South west France).

Reference: Didon-Lescot *et al.*, (1988).

Three mountain sites (watersheds) in the Cévennes National Park, each representing a typical vegetation aspect: two forest types little or no exploitation, beech (90 years) and spruce (60 years), and a grassland (pasture with *Nardus stricta*). Altitude 1200 to 1500 m. Research objectives: monitoring air quality and deposition in a remote site (background level of atmospheric load). Equipment: wet/dry and bulk collectors below the canopy and in the open. Methods for deposition assessment: comparison above and below canopy deposition; no further information.

Germany

Location: Station Postturm, Farchau/Ratheburg forest district (near Hamburg).

Reference: Michaelis *et al.* (1989).

Spruce forest in a large forest district, about 40 km north-east of the City of Hamburg (industrial conurbation). Research objectives: Assessment of deposition rates to forests for all major organic and inorganic pollutants. Equipment: for measurement of precipitation (wet/dry-collectors) and gaseous pollutants; meteorologic tower, height 48 m, with 5 platforms for collecting and measuring various properties. Methods for deposition assessment: gradient method. Eddy correlation experiments planned.

Location: Solling (Niedersachsen).

Reference: Ellenberg (1971), Matzner & Ulrich (1984), Ellenberg *et al.* (1986), Hoefken & Gravenhorst (1982), Matzner (1989).

Beech (140 years) and spruce (100 years) at 500 m above sea level; investigations began in 1966. Research objectives: Assessment of deposition rates to forests for all major pollutants. Equipment: several sorts of collectors above and below the canopy including meteorologic towers with platforms at different heights for collecting and measuring various properties. Methods for deposition assessment: for continuous monitoring since 1970 the canopy flux balance method is used; other methods, including gradient method, will be employed eventually.

Location: State of Hessen (Darmstadt - Frankfurt - Kassel).

Reference: Hanewald (1986)

Six measuring sites in spruce forests (60-80 years) distributed over the state of Hessen; part of the state measuring programme WdI (Waldschäden durch Immissionen) and of the air quality monitoring. Research objectives: monitoring and assessment of air quality and atmospheric deposition. Equipment: bulk collectors in the open and under the canopy; meteorologic towers with 3 platforms for collecting and measuring various properties: ground level, canopy surface and 5-7 m above canopy surface. Methods for deposition assessment: comparison above and below canopy deposition; no further information.

Location: Stadtwald Frankfurt.

Reference: Herrmann & Jaeschke (1989).

Forest site (spruce, Scots pine) adjacent to the City of Frankfurt (conurbation, industry, airport). Research objectives: assessment of deposition rates to forests for gaseous compounds NO, NO_2, O_3, and SO_2. Equipment: meteorologic tower, height 51 m, with 5 platforms between 21 and 50 m above ground for collecting and measuring various properties Method for deposition assessment: gradient method.

Location: Schönbuch (near Stuttgart) and Feldberg (Schwarzwald).

Reference: Bücking & Steinle (1988).

Two research sites in lower altitude (Schönbuch: 450 m) with a beech site (height 20-35 m, 80 years) and a spruce site (height 25 to 30 m, 60 years) and one site at higher altitude

(Feldberg: 1350 m) with a quasi-natural spruce forest (120-150 years) with a heterogeneous structure. Research objectives: assessment of deposition rates as dependent upon forest type and altitude, and its small-scale local variability. Equipment: bulk collectors in the open and under the canopy. Methods for deposition assessment: comparison above and below canopy deposition; no further information.

Location: Ebersberger Forst (near München).

Reference: Enders & Teichmann (1986).

Spruce forest (90 years, average height 33 m), with very uniform canopy. Research objectives: Assessment of deposition rates to forest for gaseous NO, NO_2, O_3 and SO_2. Equipment: meteorologic tower, height 41 m, with 7 platforms for collecting and measuring various properties. Methods for deposition assessment: gradient method.

Location: Grosse Ohe/Bayrischer Wald.

Reference: Enders & Teichmann (1989).

Mixed forest (70% spruce, 28% beech and others), altitude 807 m, catchment area from 770 to 1452 m above sea level. Research objectives: assessment of deposition rates to forest for major pollutants including ozone. Equipment: for measurement of precipitation (wet/dry-collectors) and gaseous pollutants; meteorologic tower, height 51 m, with 11 platforms for collecting and measuring various properties, including eddy correlation system. Methods: gradient method, eddy correlation.

The Netherlands

Location: Komgronden nature reserve (near Waardenburg).

Reference: Heil & van Dam (1988).

Two semi-natural grassland (*Arrhenaterion elatioris*) areas in a nature reserve, surrounded by agricultural land. Research objectives: assessment of atmospheric deposition in relation to grassland canopy structure. Equipment: bulk collectors above and below grass vegetation. Methods for deposition assessment: comparison above and below canopy deposition, as described by Heil & van Dam (1986).

Location: Gelderse Vallei.

Reference: Ivens *et al.* (1988).

Forest of Douglas fir and beech, surrounded by agricultural land densely stocked with cattle. Research objectives: assessment of atmospheric deposition, especially nitrogen, and flux of above ground nitrogen uptake by trees. Equipment: bulk collectors in the open and under the canopy. Methods for deposition assessment: comparison above and below canopy deposition; no further information.

Location: Speulderbos and Kootwijk (Northern Veluwe, near Appeldoorn).

Reference: Vermetten & Hofschreuder (1989).

Two forest sites of Douglas fir (28 years, height 18-20 m; 36 years, height 16-18 m) embedded in a large forest, 25 to 50 m above sea level. Investigation in the framework of the Dutch Priority Programme on Acidification. Research objectives: monitoring and assessment of air quality and atmospheric deposition, especially nitrogen, and flux of above ground nitrogen uptake by trees. Equipment: wet-only and bulk collectors in the open and under the canopy; meteorologic tower on both sites, height 30 m, with 5 platforms for collecting and measuring various parameters. Methods for deposition assessment: comparison above below canopy deposition; gradient method using the diabatic correction functions of Dyer and Hicks.

References

Adriano, D.C., & Havas, M. (eds) 1989. Acidic Precipitation, 1, *Case Studies*, Springer-Verlag, New York.

Barth, S. & Klockow, D. 1989. A contribution to the experimental quantification of dry deposition to the canopy of coniferous trees. In: H. W. Georgii (ed.), *Mechanisms and Effects of Pollutant-Transfer into Forests*. Kluwer Academic Publishers, Dordrecht, pp. 87-95.

Baumann, K. & Baumbach, G. 1989. Determination of the gaseous air pollutant uptake of a spruce branch by means of the enclosure technique. In: H. W. Georgii (ed.), *Mechanisms and Effects of Pollutant-Transfer into Forests*. Kluwer Academic Publishers, Dordrecht, pp. 169-176.

Biren, J. M., Elichegaray, C. & Vidal, J. P. 1988. Caractérisation de l'environnement atmosphérique en zone forestière. In: P. Mathy, *Air Pollution and Ecosystems*. D. Reidel, Dordrecht, pp. 480-485.

Bredemeier, M., Matzner, E., & Ulrich, B. 1988. A simple and appropriate method for the assessment of total atmospheric deposition in forest ecosystem monitoring. In: M. H. Unsworth & D. Fowler (eds), *Acid deposition processes at high elevation sites*. Kluwer Academic Publishers, Dordrecht, pp. 607-614.

Bresser, A. H. M., & Salomons, W. (eds) 1990. *Acidic Precipitation. Vol. 5, International Overview and Assessment*. Springer-Verlag, New York.

Bücking, W. & Steinle, R. 1988. Kleinräumige Verteilungsmuster der Stoffdeposition in naturnahen Waldökosystemen. In: P. Mathy, *Air Pollution and Ecosystems*. D. Reidel, Dordrecht, pp. 486-492.

Didon-Lescot, J. F., Dejean, R., Durand, P. & Lelong, F. 1988. Les Bassins versants du Mont Lozère: un observatoire du bruit de fond de la pollution atmosphérique. In: P. Mathy, *Air Pollution and Ecosystems*. D. Reidel, Dordrecht, pp. 499-507.

Dröscher, F., Nickel, J. & Mikisch, E. 1989. Measurement of dew and fog water deposition in forest stands. In: H. W. Georgii (ed.), *Mechanisms and Effects of Pollutant-Transfer into forests*. Kluwer Academic Publishers, Dordrecht, pp. 239-247.

Dyer, A. J. 1975. Measurement of turbulent fluxes by fluxation and NIFTI techniques. *Atmospheric Technology* 7, 24-29.

Ellenberg, H. (ed.) 1971. *Integrated Experimental Ecology - Methods and Results of Ecosystem Research in the German Solling Project*. Springer Verlag, Berlin.

Ellenberg, H., Mayer, R., & Schauermann, J. (Hrsg.) 1986. *Ökosystemforschung, Ergebnisse des Sollingprojekts* 1966-1986. Verlag Eugen Ulmer, Stuttgart.

Enders, G. & Teichmann, U. 1986. Gasdep - Gaseous deposition measurements of SO_2, NO_x, and O_3 to a spruce stand: conception, instrumentation, and first results of an experimental project. In: H. W. Georgii (ed.), *Atmospheric Pollutants in Forest Areas*. D. Reidel, Dordrecht, pp. 13-24.

Enders, G. & Teichmann, U. 1989. Profiles of ozone and surface layer parameters over a mature spruce stand. In: H. W. Georgii (ed.), *Mechanisms and Effects of Pollutant-Transfer into Forests*. Kluwer Academic Publishers, Dordrecht, pp. 21-35.

Fowler, D. 1980. Wet and dry deposition of sulphur and nitrogen compounds from the atmosphere. In: T. C. Hutchinson & M. Havas. *Effects of acid precipitation on terrestrial ecosystems*. Plenum Press, New York, pp. 9-27.

Fuhrer, J. & Slanina, J. 1988. Pollution Climates in Europe: Report on Session I: Deposition in Ecosystems. In: P. Mathy, *Air Pollution and Ecosystems*. D. Reidel, Dordrecht, pp. 50-55.

Galloway, J. N. & Parker, G. G. 1980. Difficulties in measuring wet and dry deposition on forest canopies and soil surfaces. In: T. C. Hutchinson & M. Havas. *Effects of acid precipitation on terrestrial ecosystems*. Plenum Press, New York, pp. 57-75.

Garland, J. A., Nicholson, K. W., & Derwent, R. G. 1988. The deposition of trace substances from the atmosphere. In: K. Grefen & J. Löbel. *Environmental Meteorology*. Kluwer Academic Publishers, Dordrecht, pp. 141-160.

Georgii, H. W. (ed.) 1989. *Mechanisms and effects of pollutant-transfer into forests*. Kluwer Academic Publishers, Dordrecht.

Glatzel, G., Kazda, M. & Markart, G. 1986. Winter deposition rates of atmospheric trace constituents in forests: assessment of total input. In: H. W. Georgii (ed.), *Atmospheric pollutants in forest areas*. D. Reidel, pp. 101-108.

Godt, J., Schmidt, M. & Mayer, R. 1986. Processes in the canopy of trees: internal and external turnover of elements. In: H. W. Georgii (ed.), *Atmospheric Pollutants in Forest Areas*. D. Reidel, Dordrecht, pp. 263-274.

Grennfelt, P. & Thörnelöf, E. (eds) 1992. Critical loads for nitrogen - a workshop report. The Nordic Council, Copenhagen, Nord 1992: 41.

Grosch, S. & Georgii, H. W. 1989. Deposition of rain and snow - dry deposition on snow. In: H. W. Georgii (ed.), *Mechanisms and Effects of Pollutant-Transfer into Forests*. Kluwer Academic Publishers, Dordrecht, pp. 51-59.

Hanewald, K. 1986. Comparison of SO_2 concentration and immission rates with bulk precipitation data. In: H. W. Georgii (ed.), *Atmospheric Pollutants in Forest Areas*. D. Reidel, Dordrecht, pp. 25-34.

Heil, G. W. & van Dam, D. 1986. Vegetation structures and their roughness lengths with respect to atmospheric deposition. In: H. F. Hartmann (ed.), *Proceedings of the 7th World Clean Air Congress 1986*. **5**, 16-21.

Herrmann, U. & Jaeschke, W. 1989. Determination of dry deposition of gases over tree tops by measured data and a numerical model. In: H. W. Georgii (ed.), *Mechanisms and*

Effects of Pollutant-Transfer into Forests. Kluwer Academic Publishers, Dordrecht, pp. 13-20.

Hoefken, K. D. & Gravenhorst, G. 1982. Deposition of atmospheric aerosol particles to beech and spruce forest. In: H. W. Georgii & J. Pankrath, *Deposition of atmospheric pollutants*. D. Reidel, Dordrecht.

Hosker, R. P. & Lindberg, S. E. 1982. Review: atmospheric deposition and plant assimilation of gases and particles. *Atmospheric Environment* **16**, 889-910.

Hough, A. M. 1988. Atmospheric chemistry at elevated sites - a discussion of the processes involved and the limits of current understanding. In: M. H. Unsworth & D. Fowler, *Acid Deposition at High Elevation Sites*. Kluwer Academic Publishers, Dordrecht, pp. 1-47.

van Hove, I. W. A., Adema, E. H. & Vredenberg, W. J. 1988. The uptake of atmospheric ammonia by leaves. In: P. Mathy, *Air Pollution and Ecosystems*. D. Reidel, Dordrecht, pp. 734-738.

Ivens,W.P., Draaijers G. P. J. & Bleuten, W. 1988. Atmospheric nitrogen deposition in a forest next to an intensively used agricultural area. In: P. Mathy, *Air Pollution and Ecosystems*. D. Reidel, Dordrecht, pp. 536-541.

Kaimal, J. C. 1975. Sensors and techniques for direct measurement of turbulent fluxes and profiles in the atmospheric surface layer. *Atmospheric Technology* **7**, 7-14.

Kreutzer, K. 1988. Experimentelle Untersuchungen zur Stoffauswaschung aus Baumkronen durch saure Beregnung. In: P. Mathy, *Air Pollution and Ecosystems*. D. Reidel, Dordrecht, pp. 242-247.,

Kroll, G., & Winkler, P. 1988. Estimation of wet deposition via fog. In: K. Grefen & J. Löbel. *Atmospheric Meteorology*. Kluwer Academic Publishers, Dordrecht, pp. 227-236.

Laitat, E. 1987. Combined effects of ozone, precipitaiton type and acidity on young spruce trees in a simulation chamber. In: J. Bervaes (ed.). *Relationships between above and below ground influences of air polutants on forest trees*. Commission European Communities, EUR 11738 EN, Brussels, pp. 245-250.

Laitat, E. & Fagot, J. 1988. Etude du déperissement forestier en Belgique - premier bilan d'une année de récolte des précipitations à la station pilote de Vielsalm. In: P. Mathy, *Air Pollution and Ecosystems*. D. Reidel, Dordrecht, pp. 554-559.

Leonardi, S. & Flückiger, W. 1988. Indirect effects of acid mist upon the rhizosphere and the leaves' buffering capacity of beech seedlings. In: P. Mathy, *Air Pollution and Ecosystems*. D. Reidel, Dordrecht, pp. 697-700.

Matzner, E. 1986. Deposition/canopy-interaction in two forest ecosystems of northwest Germany. In: H. W. Georgii (ed.), *Atmospheric Pollutants in Forest Areas*. D. Reidel, pp. 247-262.

Matzner, E. 1989. Acidic Precipitation: Cases Study Solling. In: Adriano & Havas, (eds) *Acid precipitation case studies*. Springer Verlag New York, pp. 39-83.

Matzner, E. & Ulrich, B. 1984. Raten der Deposition, der internen Produktion und des Umsatzes von Protonen in zwei Waldökosystemen. *Zeitschrift für Pflanzenernährung und Bodenkunde*, **147**, 290-308.

Mayer, R. 1985. Verfahren zur Erfassung der Schadstoffzufuhr in Waldökosystemen. *Staub, Reinhaltung der Luft* **45**, 267-292.

Mayer, R. & Ulrich, B. 1974. Conclusions on the filtering action of forests from ecosystem analysis. *Oecologia Plantarum* **9**, 157-168.

Mayer, R. & Ulrich, B. 1980. Input to soil, especially the influence of vegetation in intercepting and modifying inputs - a review. In: T. C. Hutchinson & M. Havas, *Effects of acid precipitation on terrestrial ecosystems*. Plenum Press, New York, pp. 173-182.

Meiwes, K. J., Hauhs, M., Gerke, H., Asche, N., Matzner, E. & Lammersdorf, N. 1984. Die Erfassung des Stoffkreislaufs in Waldökosystemen - Konzept und Methodik. *Berichte des Forshcungszentrums Waldökosysteme Göttingen*, **7**, 68-142.

Michaelis, W., Schönburg, M. & Stössel, R.P. 1989. Deposition of atmospheric pollutants into a North German forest ecosystem. In: H. W. Georgii (ed.), *Mechanisms and Effects of Pollutant-Transfer into Forests*. Kluwer Academic Publishers, Dordrecht, pp. 193-202.

Queirolo, F., Valenta, P. & Stegen, S. 1989. Accumulation of heavy metals in oak wood from polluted regions. In: H. W. Georgii (ed.), *Mechanisms and Effects of Pollutant-Transfer into Forests*. Kluwer Academic Publishers, Dordrecht, pp. 193-202.

Raven, J. A. 1987. Nitrogen acquisition by shots and roots occurrence and implications for acid-base-regulation. In: J. Bervaes *et al.* (eds), *Relationships between above and below ground influences of air pollutants on forest trees*. Commisssion European Communities, EUR 11738 EN, Brussels, pp. 187-195.

Roelofs, J. G. M., Boxman, A. W., van Dijk, H. F. G. & Houdijk, A. L. F. M. 1987. Nutrient fluxes in canopies and roots of coniferous trees as affected by nitrogen-enriched air-pollution. In: J. Bervaes *et al.* (eds), *Relationships between above and below gorund influences of air pollutants on forest trees*. Commission European Commun ities, EUR 11738 EN, Brussels, pp. 205-221.

Seufert, G. 1986. Effects of long-term application of SO_2, O_3 and acid rain on mineral cycling of tree seedlings in model ecosystems. In: Commission European Communities, *Air Pollution Research Report, Direct effects of dry and wet deposition on forest ecosystems - in particular canopy interactions*. pp. 51-59.

Seufert, G. 1989. Experiments on tree canopy deposition of SO_2 and resulting leaching effects. In: H. W. Georgii (ed.), *Mechanisms and Effects of Pollutant-Transfer into Forests*. Kluwer Academic Publishers, Dordrecht, pp. 305-313.

Sevruk, B. 1986. Correction of precipitation measurements. Summary Report. (Workshop on the Correction of Precipitation Measurements, Zürich, April 1985). *Zürcher Geographische Schriften* , **23**, 13-23.

Stewart, G. R., Pearson, J. & Clough, E. 1987. Characteristics of inorganic nitrogen assimilation in woody plants. In: J. Bervaes *et al.* (eds), *Relationships between above and below ground influences of air pollutants on forest trees*. Commission European Communities, EUR 11738 EN, Brussels, pp. 222-227.

Unsworth, M. H. & Fowler, D. 1988. Deposition of pollutants on plants and soils - principles and pathways. In: P. Mathy, *Air Pollution and Ecosystems*. D. Reidel, Dordrecht, pp. 68-84.

Vermetten, A. W. M., Hofschreuder, P. & Harssema, H. 1986. Deposition of gaseous pollutants in a Douglas fir forest. In. H. W. Georgii (ed.), *Atmospheric Pollutants in Forest Areas*. D. Reidel, Dordrecht, pp. 3-11.

Vermetten, A. W. M. & Hofschreuder, P. 1989. Deposition of gaseous pollutants in a Douglas fir forest: First results of the ACIFORN project. In: H. W. Georgii (ed.), *Mechanisms and Effects of Pollutant-Transfer into Forests*. Kluwer Academic Publishers, Dordrecht, pp. 61-68.

Waraghai, A. & Gravenhorst, G. 1989. Dry deposition of atmospheric particles to an old spruce stand. In. H. W. Georgii (ed.), *Mechanisms and Effects of Pollutant-Transfer into Forests*. Kluwer Academic Publishers, Dordrecht, pp. 77-86.

R. Mayer
University of Kassel
FB 13 (Landscape Ecology)
Postfach 10 13 80
D-3500 Kassel
Germany

Discussion

Question by J. Collett

I like your canopy flux balance techniques for tall canopies. What techniques would you recommend for low canopies, e.g. agricultural fields or grasslands?

Answer by R. Mayer

It is possible to apply the same technique if the canopy is not too low (it has been applied to heathland). There is, in principle, no lower limit but for short and homogeneous vegetation other methods, such as micrometeorological techniques, can very well be adapted.

Question by H. Flühler

Are there any "inert" surfaces which can be used as reference boundaries for dry deposition?

Answer by R. Mayer

Any surface has properties which may be selective for some atmospheric constituents either by physical or by chemical sorption. Therefore, dry deposition on to inert surfaces as "standard" or reference flux cannot, by defintion, solve the problem of deposition to "specific" surfaces, since boundary conditions at this surface are changing rapidly with time (e.g. stomata opening, humidity, temperature) and are very heterogeneous in space (canopy architecture, wind field).

Soil Monitoring, Monte Verità, © Birkhäuser Verlag Basel

VOLATILIZATION OF ORGANIC CHEMICALS FROM SOIL

William A. Jury

Description of the volatilization of organic chemicals from soil involves combining the model for the transport of chemical mass through the soil volume with an expression describing mass exchange with the atmosphere. The latter expression may be approximated by a zero concentration condition for extremely volatile chemicals evaporating into a well-mixed atmosphere, but in general must be described by a transfer coefficient representing the effect of a stagnant air boundary layer at the soil surface. Simple linear models have been developed to describe volatilization losses for chemicals initially placed at the soil surface, or initially placed in the subsurface with a clean layer of soil over them. These models are useful for screening, but might not include all of the processes governing chemical volatilization in the natural environment. This paper outlines the general theory and discusses specific approaches useful for screening and simulation modelling, and describes the main approaches used for monitoring volatile emissions from soil. The paper concludes with a brief discussion of the major research problems remaining in the description and measurement of volatilization from soil.

Introduction

Chemical volatilization from soil has become an important research topic in several different scientific disciplines. Agricultural researchers have studied volatilization of both pesticides and ammonia vapour, mainly from the perspective of the loss of chemical from the target zone, but recently also because of the potential threat to health posed by pesticide residues in the atmosphere (Glotfelty *et al.* 1987). Evolution of nitrous oxide and dinitrogen gases during denitrification has also been investigated extensively, principally to evaluate the loss of nitrate from the soil, but also to determine the magnitude of atmospheric loading of nitrous oxide. Volatilization of organic contaminants that are spilled or stored in the soil is now a major research interest within the engineering community, because of the potential health threat posed to people living near the point of release.

Much of the early research on volatilization was experimental. Comprehensive studies of pesticide volatilization from soil as a function of soil and environmental conditions have been performed by Spencer and co-workers (Cliath and Spencer, 1971; Spencer and Cliath 1970, 1973, 1974; Spencer *et al.* 1988). A summary of the extensive field information on volatilization has been provided by Glotfelty *et al.* (1984) and Jury *et al.* (1987).

Jury *et al.* (1983; 1984abc; 1990) developed a screening model for pesticides and volatile organics at small concentrations in which a chemical was placed in a soil compartment and was allowed to volatilize, degrade or leach under different environmental scenarios. This model differed from earlier efforts (Jury *et al.* 1980) by the inclusion of a stagnant air boundary layer at the soil surface, which is required to avoid overprediction of volatilization of compounds with fairly small Henry's constant that are moved upward by water evaporation (Jury *et al.* 1984b).

Volatile organic compounds that originate from spills of pure liquid can generate significant vapour pressures in the soil air (Devitt *et al.* 1987). Under such conditions convection of chemical by bulk air movement can potentially become important. Mendoza and Frind (1990ab) developed a transient description of water, chemical and air flow in soil. It described experimental data from a laboratory study far better than the simulations of a model that did not include mass flow. Their papers compared in detail the performance of diffusion-based and advection-diffusion simulations for several environmental scenarios.

The objective of this paper is to summarize the current knowledge on volatilization, beginning with the description of the simple diffusion-based systems and moving toward progressively more complicated descriptions. The paper closes with a discussion of current unresolved problems and areas of potential future research.

Transport Description for a Dilute Gas

The description begins with a discussion of the diffusion-based problem for a chemical that is not present in the soil as a separate, nonaqueous liquid. It is based on the assumption that the vapour pressure generated by the chemical is small enough that the air phase is stationary. The chemical mass balance equation may be written as (assuming one dimension for simplicity)

$$\frac{\partial C_t}{\partial t} + \frac{\partial J_s}{\partial z} + r_s = 0, \tag{1}$$

where J_s is the solute flux (mass of chemical flow per unit area per unit time), C_t is the total resident concentration (mass of chemical per volume of soil), and r_s is the rate of loss of mass per volume by reactions.

This equation is always valid regardless of the model used for the individual processes; it is merely a mathematical statement of all of the fates that a chemical can undergo inside of a volume element over a time interval. The exact form of the equation and the subspecies that are represented depend on the problem of interest.

The individual components of the mass balance equation are discussed below.

The General Chemical Storage Relation

The total chemical resident concentration C_t, which is expressed in terms of mass per soil volume, may be divided into different "phase concentrations" depending on where the chemical resides in the soil. Where there is no nonaqueous phase liquid present in the soil, the chemicals are found in gaseous, dissolved and adsorbed phases. The relationship between the total concentration and the phase concentrations is given in (2)

$$C_t = \rho_b C_a + \theta C_\ell + a C_g, \tag{2}$$

where ρ_b is soil bulk density, θ is volumetric water content, and a is volumetric air content. Each phase that is mobile must be represented by a flux equation in the mass balance equation. The adsorbed phase is assumed to be immobile in soil during the time of chemical contact with the adsorption site.

Vapour Phase Flux

Unless the partial pressure of the chemical vapour is very large, mass flow effects may be neglected, and vapour moves through the soil predominantly by diffusion. The gaseous diffusion mass flux J_g is assumed to be given by Fick's Law, modified to account for the tortuosity of the flow pathways:

$$J_g = -\xi_g(a) D_g^{air} \frac{\partial C_g}{\partial z} = -D_g^{soil} \frac{\partial C_g}{\partial z}, \tag{3}$$

where $\xi_g(a)$ is the gas phase tortuosity factor, D_g^{air} is the binary diffusion coefficient of the chemical in air, and D_g^{soil} is soil gas diffusion coefficient. The tortuosity factor $\xi_g(a)$ accounts for increased path length and decreased cross-sectional area. In homogeneous soil it has been modelled satisfactorily by the Millington-Quirk Equation (Millington and Quirk, 1961):

$$\xi_g(a) = \frac{a^{10/3}}{\phi^2}, \tag{4}$$

where ϕ is the porosity (volume of void space per volume of soil). If the soil is fairly dry, then other representations of the tortuosity factor such as the Penman Equation (Penman, 1940)

$$\xi_g(a) = 0.66a \tag{5}$$

may also be used.

Dissolved Phase Flux

The two mechanisms by which dissolved chemical moves through soil are convection (also called advection) with flowing water, and diffusion within the solution phase. The convection term is given simply by:

$$J_{convection} = C_\ell J_w \tag{6}$$

where J_w is the water flux.

If the water flux describes the water flow path exactly then the only other transport mechanism is diffusion within the dissolved phase. However, since the precise details of the soil geometry are never known, only a volume-averaged expression for the water flux is given. As a result, the small-scale local convection of chemical is not included in the volume-averaged expression in (6). The remaining term must be reinserted as a hydrodynamic dispersion term.

The most common expression used to describe convection, diffusion and dispersion of chemical is the convection-dispersion flux equation:

$$J_\ell = -D_e \frac{\partial C_\ell}{\partial z} + C_\ell J_w, \tag{7}$$

where D_e is the effective diffusion and dispersion coefficient, assumed to be constant in homogeneous soil. According to the assumptions of the convection-dispersion model, the dispersion process is mathematically equivalent to diffusion and therefore the coefficient D_e may be decomposed into

$$D_e = D_h + D_d, \tag{8}$$

where D_h is the hydrodynamic dispersion coefficient and D_d is the diffusion coefficient for the chemical in soil solution.

The diffusion component of the expression above may also be modelled by the Millington and Quirk tortuosity equation, this time in the form

$$D_d = \xi_\ell(\theta) D_\ell^{water}, \tag{9}$$

where $\xi_\ell(\theta)$ is the liquid phase tortuosity factor, which is related to the volumetric water content by

$$\xi_\ell(\theta) = \frac{\theta^{10/3}}{\phi^2}. \tag{10}$$

The hydrodynamic dispersion coefficient is commonly assumed to be proportional to the local water velocity

$$D_h = \lambda V = \lambda \frac{J_w}{\theta}, \tag{11}$$

where λ is the dispersivity and $V = J_w/\theta$ is the pore water velocity.

Thus, the total solute flux equation may be written in terms of the above relationships as

$$J_s = J_w C_\ell - [\lambda V + \xi_\ell(\theta) D_\ell^{water}] \frac{\partial C_\ell}{\partial z} - \xi_g(a) D_g^{air} \frac{\partial C_g}{\partial z}. \tag{12}$$

Partitioning Laws

Gaseous and Dissolved Phases: The gaseous and dissolved phases are generally assumed to be in equilibrium at all times, with Henry's Law describing the partitioning. In dimensionless form, this law may be written as

$$C_g = K_H C_\ell, \tag{13}$$

where K_H is Henry's constant.

This relationship has been shown to hold over the entire range of saturation for many compounds (Spencer and Cliath, 1970) so that Henry's constant may be estimated from the saturated vapour density and the water solubility, as

$$K_H = \frac{C_g^*}{C_\ell^*}, \tag{14}$$

where * refers to saturation. The relationship (14) has been shown to hold over the entire range of saturation for many compounds (Spencer and Cliath, 1970)

Adsorbed and Dissolved Phases: The adsorbed and dissolved phases are characterized at equilibrium by the adsorption isotherm

$$C_a = f(C_\ell), \tag{15}$$

which may be postulated in a number of nonlinear forms using the Langmuir or Freundlich models.

A special case of the Freundlich isotherm is the linear isotherm

$$C_a = K_d C_\ell \tag{16}$$

where K_d is the distribution coefficient. Equation (16) holds approximately at low concentrations for many nonionic organic compounds binding to soil minerals and organic matter (Hamaker and Thompson, 1972). Adsorption occurs primarily to organic matter surfaces unless the organic matter content is very small.

As a means of devising an adsorption index for a given compound that is independent of the soil, the distribution coefficient may be decomposed into the product of the organic C fraction f_{oc} and an adsorption coefficient per unit organic matter content K_{oc}:

$$K_d = f_{oc} K_{oc}. \tag{17}$$

The organic C partition coefficient K_{oc} is generally less spatially variable than the distribution coefficient K_d (Hamaker and Thompson, 1972).

The partitioning laws allow relations to be developed between the total concentration and the concentration in each of the phases. For example, the total concentration expressed in (2) may be written in terms of the mass stored in the other phases as follows:

$$C_t = (\rho_b + \frac{\theta}{f_{oc}K_{oc}} + \frac{aK_H}{f_{oc}K_{oc}})C_a = R_a C_a \tag{18}$$

$$C_t = \ (\rho_b f_{oc} K_{oc} + \theta + aK_H)C_\ell \ = R_\ell C_\ell \tag{19}$$

$$C_t = \ (\frac{\rho_b f_{oc} K_{oc}}{K_H} + \frac{\theta}{K_H} + a)C_g \ = R_g C_g \tag{20}$$

where R_a, R_ℓ, R_g are the adsorbed, dissolved, and gaseous phase partition coefficients, respectively.

Chemical and Biological Reaction Processes

Chemical and biological decomposition of organic compounds is affected by many soil factors such as temperature, water content, and organic carbon, and by various biological factors such as microbial species and population density (Valentine, 1986). When microbial biomass is not limiting it is common to express biological breakdown of a concentration C of a compound as a first-order reaction. If the processes of degradation are largely unknown, the overall reaction rate r_s is generally divided into first-order processes assumed to be occurring within the dissolved and sorbed phases

$$r_s = \mu_a C_a + \mu_\ell C_\ell. \tag{21}$$

Using the partition laws above, this can be expressed in terms of the total concentration as

$$r_s = (\frac{\mu_a}{R_a} + \frac{\mu_\ell}{R_\ell})C_t = \mu_E C_t, \tag{22}$$

where μ_E is the effective rate coefficient of the dissolved and sorbed phase degradation. It is related to the overall degradation half life $\tau_{1/2}$ by

$$\tau_{1/2} = \frac{\log(2)}{\mu_E} \tag{23}$$

Upper Boundary Condition

Jury *et al.* (1983) assumed that volatilization from a bare soil surface to the atmosphere was limited by gaseous diffusion through a stagnant boundary layer at the soil-air interface. The mathematical expression of this diffusion-limited transport to the atmosphere may be represented in terms of the following boundary condition:

$$J_s(0,t) = h(C_g(0,t) - C_{air}), \tag{24}$$

where h is the transfer coefficient, and C_{air} is the concentration of the gas in the free air above the stagnant boundary layer. The transfer coefficient h is actually equal to D_g^{air}/d, where d is the thickness of the air boundary layer. Jury *et al.* (1983) indicated that a thickness of 5 mm was consistent with volatilization occurring from bare soil surfaces in the known field experiments where sufficient information was available to make an estimate.

Effective Coefficients

When the partitioning laws for equilibrium processes are used to eliminate unknowns from the transport equations the entire transport and reaction process can be written in terms of the total concentration C_t. Therefore, the solute flux law (12) may be written as

$$J_s = V_E C_t - D_E \frac{\partial C_t}{\partial z}, \tag{25}$$

where

$$V_E = \frac{J_w}{R_\ell} \tag{26}$$

is the effective solute velocity, and

$$D_E = \frac{D_e}{R_\ell} + \frac{D_g^{soil}}{R_g} \tag{27}$$

is the effective liquid and gas phase diffusion-dispersion coefficient. Similarly, the mass balance can be combined with the reaction and storage laws to produce the final form of the transport equation:

$$\frac{\partial C_t}{\partial t} = D_E \frac{\partial^2 C_t}{\partial z^2} - V_E \frac{\partial C_t}{\partial z} - \mu_E C_t. \tag{28}$$

Equation (28) was used by Jury *et al.* (1983, 1990, 1992) as an environmental screening model to evaluate pesticide behaviour in agricultural scenarios. The screening model philosophy and approach are described in the next section.

Screening Models of Chemical Volatilization from Soil

Screening models for determining the fate of chemicals in the environment use idealized scenarios and simplify the processes to examine the transport and transformations of many chemicals subjected to the same conditions. They are designed to be used with limited information about the chemicals and the environmental conditions. Their primary purposes are to sort chemicals into behaviour groups and to determine their relative behaviour.

Two different volatilization screening models have been developed from this philosophy. The first is intended to evaluate pesticide fate under agricultural conditions, and the second is to determine the volatilization potential of buried chemicals. These are briefly discussed below.

Behaviour Assessment Model

The Behaviour Assessment Model published by Jury *et al.* (1983; 1984abc) was developed to assist in preliminary evaluation of pesticide losses by leaching, volatilization, and degradation under various conditions in soil, using only the basic chemical information required during pesticide registration. The specific assumptions were as follows:

- Chemical vapour obeys Fick's law (3) and dissolved chemical obeys the convection-dispersion flux equation (7).
- Chemical is placed at uniform initial concentration in a soil layer of thickness L at $t = 0$.
- Water flow J_w is steady and either upward, downward, or 0.
- Chemical volatilizes through a stagnant air layer above the surface according to (24).
- Chemical undergoes first-order degradation (22) and equilibrium linear adsorption (16) during transport.
- Environmental fate is summarized in terms of the fraction volatilized, degraded, or leached.

With these assumptions, the three phases of the chemical can be expressed in terms of the total chemical concentration, and the single transport equation (28) describes the process.

The volatilization condition (24) at the surface boundary layer may also be expressed in terms of the total concentration C_t as

$$J_s(0,t) = \left(-D_E \frac{\partial C_l}{\partial z} + V_E C_t\right)_{z=0} = H_E\left(C_l(0,t) - R_g C_{air}\right), \tag{29}$$

where $H_E = h/R_g$ is the effective transfer coefficient of the surface boundary layer.

The effective dispersion coefficient D_E, solute velocity V_E, transfer coefficient H_E, and rate coefficient μ_E are functions of the soil conditions and the pesticide chemodynamic properties K_H, K_{oc}, and $\tau_{1/2}$. Screening model calculations may then be conducted under standard soil and environmental conditions for many chemicals simultaneously to determine their relative behaviour.

Table 1 summarizes the chemodynamic properties of five common pesticides for which the screening model calculations will be demonstrated.

Two different scenarios are considered: zero water flow and steady state upward flow at 0.5 cm d^{-1}, from a sandy soil ($\rho_b = 1.5$ g cm^{-3}, $\theta = 0.25$, $f_{oc} = 0.0075$) in which the pesticide is located in a layer of 10 cm thickness at $t = 0$. This might correspond to a farming operation in which the chemical was incorporated into the soil right after application, or if it was applied in solution during an irrigation.

Tables 2 and 3 summarize the mass balance components after 10 days, expressed in terms of the percent volatilized, degraded, or remaining in the soil.

The values in Tables 2 and 3 illustrate the kinds of information provided by the behaviour assessment model. First, strongly adsorbed compounds such as DDT scarcely volatilize under either condition, whereas EDB and DBCP are quite volatile. Moreover, atrazine and 2,4-D do not volatilize when water is not flowing, but lose a significant amount to the atmosphere under continuous evaporation, because they concentrate at the soil surface and increase the vapour density there. Thus, under the same conditions DBCP can actually lose 50% of its mass to the atmosphere while DDT has almost no volatile losses, even though their Henry's constants are similar.

Table 1: Chemodynamic properties of five pesticides (Values taken from Jury *et al.* 1984b)

Pesticide	K_{oc}	K_H	$\tau_{1/2}$
	cm^3 g^{-1}	-	days
Atrazine	160	2.5 x 10^{-7}	71
2,4-D	20	5.6 x 10^{-9}	15
DBCP	130	4.0 x 10^{-3}	1000
DDT	240000	2.0 x 10^{-3}	3840
EDB	44	3.5 x 10^{-2}	3650

Table 2: Fate summary for the pesticides in Table 1 under a scenario of ten days without water evaporation

Pesticide	% Volatilized	% Degraded	% Remaining
Atrazine	0.10	9.30	90.6
2,4-D	0.00	37.0	63.0
DBCP	7.40	0.40	92.2
DDT	0.30	0.20	99.5
EDB	5.70	19.9	74.4

Table 3: Fate summary for the pesticides in Table 1 under a scenario of ten days with water evaporation at 0.5 cm d^{-1}

Pesticide	% Volatilized	% Degraded	% Remaining
Atrazine	6.00	9.10	84.9
2,4-D	5.00	36.5	58.5
DBCP	50.0	0.60	49.4
DDT	0.30	0.20	99.5
EDB	19.7	18.9	61.4

Compounds such as atrazine with smaller Henry's constants are not normally expected to volatilize significantly, and the prediction made by the behaviour assessment model of enhanced volatilization when water is evaporating required testing to confirm the phenomenon. Spencer *et al.* (1988) performed laboratory tests on prometon ($K_H = 10^{-7}$) under water evaporation and measured continually increasing volatilization fluxes over time during the 20 days of the experiment. At the conclusion of the study they found that the concentration of the chemical in the top 1 mm of soil had increased to more than 50 times the initial concentration. Under the same conditions, lindane, with greater adsorption and a much larger Henry's constant ($K_H = 1.3 \times 10^{-4}$) had a decreasing volatilization flux with time and a surface concentration at the end of the study that was near zero.

The dependence of chemical volatilization on the value of Henry's constant under water evaporation conditions had been predicted by Jury *et al.* (1984a) using the following analysis. Transport of the chemical mass to the soil surface from the lower soil profile is approximately equal to the convective flux $J_w C_\ell$, and transport away from the surface is given by hC_g, assuming that the concentration in the free air is negligible. Those compounds for which $J_w C_\ell >> hC_g$ will accumulate at the soil surface, which will increase the vapour concentration and cause the volatilization flux to increase over time. Jury *et al.* (1984a) showed that this condition is equivalent to the inequality $K_H << 10^{-4}$, so that the Henry's constant could be used to place organic chemicals in different volatilization categories.

Buried Chemical Volatilization Model

The Buried Chemical Volatilization Model, published by Jury *et al.* (1990, 1992) is a modification of the Behaviour Assessment Model, wherein the initial condition consists of a uniform concentration of contaminated soil under a layer of chemical-free soil. The contaminated layer may be either of finite or infinite thickness. The upper boundary condition (29) and the transport equation (28) are the same as in the Behaviour Assessment Model.

Several screening tests were performed on volatile organic chemicals by Jury et. al. (1990, 1992). Among them was a simple evaluation of the thickness of uncontaminated soil required to minimize volatilization losses (which they defined as $< 0.7\%$ of initial mass) under conditions of upward diffusion and simultaneous degradation without water flow.

Table 4 summarizes the environmental fate properties for several volatile organics that will be used to illustrate the screening tests. The thickness of soil L required to restrict volatilization losses to insignificant amounts depends on the relationship between the diffusive travel time to the surface and the degradation time. The formula for the cumulative volatilization loss V_∞ from a buried deposit of chemical mass M_0 (derived in Jury *et al.* (1990)) is given by

$$\frac{V_\infty}{M_0} = \exp[-L(\mu_E/D_E)^{1/2}]. \qquad (30)$$

Table 5 shows the calculated cover thicknesses required to restrict volatilization losses of the five compounds in Table 4 for two different soil types. As seen in this Table, EDB requires a significant soil cover, because of its persistence in the soil. In contrast, other compounds such as benzene degrade so rapidly that only thin layers of soil are necessary to delay the travel time sufficiently for the compound to be converted.

This calculation neglects many important effects such as upward water flow and variable water content that can influence the migration of chemical to the surface. However, it does

Table 4: Chemodynamic properties of five volatile organic compounds (Values taken from Jury *et al.* 1990 and 1992)

Compound	K_{oc}	K_H	$\tau_{1/2}$
	$cm^3\ g^{-1}$	-	d
Benzene	80	0.22	5-16
Chloroethene	56	0.44	30-180
EDB	44	0.035	3650
Toluene	98	0.28	4-22
Xylene	50	0.28	7-30

Table 5: Soil Cover Thickness (cm) Required to Restrict Volatilization Losses to < 0.7% of Mass Incorporated into Soil

Compound	Sandy Soil	Clayey Soil
Benzene	0.1 - 1.6	≈ 0
Chloroethene	16.3 - 44.5	0.1 - 0.7
EDB	1087	218
Toluene	0.1 - 3.5	≈ 0
Xylene	1.0 - 10.8	≈ 0

provide a means for identifying compounds such as EDB that pose a threat to the quality of the atmosphere above the surface, even when buried well below the surface or perhaps even dissolved in ground water. Compounds that are identified as requiring a thick cover to prevent significant loss by volatilization should be regarded as posing a serious risk if such covers are absent.

There are additional transport mechanisms that can become important when the gas is not at a dilute concentration in the soil.

Transport Description for a Dense Gas

The derivations in the previous sections assumed that the gas in the soil air was so dilute that the air pressure and density differed only insignificantly from their values in the absence of the diffusing gas. When highly volatile compounds enter the soil at concentrations near saturation, however, this assumption is not valid and a more general description is needed. The major changes required and their effects are briefly discussed below.

Mass Flow Effects

Steady Evaporation of a Gas into Air: The form of the gas flux law presented in (3) is valid only if the partial pressure of the diffusing gas is so small that it can be neglected. The most general form of Fick's Law of diffusion for gas A diffusing through gas B may be written for one dimension, using molar fluxes to simplify the subsequent derivations, as

$$N_A - X_A(N_A + N_B) = -cD_A^B \frac{dX_A}{dz} \tag{31}$$

where N_i, $i = A, B$ is the molar flux of gas i relative to stationary coordinates, $c = c_A + c_B$ is the molar concentration, D_A^B is the binary diffusion coefficient, and $X_A = c_A/c$ is the mole fraction. The second term on the left side of (31) accounts for the transport by bulk fluid motion caused by the concentration gradient of the background gas B. This equation has the same mathematical form as the pure diffusion equation (3) when it is written in a coordinate system that is referenced to the fluid motion (Bird *et al.* 1960).

An example taken from the book by Bird *et al.* (1960) is useful in illustrating the importance of the new term in the diffusion equation above. Imagine a liquid pool of substance A evaporating at the fixed location $z = 0$ into a chamber containing gas B, maintaining the equilibrium mass fraction X_{A0} there. At height L at the top of the chamber, a fixed mixture of A and B is passed over the cylinder, so that the constant mole fractions X_{AL}, X_{BL} are maintained at $z = L$. If the gases behave as an ideal mixture, and constant pressure and temperature are maintained within the chamber, then when steady flow develops in the system, c = constant everywhere. Further, in steady state the total molar flux $N_B = 0$ of the background gas must be zero everywhere, since it must vanish at the interface with the pure liquid A.

Under these conditions, the steady flux of gas A becomes

$$N_A = -\frac{cD_A^B}{1 - X_A}\frac{dX_A}{dz}, \tag{32}$$

which, for constant c, D_A^B is equivalent to

$$\frac{d\log(1 - X_A)}{dz} = \phi_1, \tag{33}$$

where ϕ_1 is a constant.

Applying the two boundary conditions

$$X_A(0) = X_{A0} \tag{34}$$

$$X_A(L) = X_{AL} \tag{35}$$

and integrating (33), we obtain finally

$$\frac{(1 - X_A)}{(1 - X_{A0})} = \left(\frac{(1 - X_{AL})}{(1 - X_{A0})}\right)^{(Z/L)}. \tag{36}$$

The concentration profile of X_A as a function of z is given in Fig. 1a as a function of the mass fraction X_{A0}, assuming that $X_{AL} \approx 0$.

The molar flux corresponding to the profile in (36) is given by

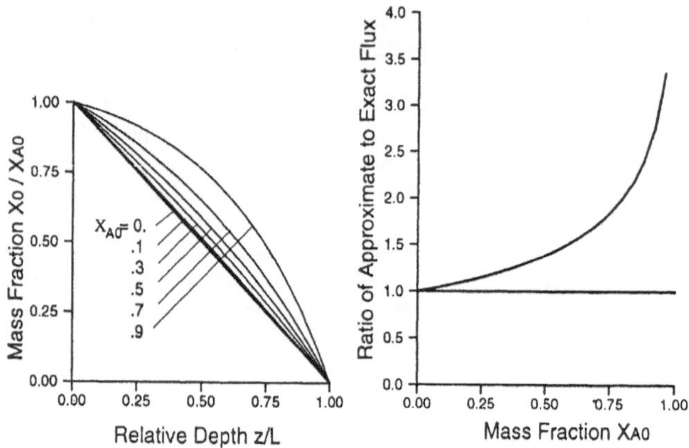

Figure 1: Concentration profiles (left) and gas flux (right) of the diffusing gas A in steady flow from a source of pure liquid into a medium of gas B. Right figure is scaled by the value of the approximate flux obtained by ignoring partial pressure effects.

$$N_A = \frac{cD_A^B}{L} \log\Big(\frac{(1 - X_{AL})}{(1 - X_{A0})}\Big), \tag{37}$$

whereas the approximate flux equation predicted by the approximate form (3) of Fick's law which neglects the concentration gradient in gas B, is given by

$$N_A \approx \frac{cD_A^B}{L}(X_{A0} - X_{AL}). \tag{38}$$

For the case $X_{AL} \approx 0$, the ratio of the exact flux (37) to the approximate flux (38) is given by

$$\frac{N_{A_{exact}}}{N_{A_{approx}}} = -\frac{\log(1 - X_{A0})}{X_{A0}}. \tag{39}$$

Equation (39) is plotted in Fig. 1b as a function of X_{A0}. This figure shows clearly that the approximate equation (38) has negligible error at small concentration of the diffusing gas, but seriously underestimates the flux when the partial pressure of the diffuser is significant.

Density Effects on Gas Transport

Several volatile organic compounds are denser than pure air. Consequently, following a spill of nonaqueous liquid, or the application of a fumigant like methyl bromide, forced convective motion arising from density gradients within the soil air will cause contaminant to be moved into the uncontaminated region, a process that may dominate diffusive transport during the early stages.

Simulation of this transient stage of contaminant transport is quite complex, because separate equations are required to describe the flow of each fluid in the soil. Sleep and Sykes (1989) and Mendoza and Frind (1990ab) have addressed this problem with numerical calculations of volatile contaminant migration, using model equations that account for the effect of the contaminant on the air density. Mendoza and Frind (1990a) define a relative density

$$\rho_{rv} = \frac{\rho_A + \rho_B}{\rho_A} = \frac{X_A M_A + X_B M_B}{M_A}, \tag{40}$$

where M_i, $i = A, B$ is the molecular weight of the pure compound. This concentration-dependent density is used in the equation to calculate the total gas flow in response to pressure gradients, which requires a pressure-driven gas transport equation of the form (Bear 1972):

$$\frac{\partial}{\partial x_i}\left[\frac{\rho_f k_{ij}^0 k_{rf}}{\mu_f}\left(\frac{\partial p_f}{\partial x_j} + \rho_f g \frac{\partial z}{\partial x_j}\right)\right] + q_f = \frac{\partial}{\partial t}(\phi \rho_f S_f), \tag{41}$$

where the subscript f represents the fluid phase of interest, x_i is spatial coordinate i , k_{ij}^0 are components of the intrinsic permeability tensor, k_{rf} is relative permeability, μ_f is the kinematic fluid viscosity, p_f is fluid pressure, q_f is a source or sink term, and S_f is the volumetric fluid saturation. Mendoza and Frind (1990ab) show that during the transient phase density-driven flow can contribute in a dominant manner to contaminant vapour transport in porous media if the concentrations of contaminant are large enough to cause substantial density gradients in the system. Figure 2, taken from their first paper, shows comparative simulations of TCE transport following a hypothetical spill of 200 L of pure liquid using models that neglect (top graph) and include (bottom graph) the advective flow contribution to vapour movement. Clearly, under these conditions the diffusion model alone is not adequate to describe the transport and fate of this environmental pollutant in soil.

Monitoring Gas Flux from Soil Surfaces

The methods for monitoring volatile emissions from the soil surface may be grouped into two categories: flux chamber methods and atmospheric flux estimate methods.

Flux Chamber Methods

A flux chamber is simply a box placed over the soil surface to trap gas as it enters the atmosphere (Rolston, 1986). The simplest type is the so-called closed chamber (Matthias *et al.* 1980), which allows the gas concentration to accumulate over time. By measuring the

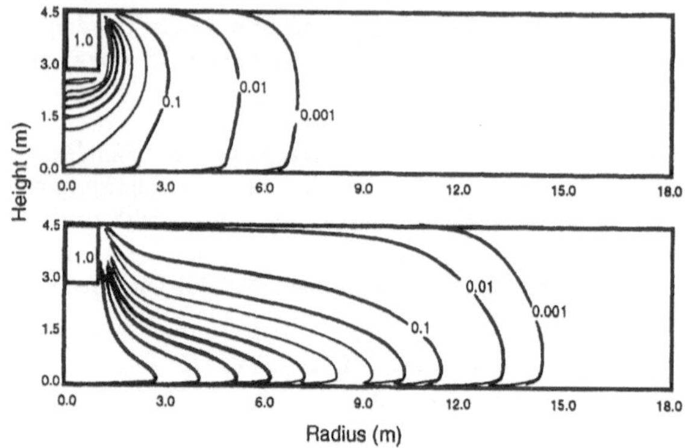

Figure 2: Calculations of TCE migration following a hypothetical spill using models that assume just diffusion and both advection-diffusion transport mechanisms (Taken from Mendoza and Frind (1990a).

chamber concentration at several times the mass entering the known air volume of the system can be calculated. The device cannot be used for long times, because the gas accumulation will cause a decrease in the gas entering the chamber compared to that escaping through the uncovered soil. In addition, the box perturbs water evaporation, wind, and energy exchange at the surface, processes which can influence volatilization. Errors can also result from incomplete mixing of the gas in the chamber.

The problem of gas accumulation in the air space can be avoided by the flow-through method, which consists of a chamber with a continuous air flow through the system. By measuring the concentration at the outlet end of the air stream and multiplying it by the flow rate, the mass loss through the surface can be estimated. The surface conditions are disturbed from their natural state as with the other method, however, and in addition the flow rate across the surface might create an added effect. Problems can be minimized by maintaining pressure and temperature close to natural levels using a large air space. Clendening and Jury (1990) used flow-through flux chambers to measure pesticide volatilization in a mass balance study conducted in the field.

Atmospheric Flux Estimates

The methods of estimating atmospheric flux all attempt to calculate the gas flux leaving the soil surface from the turbulent flux equation

$$J_g = -K_g \frac{\partial c}{\partial z}, \tag{42}$$

where J_g is the instantaneous flux rate in the upward direction, K_g is the turbulent mass transfer exchange coefficient, c the gas concentration in the air, and z is vertical position above the surface. Because J_g and K_g cannot be measured directly, indirect ways of applying (42) have been proposed.

1. **Aerodynamic Method** In the aerodynamic method, the transfer coefficient K_g is related to measurable quantities by assuming that the wind profile is logarithmic, and using empirical factors to correct for atmospheric stability. In the modified form the equation requires time-averaged measurements of gas concentration and horizontal wind speed at several heights above the surface.

2. **Energy Balance** In the energy balance method the transfer coefficient for sensible heat is measured by using a surface energy balance and the Bowen ratio method (Majewski *et al.* 1990). This coefficient is then assumed to be equal to K_g, and the volatile flux is estimated from (42) by measuring the concentration gradient in the atmosphere above the surface.

3. **Eddy Correlation** Here high frequency measurements of the product of vertical wind speed and temperature fluctuation about their mean values are made to determine a time-averaged estimate of the sensible heat flux. This is used to calculate the heat transfer coefficient which is assumed to be equal to the gas transfer coefficient. Then (42) is used as in the Energy Balance method to calculate J_g.

4. **Integrated Horizontal Flux** This method makes a downwind measurement of the product of the average horizontal wind speed and concentration over the entire vertical profile where the concentration plume is passing. The principle involved in the measurement is that all evolving vapour will leave the field laterally over a height where the sensors are located.

Each of these methods was tested in a field study by Majewski *et al.* (1990), and was found to give statistically identical estimates of the flux of four pesticides during an experiment. The authors cautioned, however, that no absolute standard of comparison was used for the flux estimates, and all methods might have been in error. They particularly questioned the assumption of identical transport coefficients for pesticide and heat, which had been shown in the past to be inaccurate at certain times in the day.

Areas of Future Research

This overview has presented both simple and complex models for transport and volatilization of organics in soil, and has described the monitoring procedures. Despite significant progress in computational representations of gas advection and multiphase flow, there are still many complications present in actual contamination events that are not represented by any models and limitations to the interpretations of monitoring signals. Significant among these limitations are:

- *Structural Effects on Diffusion.* All of the tortuosity models assume that diffusion occurs through a homogeneous soil matrix. In reality, many soils contain void spaces in the surface layer that can act as conduits for rapid release of compound. Representation of this effect may require separate characterization of flow in the two regions, and mass exchange between them. Such a model may be fairly straightforward to formulate and solve, but will contain parameters that cannot be measured independently.

- *Unstable Flow of Gas Mixtures.* The multiphase flow models do not allow for unstable flow, which can occur under a variety of conditions in real soil. With respect to the problems discussed in this paper, the presence of a dense gas overlying a less dense air phase invites unstable streaming of the dense gas through the air. Such a mechanism can dominate transport, but will produce a concentration distribution that defies prediction.

- *Nonequilibrium Dissolution of Gas into Solution.* Henry's law might not apply to those parts of the soil matrix where solution is held in isolated regions. The interphase dissolution under such conditions will be limited by diffusion, and the speed of the transfer will depend strongly on details of the local geometry that are not observable. The same restrictions may apply to the dissolution and volatilization of nonaqueous liquid trapped in soil pores after a contaminant spill.

- *Microbial Degradation of Volatile Organics.* Experimental information on the biodegradation of volatile organics in soil is very limited, and has usually been obtained in the laboratory under conditions that differ considerably from those in the natural environment. A host of problems, such as modelling microbial growth, representing degradation in different phases, and identifying diffusive limits to microbial attack on compounds, must be solved before the transport and degradation problem can be simulated quantitatively.

- *Aerodynamic Mass Transfer Measurements.* The prevailing methods all have limitations, either because of inadequate similarity in the turbulent transport of heat and mass through the air layer above the surface, or because of flux divergence or instrument response time. If the frequency with which gas concentration measurements are taken can be increased, then methods such as Eddy Correlation can be used directly on the gas flux equation.

References

Bear, J. 1972. *Dynamics of fluids in porous media.* Elsevier, N. Y.

Bird, R. B. ,Stewart, W. A. and Lightfoot, E. N. 1960. *Transport phenomena.* John Wiley & Sons, N. Y. , 780 p.

Clendening, L. D. and Jury, W. A. 1990. A pesticide field mass balance study. In: D. Kurtz (ed) *Long range transport of pesticides.* Lewis Publishers, Chelsea, Michigan.

Cliath, M. M. and Spencer, W. F. 1971. Movement and persistence of dieldrin and lindane in soil as influenced by placement and irrigation. *Soil Science Society of America Proceedings* **35,** 791-795.

Devitt, D. A., Evans, R. B. ,Jury, W. A., Starks, T. S. , and Van Ee, J. 1987. *Soil gas sensing for detection and mapping of volatile organics.* National Water Well Association. Dublin, OH, 270 p.

Glotfelty, D. E., Taylor, A. W., Turner, B. C., and Zoller, W. H. 1984. Volatilization of surface-applied pesticides from fallow soil. *Journal of Agricultural and Food Chemistry* **32,** 638-643.

Glotfelty, D. E., Seiber, J. N., and Liljedahl, L. A. 1987. Pesticides in fog. *Nature* **325,** 602.

Hamaker, J. W. and Thompson, J M. 1972. Adsorption. pp. 49-144. In C. A. I. Goring and J. W. Hamaker (eds.) *Organic chemicals in the soil environment.* Marcel Dekker. Inc., New York.

Jury, W. A., Grover, R., Spencer, W. F., and Farmer, W. J. 1980. Predicting vapor losses of soil-incorporated triallate. *Soil Science Society of America Journal* **44,** 445-450.

Jury, W. A., Spencer, W. F., and Farmer, W. J. 1983. Model for assessing behavior of pesticides and other trace organics using benchmark properties. I. Description of model. *Journal of Environmental Quality* **12,** 558-564.

Jury, W. A., Farmer, W. J, and Spencer, W. F. 1984a. Model for assessing behavior of pesticides and other trace organics using benchmark properties. II. Chemical classification and parameter sensitivity. *Journal of Environmental Quality* **13,** 567-572.

Jury, W. A., Spencer, W. F., and Farmer, W. J. 1984b. Model for assessing behavior of pesticides and other trace organics using benchmark properties. III. Application of screening model. *Journal of Environmental Quality* **13,** 573-579.

Jury, W. A., Spencer, W. F., and Farmer, W. J. 1984c. Model for assessing behavior of pesticides and other trace organics using benchmark properties. IV. Review of experimental evidence. *Journal of Environmental Quality* **13,** 580-585.

Jury, W. A., Winer, A. M., Spencer, W. F., and Focht, D. D. 1987. Transport and transformations of organic chemicals in the soil-air-water ecosystem. *Reviews of Environmental Contamination and Toxicology* **99**, 120-164.

Jury W. A., Russo, D., Streile, G., and Elabd, H. 1990. Evaluation of volatilization by organic chemicals residing below the soil surface. *Water Resources Research* **26**, 13-20.

Jury W. A., Russo, D., Streile, G., and H. Elabd, H. 1992. Correction to: Evaluation of volatilization by organic chemicals residing below the soil surface. *Water Resources Research* **28**, 607-608.

Majewski, M. S., Glotfelty, D. E., Paw U, K. T., and Seiber, J. N. 1990. A field comparison of several methods for measuring pesticide evaporation rates from soil. *Environmental Science and Technology* **24**, 1490-1497.

Matthias, A. D., Blackmer, A. M., and Bremner, J. M. 1980. A simple chamber technique for measurement of nitrous oxide emissions from soil. *Journal of Environmental Quality* **9**, 251-256.

Mendoza, C. A. and Frind, E. O. 1990a. Advective-dispersive transport of dense organic vapors in the unsaturated zone. I. Model development. *Water Resources Research* **26**, 379-387.

Mendoza, C. A. and Frind, E. O. 1990b. Advective-dispersive transport of dense organic vapors in the unsaturated zone. II. Sensitivity analysis. *Water Resources Research* **26**, 388-398.

Millington, R. J. and Quirk, J. M. 1961. Permeability of porous solids. *Transactions of the Faraday Society* **57**, 1200-1207.

Penman, H. L., 1940. Gas and vapour movements in the soil. *Journal of Agricultural Science* **30**, 437-462.

Rolston, D. E. 1986. Gas Flux. p. 1103-1119. In:A. Klute (ed) *Methods of soil analysis Vol 1. 2nd Ed.*, American Society of Agronomy Monograph 9, Madison, Wi.

Sleep, B. E. and Sykes, J. F. 1989. modelling the transport of volatile organics in variably saturated media. *Water Resources Research* **25**, 81-92.

Spencer, W. F. and Cliath, M. M. 1970. Desorption of lindane from soil as related to vapour density. *Soil Science Society of America Journal* **34**, 574-578.

Spencer, W. F. and Cliath, M. M. 1973. Pesticide volatilization as related to water loss from soil. *Journal of Environmental Quality* **2**, 284-289.

Spencer, W. F. and Cliath, M. M. 1974. Factors affecting the loss of trifluralin from soils. *Journal of Agricultural Food Chemistry* **22**, 987-991.

Spencer, W. F., Cliath, M. M., Jury, W. A., and Zhang, L.Z. 1988. Volatilization of organic chemicals from soil as related to their Henry's Law constants. *Journal of Environmental Quality* **17**, 504-509.

Valentine, R. L. , 1986. Nonbiological dcgradation in soil. p. 223-243. In: S. M. Hern and S. Meloncon (Eds.) *Vadose zone modelling of organic pollutants.* Lewis Pub., Chelsea, MI.

W. A. Jury, Department of Soil and Environmental Sciences
University of California Riverside, CA 92521, USA

Discussion

Question by A. Müller
Pesticide is usually applied to a crop or other surface vegetation rather than on bare soil. The volatilization from the plant should also be considered.

Answer by W. Jury
The volatilization from the soil is much easier to model or measure than it is from the plant, because the surfaces of the latter are not well defined. I agree that it should be considered, but it would be very difficult to put into a model of the type we use for screening.

Question by S. van der Zee
Since you can distinguish between compounds for which resistances are mainly in the soil or in the boundary layer, does that mean if the soil predominates, that monitoring should be done in the soil rather than above ground?

Answer by W. Jury
No it doesn't, because the soil resistance causes the chemical to have very steep concentration gradients near the surface which would be very difficult to measure accurately enough to estimate the flux. Also, the diffusion coefficient would have to be calculated or measured, which would create additional errors. When soil resistance predominates, the concentration falls to zero at the surface, which allows the boundary layer to be neglected for modelling purposes. That is the major simplification.

Question by R. Frossard
Do you have experience with the volatilization of PCBs? What do you know about them?

Answer by W. Jury
I haven't worked with them, but I do know that they have been found at considerable distances from where they entered the soil, so they do volatilize. Volatilization is difficult to predict because they are a mixture and do not have a single vapour pressure.

Question by J. Collett
Are there now, or will there be developed soon, instruments capable of measuring the compounds you are interested in with a time resolution sufficient for application in the eddy correlation method?

Answer by W. Jury
Such instruments do not currently exist for these compounds. It is uncertain whether they might ever be available, since very small time resolution is required. Development of such instruments is a topic for future research, as I indicated in my summary.

Soil Monitoring, Monte Verità, © Birkhäuser Verlag Basel

MONITORING OF LEACHING OF REACTIVE SOLUTES IN HETEROGENEOUS SOILS: PROBLEMS EXPECTED FROM A MODELLING PERSPECTIVE

Sjoerd E.A.T.M. van der Zee, Willem Jan P. Bosma & Hans C.L. Meeussen

Soils are intrinsically heterogeneous with regard to physical, chemical, and biological properties. To deal with the random nature of this heterogeneity stochastic approaches are necessary. Recently developed stochastic models may be adequate for predicting the average leaching rate. These models do not reveal at what places in the field leaching occurs. Hence, monitoring of leaching requires a measurement instrument with sufficient size to sample a representative volume of soil. For developing a monitoring strategy it appears necessary to give more attention to data assessment and transient modelling, as well as modelling of coupled physical, chemical and biological transport in heterogeneous domains. In particular, measurements that are relevant to quantify preferential flow may require innovative concepts.

Introduction

Reactive solute transport in soil is a complex topic of research. Because solute displacement is subject to water flow, which is a highly nonlinear and dynamic process, the mass balance for water has to be resolved first. This modelling is in itself complicated by the interactions with vegetation and evaporation which couple the soil with the atmosphere. Furthermore, soil is spatially variable in depth and in the horizontal plane with regard to, for example, the hydraulic properties.

In addition, most solutes of practical interest (e.g. nutrients, contaminants) are chemically reactive. They can sorb on to the solid phase by adsorption and precipitation, and they can be released by desorption and dissolution. In the soil solution different solutes may interact by forming dissolved or sorbing complexes. These processes, comprehensively termed speciation processes, often appear to be sensitive to conditions such as redox potential, pH, ionic strength and composition of the solution. Hence, solute transport is generally a multicomponent phenomenon. Whereas the speciation processes can be accounted for using chemical equilibrium computer packages, many of these packages do not consider chemical kinetics or physical non-equilibrium aspects. Nevertheless, these "kinetics" often dominate in unsaturated soil.

A further complication is caused by biological activity in soil. We do not have to deal only with uptake or release of solutes by plant roots. Microbial activity, e.g. denitrification, is still a poorly understood process. Generally, plant root and microbial activity is sensitive to various physical and chemical conditions. Moreover, perhaps because of the heterogeneity of these conditions and the growth of roots and microbial populations, biological activity is highly spatially variable. Because of this complexity and complications caused by initial and boundary conditions, the fate of solutes in soil has many unresolved problems. In this paper, we show that the development of a monitoring strategy can benefit from modelling. Also, we show by modelling that monitoring of fluxes (e.g. leaching) requires consideration of heterogeneity.

Monitoring total contents

The aim of soil monitoring may be restricted to the assessment of changes in the total content of one or several compounds as a function of time but improvement in the understanding of soil processes may also be desired. Usually the first aim is easier to achieve than the second as the problem posed is simpler.

Nevertheless, the assessment of temporal changes in total amounts may not be straightforward. A single approach for many compounds is rarely advised because of the difference in dynamics. Consider for instance a plough layer of 30 cm thickness and atmospheric deposition of a linearly adsorbing solute (sorption, $s = K_d \cdot c$). Then, the accumulation rate equals (Boekhold & Van der Zee, 1991)

$$\frac{dT}{dt} = A - kT \tag{1}$$

where T is the total content, t is time, A is the deposition rate and k is a parameter related to the adsorption coefficient as well as the water flux. When environmental policymakers wish to know how much is deposited (each year or decade), the answer depends on A and k. For a non-reacting compound (e.g. chloride or the radionuclide species technetate) a considerable fraction may be leached within a year (small K_d, large k). A very frequent measurement or a very thick monitored soil layer is needed in this case. But if a compound is very reactive and almost completely immobilized in soil (k small) the accumulation rate may be almost equal to the deposition rate (A). The total amount (T) as a function of time is found by integrating equation (1). This yields

$$T = \frac{A}{k}[1 - \exp(-kt)] + T_o \exp(-kt) \tag{2}$$

where T_0 is the amount at t=0 (start of monitoring). For small k values (i.e. $kt \approx 0$ the change of T may be hard to detect because equation (2) reduces to $T \approx (A/k)\,[1\text{-}1] + T_0 = T_0$. In such cases, either a smaller soil layer thickness or a larger time interval of measurement is required. This simple example shows that a sound monitoring strategy depends on relationships between input rates (A), soil as well as compound properties (all lumped in k), and the monitoring layer thickness (as depth affects the values of T, A and k). Because input rates and soil and compound properties may differ between sites, the monitoring strategy (i.e. depth of measurement, time interval) also has to differ between sites. On one site the strategy has to differ between compounds too.

A dominant problem that complicates monitoring is spatial variability (or heterogeneity). To avoid duplicating this aspect is discussed in the next section, even though it is relevant also for monitoring total contents.

Monitoring processes and transport

Monitoring soil processes intended to increase understanding of mobility (fluxes), biological availability and effects on soil (e.g. as an ecosystem or resource) is usually far more demanding than assessing changes in total contents. Generally, several physical, chemical and biological relationships are simultaneously studied. Because one of the main complications of monitoring is spatial variability, this aspect is given emphasis. The solute mass balance is given by

$$\frac{\partial \theta c}{\partial t} + \frac{\partial \rho s}{\partial t} - \nabla . (\theta D . \nabla c - \theta \upsilon \nabla c) = \Phi_c \tag{3}$$

where θ is the volumetric water fraction, ρ is the dry bulk density, c is the concentration, s is the sorbed amount, D is the dispersion tensor, υ is the flow velocity and Φ_C is a function describing uptake by biota or degradation. This mass balance equation may be solved numerically by taking transient flow and spatial variability of chemical interactions (s), flow (θ, υ) and biological interactions (Φ_C) into account. Generally, this involves a tremendous experimental parameter assessment and numerical effort. Consequently, the transport problem has been simplified considerably in most cases.

During transport solute contacts surfaces (soil matrix, roots, organisms) that have a particular capacity for sorption, uptake or degradation of the solute. Depending on the kinetics, solute loss (either temporary or irreversible) occurs. Because of this loss of solute, the concentration front may be retarded (by a factor R), the concentration may be buffered at a particular value and the solute front shape may be distorted (Van der Zee & Destouni, 1992). In particular, front retardation and concentration buffering can affect the fluxes through a control plane, e.g. at depth $z=L$. Retardation can cause the solute to arrive late at the depth, L, whereas concentration buffering limits the flux, equal to $\theta\ v\ c(L,t)$. As an example let us consider pesticide transport, where sorption can be described by the Freundlich equation ($s = K_F\ c^n$, $0<n \leq 1$) and biodegradation obeys $\Phi_C = -\mu c$, where K_F, n and μ are parameters. The retardation factor in that case equals

$$R = 1 + \frac{\rho}{\theta} K_f c^{n-1} \tag{4}$$

which is concentration dependent. Assuming that a constant concentration enters the soil surface this yields the results in Figure 1.

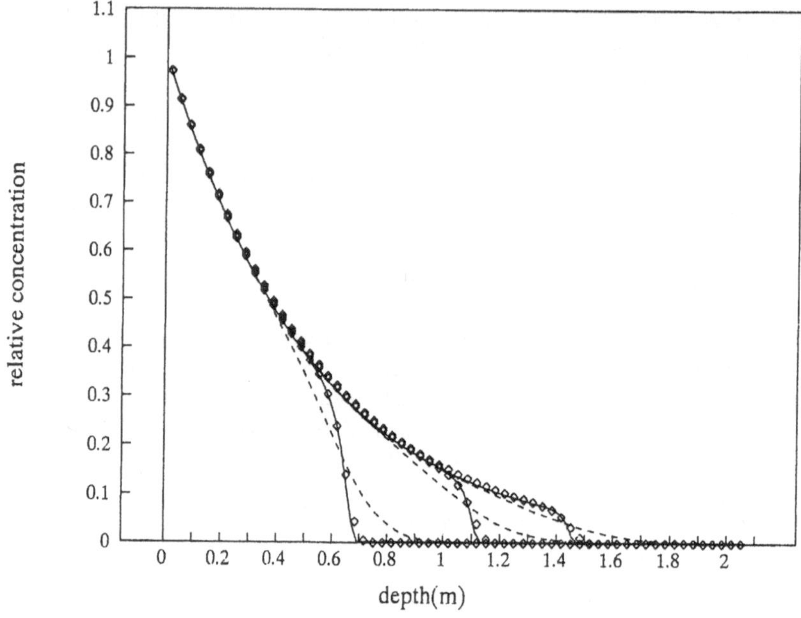

Figure 1: Fronts for three times for a solute with Freundlich sorption ($n = 0.65$) and first-order degradation. Solid line: analytical approximation, dashed line: analytical solution for $n = 1$ (at three different times as for $n = 0.65$!), symbols: numerical solution.

We see that the concentration decreases with increasing depth because of degradation. Because R depends on concentration according to equation (4), the front ($v_f = v/R$) has a decreasing velocity: during a particular time interval the downstream part of the front moves initially 0.63 m; during the next interval only 0.37 m. The effect of nonlinear sorption is apparent from the differing curves for $n = 1$. Another (hidden) effect is that the linearized solution is forced to have the same depth of the mean front as the nonlinear curves. This is only possible by calculating the linear fronts for different times as the nonlinear fronts, because the retardation factor for linear sorption is independent of concentration: if $n = 1$ the front does not decelerate in case of biodegradation.

In the field spatial variability of hydraulic properties (regulating both v and θ) and of (bio)chemical properties given by s and Φ_c may dominate transport behaviour. In the case of preferential flow, i.e. variability of v and θ we have two related mechanisms: (i) water and solute are channelled through a limited part of soil and (ii) because of the relatively fast movement through channels, most of the sorption, uptake and degradation capacity is bypassed.

The effect of preferential flow depends on how continuous preferential channels are between the soil surface and freatic water, as well as whether mass transfer into immobile zones is fast or slow. The latter case would counteract bypass flow.

In preferential flow we have a large experimental problem because it is often not known where the bypass flow will occur and how fast mass transfer into immobile zones will be. Hence, we have little *a priori* information on where to install our measurement equipment so that we obtain good approximations of the real fluxes. In addition to hydraulic variability, variability of boundary conditions and chemical and biological properties are usually important.

To understand better why measurements in the field and modelling results may be dissimilar various types of models have been developed. The most empirical is the deterministic dual porosity model used by Schulin *et al.* (1987). It distinguishes mobile and immobile zones which are related through a first-order mass transfer relation.

A type of model that is better for horizontal variability was proposed by Dagan & Bresler (1979), Jury (1983), Van der Zee & Van Riemsdijk (1987), and Destouni & Cvetkovic (1991). These models account for preferential flow by allowing infinite (vertical) correlation length for spatially variable soil properties, where soil is modelled as a bundle of parallel stream tubes. Dagan & Bresler (1979) accounted for random recharge and non-reactive transport, Jury (1983) considered linear adsorption with arbitrary correlation of K_d and the hydraulic conductivity K, Van der Zee & Van Riemsdijk (1987) took non-linear adsorption into account, and Destouni & Cvetkovic (1991) considered linear kinetic adsorption/desorption as well as heterogeneity. The more versatile approach of Destouni & Cvetkovic (1991), that predicts among others things double peaks in breakthrough, was compared by Destouni & Van der Zee (1992) with results of Van der Zee & Boesten (1991). In most cases the simple pragmatic approach by Van der Zee and Boesten (1991), to describe the fraction of applied contaminant leached into ground water for a single, instantaneous injection (of nitrate) at the soil surface, appeared to be useful (Figure 2). Van der Zee and Boesten (1991) assumed, as did the others, infinite vertical correlation lengths for the spatially variable properties. In which case the leached nitrate fraction (m_F) at depth (L) can be approximated with

$$m_F = \exp\left\{-\frac{1}{2}P_e\left[\sqrt{\left(1+\frac{4\mu L}{\upsilon P_e}\right)}-1\right]\right\} \tag{5}$$

assuming zero retardation and first order rate laws for nitrate uptake and denitrification. Variability of K, of K_d, and of the first order degradation rate constant were accounted for with a Peclet number (P_e) . The Peclet number is related to the variation coefficient v of $x = \mu RL/\upsilon$ according to

$$\text{v}^2 = 2[P^e - 1 + \exp(-P^e)]/P^{e^2}.$$

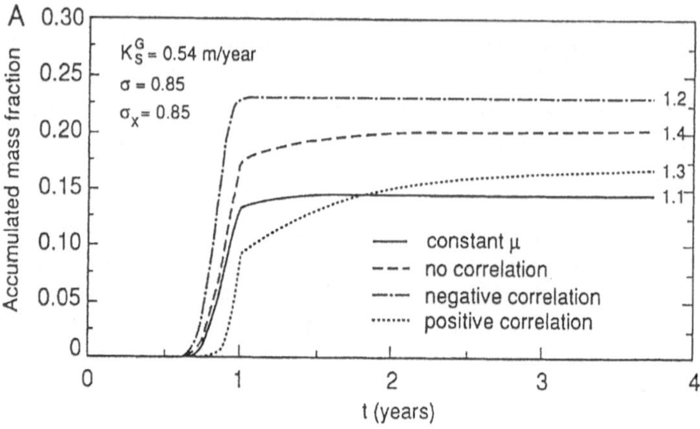

Figure 2: Leached mass fraction of nitrate as a function of time for correlated or uncorrelated random degradation rate parameter (μ) and K_S [From: Van der Zee & Destouni, 1992]. The final fractions of cases 1.1, 1.3 and 1.4 were accurately predicted also with the model of Van der Zee & Boesten [1991].

This approach for the field scale appeared to be very useful and is based on the assumption of parallel non-interacting uniform stream tubes. Although it might capture the main features for the field scale extending horizontally over many correlation scales it has its limitations. One limitation is the assumption of vertical uniformity. In reality soils are layered with mainly small vertical correlation lengths. In such cases a solute plume moving deeper might reveal a compression and expansion behaviour of the solute plume thickness (or front variance) as a function of time (Russo, 1991). Considering such a situation for K spatially variable with depth Destouni (1992) showed how such behaviour can be incorporated in the parallel stream tube model. Although multidimensional models, such as given by Bellin *et al.* (1992) for spatially variable K and K_d, or simplifications as described above, might be adequate for mean breakthrough, they do not enable us to predict where leaching will occur. This prediction is possible only when the spatial correlation length *(l)* of transport is large compared with the distances between measurements. For illustration, we calculated three dimensional vertical transport for a solute adsorbing according to the Freundlich equation, assuming only K_F is spatially variable in the horizontal plane with $l = l_T$. We generated K_F-fields of different sizes (Figures 3a, 4a) but the same degree of statistical homogeneity by keeping l_T/L constant at 0.1 (L is the size of each field).

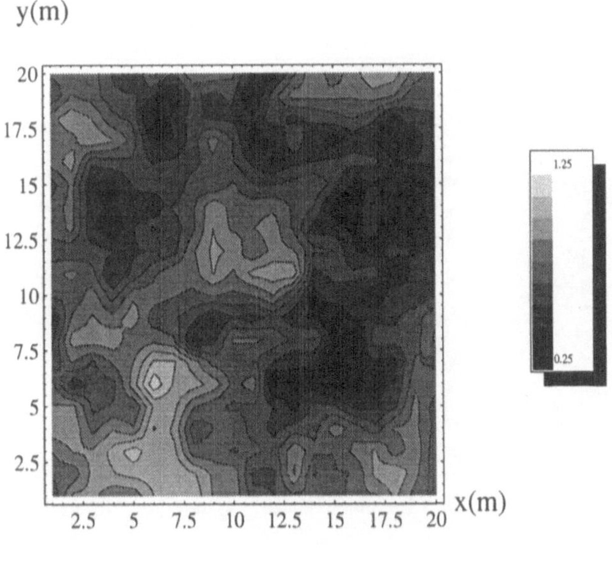

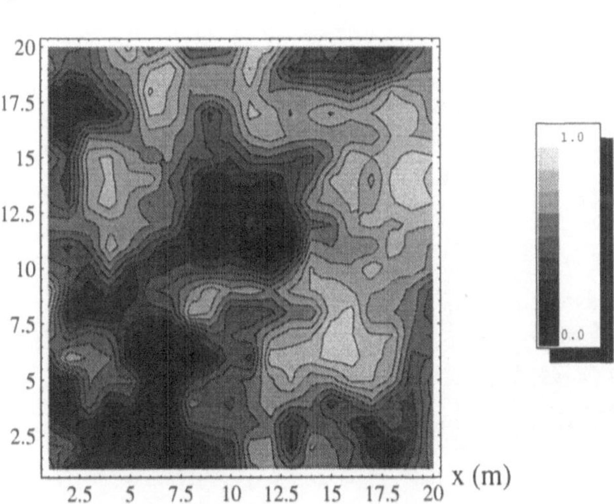

Figure 3. Contour maps of sorption coefficient (3a) and concentration at mean breakthrough at 1 m depth (3b) for A_T/l_T small (large site: distances in meters)

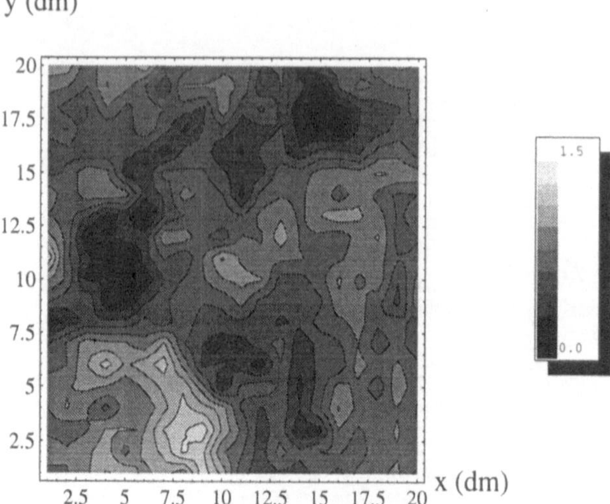

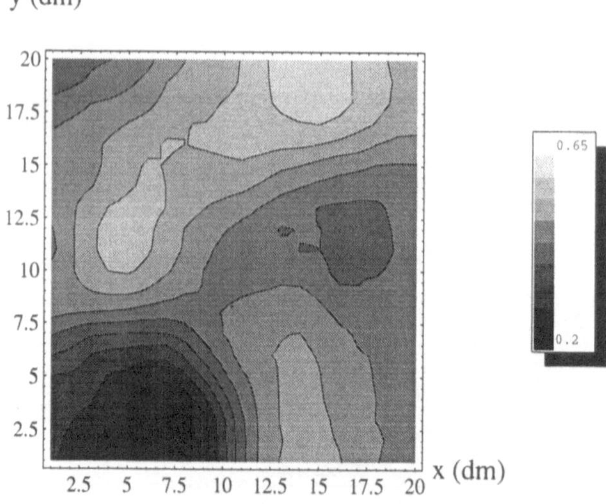

Figure 4: Contour maps of sorption coefficient (4a) and concentration at mean breakthrough at 1 m depth (4b) for A_T/l_T large (small site: distances in dm).

In both cases the transversal dispersivity was the same (A_T = 0.1 m). Transversal mixing resulted in a smoother concentration distribution at phreatic level (1 m depth) at the time of mean breakthrough for the small (L = 2 m) then for the large field (L = 20 m), as shown in Figures 3b, 4b. The breakthrough curves are almost the same as shown in Figure 5a, although the differences of the front variances (Figure 5b) indicate more dispersion in the vertical direction of flow for the larger field (A_T/l_T = 0.005). The relative correlation scale of the concentration distribution (l_T^c / L) measures where the autocorrelation (Figure 6) becomes negligible. Figure 6 suggests that fewer point measurements are needed to find the "preferential flow" paths of the small field than for the larger field for this simplified case. This is because the relative size (compared to L) of rapidly solute conducting areas is generally larger for fields with a large ratio A_T/l_T *or* l_T/L or both.

Multicomponent Transport

The previous section mentioned some of the complexity due to spatial variability of soil properties with regard to transport modelling. However, solutes are subject to a more complicated chemistry than the simple monocomponent equations mentioned so far. In multicomponent transport, solutes are considered that react with each other as well as with the solid phase. In such cases a transport equation has to be resolved for each component, taking interactions between components into account.

Transport of iron cyanide, ($Fe(CN)_6{}^{3-}$ and $Fe(CN)_6)^{4-}$, a common soil contaminant on several kinds of industrial sites, involves chemistry more intricate than that considered above. The main result of interaction between this cyanide and the soil solid phase, in acid soils, is the precipitation of iron cyanide in the form of Prussian blue, $Fe_4(Fe(CN)_6)_3(s)$, together with simultaneous dissolution of iron (hydr)oxides (Meeussen *et al.* 1990, 1992). The equilibrium between dissolved and precipitated iron cyanide can be represented as follows:

$$\mathrm{Fe_4(Fe(CN)_6)_3(s) + 12H_2O \leftrightarrows 4Fe(OH)_3(s) + 3Fe(CN)_6^{3-} + 12H^+ + 3e} \tag{7}$$

or, under more reduced conditions:

$$3Fe_4(Fe(CN)_6)_3(s) + 32H_2O \leftrightarrows 4Fe_3(OH)_8(s) + 9Fe(CN)_6^{3-} + 32H^+ + 5e \tag{8}$$

These reactions show that the solubility of Prussian blue is closely related to the chemistry of iron, and depends on both pH and redox potential due to the protons and electrons involved. Precipitation or dissolution of this mineral also influences pH and redox potential. Therefore, it is possible to use the precipitation reactions of iron cyanide and iron as a simple chemical model to describe pH, redox potential and concentrations of dissolved of iron cyanide using total amounts of iron minerals $(FeOH)_3$ and $Fe_3(OH)_8$ and Prussian blue, and initial pH as input. This chemical model was incorporated in a multicomponent transport algorithm and used to

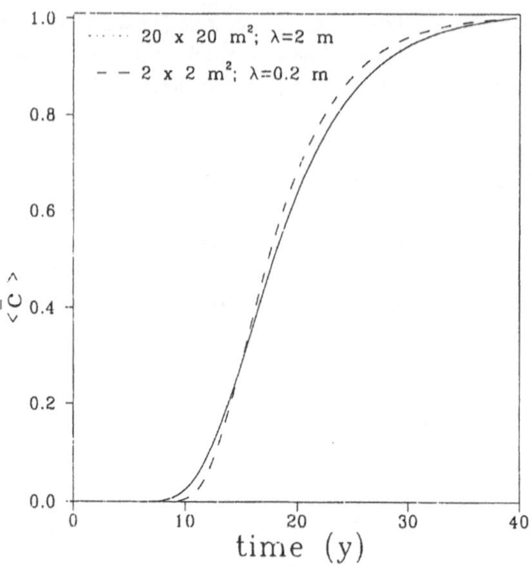

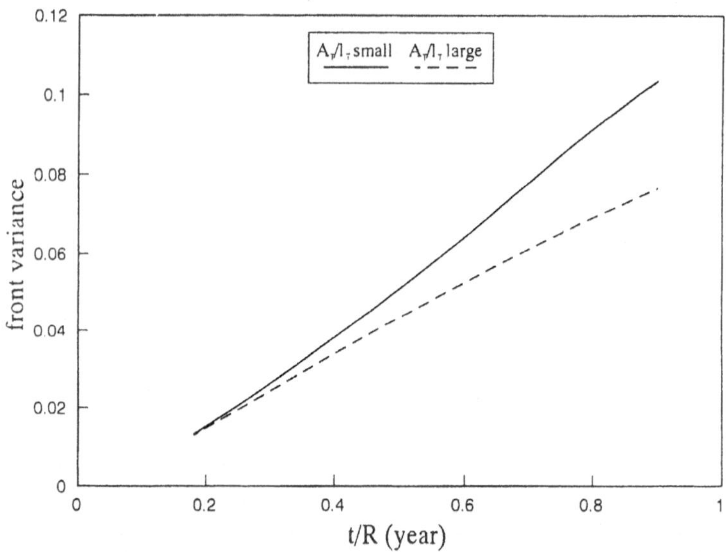

Figure 5: *Breakthrough curves (5a), averaged at 1 m depth for the horizontal plane for the case A_T/l_T is small (solid) and large (dashed) and front thickness variance which is a measure of dispersion in the mean flow direction (5b).*

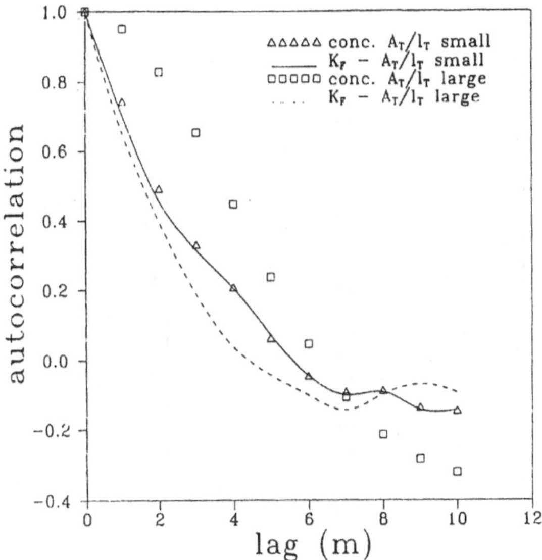

Figure 6: *Autocorrelation of concentration and K_F for the cases of Figure 3 & 4 at mean breakthrough time. For A_T/l_T large, lags should be multiplied by 0.1.*

model the transport of iron cyanide leaching from waste material at the surface in a contaminated sandy soil. The soil consisted of an oxidized layer containing only $Fe(OH)_3$ and $Fe_3(OH)_8$. The results of these calculations are shown in Figure 7. The calculated redox and pH values are close to those measured at field sites, showing accumulation of iron cyanide in the upper part of the reduced layer. This suggests that we have considered the essential chemical aspects correctly, although for example electrochemical constants are only approximately known. So this example shows that, to describe cyanide redistribution we require knowledge of pH and redox buffering, solid phase reactivity and solution composition as a function of space and time.

Likewise, Boekhold *et al.* (1990, 1993) showed that for Cd-transport organic matter content and pH are dominating properties in sandy soil. The implications are that realistic predictions of leaching require the assessment of the covariances of properties, boundary and initial conditions. Even then, only the statistical structure has been determined. It is still a problem to predict as it is *a priori* unknown which realisation of the ensemble of possible realisations is the true one.

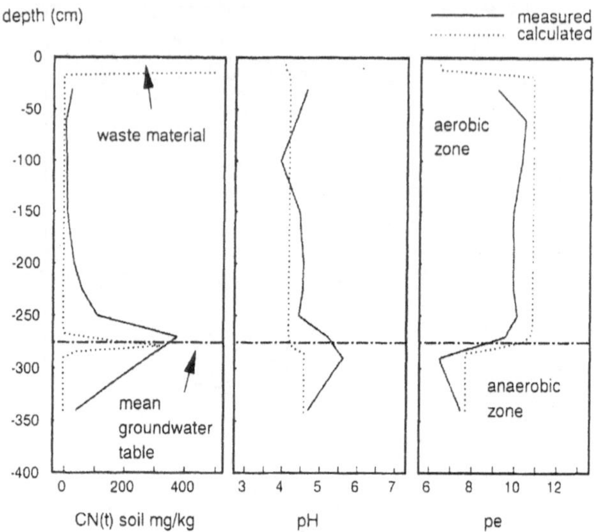

Figure 7: Distribution in depth of cyanide concentration, pH and redox potential as calculated and as measured at a representative field site.

Concluding Remarks

Soil is spatially and temporally variable. The complications arising in the development of a sound monitoring strategy are due to the complexity of the transport phenomena that underlie changes in soil composition. Spatial variability and retarding processes can obscure temporal changes in composition. Additionally, spatial variability generally necessitates either a dense network of point measurements or measurement equipment that samples large soil volumes, otherwise data may be very inaccurate. We have given examples to illustrate how mathematical modelling can help us develop a monitoring system. In particular, we showed that the spatial and temporal variability of many physical, chemical and biological properties might have to be monitored for a sound assessment of fluxes and processes in soil.

References

Bellin, A., Rinaldo, A., Bosma, W J.P. & Van der Zee, S.E.A.T.M. 1992. A study on transport of solutes in heterogeneous porous media. *Computational Methods Water Resources IX*, **2**: Mathematical modeling in water Resources (T.F. Russel *et al.*, eds), Elsevier/Computational Mechanics Publisher, 147-154.

Boekhold, A.E., Temminghoff, E.J.M. & Van der Zee, S.E.A.T.M. 1993. Influence of electrolyte composition and pH on cadmium sorption by an acid sandy soil, *Journal of Soil Science* **44**, 85-96.

Boekhold, A.E., Van der Zee, S.E.A.T.M. & de Haan, F.A.M. 1990. Prediction of cadmium accumulation in a heterogeneous soil using a scaled sorption model; In: IAHS publication no 1995, pp. 211-220.

Boekhold, A.E. & Van der Zee, S.E.A.T.M. 1991. Long term effects of soil heterogeneity on cadmium behaviour in soil. *Journal Contaminant Hydrology* **7**, 371-390.

Dagan, G. & Bresler, E. 1979. Solute dispersion in unsaturated heterogeneous soil at a field scale, 1, Theory. *Soil Science Society of America Journal* **43**, 461-467.

Destouni, G. 1992. The effect of vertical soil heterogeneity on field scale solute flux. *Water Resources Research*, **28**, 1303-1310.

Destouni, G. & Cvetkovic, V.D. 1991. Field scale mass arrival of sorptive solute into the ground water. *Water Resources Research* **27**, 1315-1325.

Jury, W.A. 1983. Chemical transport modeling: Current approaches and unresolved problems. In: *Chemical Mobility and Reactivity in Soil Systems;* Soil Science Society of America, Madison, Wisconsin. 1983.

Meeussen, J.C.L., Keizer, M.G.& van Riemsdijk, W.H. 1990. The solubility of iron cyanide complexes in soils. In: *Contaminated Soil* '90. Kluwer Academic Publishers, Dordrecht, pp. 367-374.

Meeussen, J.C.L., Keizer, M.G., van Riemsdijk, W.H. & de Haan, F.A.M. 1992. Dissolution behaviour of iron cyanide (Prussian Blue) in contaminated soils. *Environmental Science & Technology* **26**, 1832-1838.

Russo, D. 1991. Stochastic analysis of simulated vadose-zone solute transport in a vertical cross section of heterogeneous soil during unsteady flow. *Water Resources Research*, **27**, 267-283.

Schulin, R., van Genuchten, M.Th., Flühler, H. & Ferlin, P. 1987. An experimental study of solute transport in a stony field soil. *Water Resources Research* **23**, 1785-1794.

Van der Zee, S.E.A.T.M. & Destouni, G. 1992. Transport of inorganic solutes in soil. *Advances in Soil Science,* 95-146.

Van der Zee, S.E.A.T.M. & Boesten, J.J.T.I. 1991. Effects of soil heterogeneity on pesticide leaching to ground water. *Water Resources Research* **27**, 3051-3063.

Van der Zee, S.E.A.T.M. & van Riemsdijk, W.H. 1987. Transport of reactive solute in spatially variable soil systems. *Water Resources Research* **23**, 2059-2069.

Van Genuchten, M.Th. 1980. A closed-form equation for predicting the hydraulic conductivity of unsaturated soils. *Soil Science Society of America Journal* **44**, 892-898.

S.E.A.T.M. Van der Zee, W.J.P. Bosma & J.C.L. Meeussen
Department of Soil Science and Plant Nutrition,
Agricultural University
P.O. Box 8005
6700 EC Wageningen
The Netherlands.

Discussion

Question by R. Schulin

Do you think that transport models can be used to determine contaminant fluxes out of the root zone into the groundwater zone in the same way as empirically calibrated models are used in hydrology?

Answer by S. van der Zee

I think in principle yes. However, it would require many data for calibration. This is probably still a research problem which should be investigated. Generally, I see mechanistic transport models as tools for summarizing and increasing basic understanding, rather than as experimentally useful in natural (real world) conditions. However, such a fundamental mechanistic basis may subsequently be used to develop inverse models as given by Rinaldo and co-workers and these have many similarities with Jury's Transfer function model. Rinaldo's modelling approach shows the enormous potential of calibrated models of the type used in hydrology, with regard to soil, ground and surface water monitoring strategies.

Soil Monitoring, Monte Verità, © Birkhäuser Verlag Basel

THE ROLE OF SOIL EROSION IN THE MOVEMENT OF POLLUTANTS

J.N. Quinton & R.J. Rickson.

An overview is given of the changing emphasis in studying the impacts of soil erosion, from those concerned with the effects of soil erosion on soil productivity to those linking soil erosion with environmental damage. Eroded sediment can be perceived as a pollutant, but it is the chemicals adsorbed onto erodible particles that cause most damage both on-site and off-site. Knowledge of erosion processes and sediment delivery relationships gives an insight into pollutant removal, transport and relocation or deposition. Pollutants associated with eroded sediment include agro-chemicals and heavy metals, which may originate from agricultural, industrial or municipal sources. The management of these sources can thus affect the potential environmental impacts off-site. Data from erosion studies are used to illustrate the selective removal of chemically active sediment and runoff by erosion processes. The way forward is outlined, concentrating on modelling as the most efficient methodology for further studies. Physical, rather than empirical soil erosion prediction models are able to simulate mathematically the processes by which soil and any associated pollutants are detached, transported and re-located off-site.

Introduction

Since the first erosion plots were established in the US during the early 1900s (De Ploey & Gabriels, 1980) the majority of soil erosion research has concentrated on measuring the amounts of soil lost from agricultural land under different farm practices. Quantifying the physical loss of the soil resource was seen as a key step in understanding the role of soil erosion in the reduction of land capacity to support vegetation, specifically crop growth for food and fibre production. Recently it has been recognised that it is not simply the amount of soil lost that is important, but the detrimental effect of soil erosion on soil water holding capacity, nutrient status, structure and depth. This perception of soil erosion as having primarily "on-site" consequences is still prevalent today, as recent papers by Rubio *et al.* (1990) and Fullen (1991) illustrate. In some developing countries where soil erosion can have catastrophic effects on farmers' livelihoods this approach is understandable. However, in many countries there is increasing recognition that the problems of soil erosion are not confined to the source of the erosion (i.e. in the farmers fields), but are also important at sites at which any eroded material is finally deposited (Klaine *et al.*, 1988, Catt *et al.*, in press & Harrod, in press). Here the products of soil erosion, namely sediment and any soil component eroded with the sediment, may become a source of pollution. The pollution of land or water away from the eroded sites is termed "non point source pollution".

This paper aims to illustrate the role played by soil erosion in redistributing pollutants within the environment, drawing on examples from the literature and from the Woburn erosion experiment which is run jointly by Silsoe College and Rothamsted Experimental Station. Emphasis is placed on soil erosion by water, rather than that by wind or mass movement.

Source of Pollutants

The source of the pollutants moved by soil erosion is the soil itself. Indeed, eroded sediment can be viewed as a pollutant if it reaches water supplies such as reservoirs, where sedimentation reduces water quality and water storage capacity. However, pollutants are contained within the soil, and the types and amounts of pollutants will depend not only on the natural processes which give each soil its physical, chemical, morphological, biological and spatial characteristics, but also on the additions made to it by man.

In an attempt to increase crop production from our fixed land resources farmers have increasingly turned to yield-enhancing agro-chemicals. These may be mineral or organic fertilisers. There has been substantial research into the role of soil erosion in the losses of nutrients, especially N, P and K. Paradoxically, these are the chemicals which are often added to replace nutrient depletion due to erosion. Prato *et al.* (1989) state that nationally in the United States 46% of sediment, 47% of total P, and 52% of total N discharged into waterways come from agricultural sources.

Nitrogen is often lost when nitrate in the soil solution leaches. Although nitrate occurs on the surfaces of soil particles and organic matter, it is only weakly adsorbed, and it will readily move into solution. Thus most nitrate is lost by leaching or in surface flow rather than with the erosion of sediment. However, N is found as part of the soil organic matter, essential for the functioning of the soils microbial biomass. Because of this association erosion losses of nitrogen can be considerable. Table 1 illustrates the relative proportions of nitrogen moving in the solid and solution phases using the data of Catt *et al.* (in press). For both treatments (across-slope and downslope cultivation), the vast proportion of the nitrogen goes with the sediment, which Catt and his co-workers believe is mostly in the form of particulate organic matter.

Phosphorus is easily and strongly adsorbed onto clay particles and organic matter as phosphate ions (Barrow, 1978). Thus it is phosphate losses which are associated with erosion of soil particles. Kronvang (1990) quotes 85% of total P losses from a catchment in sediments on which the P was adsorbed.

Table 1. Comparison between the mean nitrogen losses (kg/ha) in runoff and transported sediment under crops of potatoes, winter wheat, and winter barley, for different cultivation directions at the Woburn erosion experiment (from Catt *et al.*, in press).

Cultivation Treatment	N in runoff kg/ha	N in sediment kg/ha
Across slope	0.42	17.54
Down slope	0.65	15.28

Other agro-chemicals which may become pollutants include herbicides and pesticides. These agro-chemicals are perceived to have a beneficial effect on crop yields with little added cost to the farmer, given the increase in revenue due to higher tonnage of crops brought about by these artificial additives. These yield-enhancing chemicals form one source of potential pollutants. When they are re-distributed by erosion, contamination may result elsewhere. The British National Rivers Authority (NRA) (1992) has highlighted soil erosion as an important agent in the movement of the pesticide aldrin, which has been found in concentrations exceeding the Environmental Quality Standard, identified in the EC Dangerous Substances Directive, down stream of bulb fields in West Cornwall. Because of the slow degradation of aldrin to dieldrin in this environment, the NRA expects that "the risk of chronic and diffuse pollution will remain for many years in this area".

Other sources of pollution are farm, municipal and industrial wastes. These may be used by farmers and land managers as means of increasing organic matter and hence soil fertility and structural stability. Albaladejo & Diaz (1990) report experiments which used municipal refuse spread on the surface of the soil to combat erosion. However, such wastes often contain lage concentrations of heavy metals, including copper, mercury, lead, nickel, vanadium, cadmium and cobalt, depending on the source. For example, livestock slurry can contain much copper, which is present in some animal feedstuffs. It is known that earthworms, which play an important role in improving soil, are sensitive to copper in the soil. Such detrimental impacts led Barrow (1991) to conclude that "the use of organic wastes for soil improvement is anything but a simple, trouble free process".

Potential pollutants are sometimes added accidentally, perhaps best illustrated by the catastrophic release of radioactive nuclei into the environment by the Chernobyl reactor. This disaster has subsequently led to the pollution of many soils, which have now become virtually useless for many generations to come, in terms of agricultural production. Other contamination arises from the wet, dry or occult deposition of pollutants from industry, see for example Collett in this volume.

Toxic elements such as boron and selenium occur naturally, but may be subject to redistribution by erosion when they are brought to the soil surface as a result of human activity. One example of this is where ground water rich in with these elements is used for surface irrigation. These elements can be precipitated at the soil surface where they are susceptible to re-distribution by erosion.

Impacts on the environment

Prior to the 1940s contamination by agro-chemicals was not perceived to be an environmental problem because there were few such products (Barrow, 1991). However, the last 50 years has seen an increase in the use of these potential pollutants, as well as an increased awareness of the potentially detrimental effects these substances have on our environment. Even so, some researchers claim we still do not fully understand the long term effects of agro-chemical use and mis-use. There is also the problem that the wide variety of agro-chemicals currently being used make the monitoring and detection of their fate difficult (Barrow, 1991). Bearing in mind these possible shortcomings, the impacts of the removal and redistribution of pollutants can be classified as those occurring "on-site" and those occurring "off-site".

On-site impacts

Removal of soil nutrients (either natural or artificial) by erosion reduces soil fertility and productivity. Ironically, selective removal of nutrients by erosion means that more fertilizer must be applied in the following season to maintain yields. For convenience and economy artificial, chemical fertilisers predominate. Application of more fertilizer may result in larger concentrations of these chemicals in the soil, and thus the likelihood of greater amounts of nutrients being carried away by erosion. Hadas *et al.* (1989) have shown that unlike organic fertilisers, chemical fertilisers have no beneficial effects on soil structure or aggregation, making the soils susceptible to further erosion. Hence a vicious circle is set up in which erosion leads to less production, artificial fertilizers have no effect on the soil structure, and further erosion and loss of productivity ensues. Hence even larger applications of nutrients are needed to mask these losses.

Off-site impacts

When pollutants are transported by runoff or stream flow, they may have detrimental effects on water quality, and once deposited, they may contaminate the surrounding environment. The increased concern for the former can be reflected by the issuing of EC guidelines as to acceptable levels of nitrate and phosphate in water supplies, for example.

Probably the most documented off-site impact of nutrient loss is that of eutrophication of water courses, lakes and reservoirs. This is especially true of phosphorus and it is often the limiting factor in the process (Sharpley & Smith, 1990). Particulate P is often transported to these destinations with eroded sediments, on to which it is strongly adsorbed. This strong adsorbtion of P means it is released only slowly into solution, and therefore provides a long term source of P for algae and other aquatic flora. The amount directly accessible is determined by whether any re-suspension of P occurs, as well as the ways in which sedimentation occurs in the water. The amount of bioavailable P also depends on an interaction between rainfall, runoff and sediment. With large concentrations of bioavailable P there is a growth of undesirable algae and other aquatic weeds. The excessive biological activity leads to eutrophication of the water and affects its potential for fisheries, recreation, and industrial and domestic use of the water.

Pesticides and herbicides may cause the insidious loss of vital soil organisms if they are concentrated away from their intended target application site. Thus they may contribute to disruption of N and sulphur recycling, as well as leading to a reduction in the symbiotic relationships between soil organisms and crops (for example the beneficial role of mycorrhizal fungi or soil arthropods). These agrochemicals are known to have affected earthworm, termite and pollinating insect populations.

Once eroded, pollutants may be transported into streams, rivers, lakes and reservoirs, depending on the delivery ratios of the catchment (see Delivery of pollutants). As most pollutants are adsorbed on to highly erodible particles, eroded sediment often contains larger concentrations of the pollutants than present in the uneroded soil. This discrepancy is often expressed as an "enrichment ratio", i.e. the ratio of pollutant in the eroded sediment compared with the concentration of the same pollutant in the parent (uneroded) soil. Such concentration of pollutants is further increased if deposits are very local. Thus pollutants may be very concentrated where deposition takes place, even if their origins are quite disparate. Harrod (in press) illustrates this for the two pesticides, aldrin and dieldrin (see Table 2), when he compares concentrations in the middle of the field with those measured in colluvium. It is interesting to note that the concentration factor for aldrin is much higher than for dieldrin. Harrod suggests that this may be related to the soil components with which they are associated having different erosional or depositional properties.

Table 2. Aldrin and dieldrin (μg/kg) in soil and colluvium in thirteen Cornish fields (from Harrod, in press).

	Mean	Maximum	Minimum
Aldrin			
Field centre	88	300	2
Colluvium	685	2475	2
Concentration factor[1]	14.4	51.5	0.08
Dieldrin			
Field center	264	1100	10
Colluvium	358	710	20
Concentration factor	3.6	15.8	0.1

Another important aspect is transport to unprotected sites. Although not well documented, we have observed the transport of mine spoil into a nature reserve.

[1]The concentration factors are based on data from specific sites and are not derived from the information in the table.

The factors affecting movement of pollutants by erosion

Before we begin to assess the contribution that soil erosion processes play in the movement of pollutants we must understand how this movement takes place. Depending on the pollutant - soil interaction we can classify chemicals as those : (i) primarily transported with the sediment; and (ii) transported with surface runoff (Baker & Johnson, 1983).

This distinction needs to be quantified before a control strategy can be designed and implemented. Measures which combat the movement of sediment and any associated particulate pollutants (such as the use of vegetative filter strips) may not be effective in controlling runoff and thus the movement of any pollutants in solution. Unfortunately, the division between particulate and solute forms of the pollutants is not fixed. For example, solute P may transform to particulate P depending on the concentration and adsorption capacity of any sediment in the stream flow, as well as the existing solute level in the flow (Sharpley & Smith, 1990). These in turn are determined by complex rainfall/runoff/sediment interactions (Sharpley, 1985).

Baker & Johnson (1983) explore the relationship between the equilibrium concentrations of soil-adsorbed chemicals and those in solution, and the proportion of chemicals lost with the sediment. They allot a value (K) to potential pollutants which reflects the degree of adsorption of a certain chemical in the soil. Figure 1, adapted from their paper, illustrates the qualitative effect of K on chemical losses associated with sediment. As K increases so does the proportion of the chemical adsorbed onto the soil and, not surprisingly, the proportion lost with the eroded sediment.

The solubility of the pollutant in question also plays an important role in determining the partitioning between the solid and liquid phase. In Wauchope's (1978) review he concludes that the largest concentrations of pesticides in runoff are associated with the most soluble pesticides. However, these concentrations can be drastically reduced by the movement of such pesticides into the soil interior. When this occurs, erosion of soil particles is then needed for transport and re-location to take place.

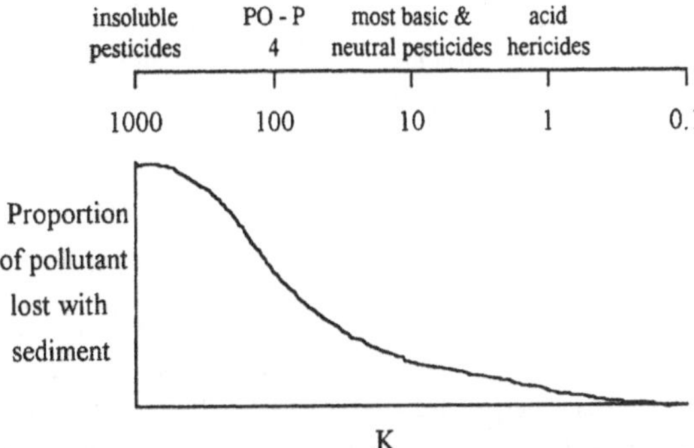

Figure 1. Relationships between the adsorption coefficient K and losses of chemicals with sediment (from Baker & Johnson, 1983).

The persistence of the chemical in question also affects greatly the amount of chemical lost from the soil. This is particularly true of pesticides where persistence (the time taken for concentrations to fall to 10% of the original value) may range from a few days to several years. However, with the banning of most persistent organo-chlorine insecticides, most pesticides now have a persistence of less than one year (Wauchope, 1978). As most pesticides degrade exponentially, the critical period as far as losses by erosion are concerned is during the first few days after application. It is then that concentrations are at their largest.

The persistence of added elements, such as N, P and K, might be considered to be indefinite, but all are subject to losses. Depending on the element in question, these may be through plant uptake, leaching or volatilisation, giving the elements only a limited time span during which they are susceptible to movement by erosion.

The timing and magnitude of any rainfall event after the application of the pollutant is crucial in determining its transport. Hudson (1981) found that 50% of annual erosion came from only 2 rainstorms, and that in one year, 75% of the erosion took place in just 10 minutes. Morgan (1986) quotes similar frequency/magnitude relationships for soil erosion in temperate areas. If one of these events coincides with the period when pollutant concentrations are at their largest, there is likely to be a significant loss of chemicals with any eroded sediment (Wauchope, 1978).

Thus the nature of the relationship between the chemical and the soil helps us to understand how pollutants are associated with both soil particles and organic matter. To understand the redistribution and movement of these pollutants, it is also necessary to appreciate the processes of soil erosion. The impact of raindrops on an unprotected, bare soil may break soil aggregates into smaller particles. This detachment from the soil mass makes the disturbed soil particles more transportable by rain splash. It is, however, overland flow which is the main agent for the transport of detached materials. The smaller size of the detached particles and their discreet nature makes them more easily entrained in and transported by overland flow, whether it is in the form of sheet flow or concentrated in a rill or gully channel. These smaller size fractions (clays, silts and organic matter) have larger concentrations of adsorbed pollutants because of their greater surface area and charge densities. The small mass of these particles means that they have low settling velocities and will therefore travel long distances before being deposited.

Delivery of pollutants

Given that the site of deposition of eroded material (and any adsorbed pollutants) is rarely close to its source, techniques are required to indicate these sources and the likely destination of the pollutants. As the pollutants are intimately associated with sediment and runoff, the monitoring of soil erosion could be one way to assess this transport. Nearly all soil erosion monitoring has taken place at a plot scale, often on plots measuring 22 x 1.6m, following the standard USDA plot size. Although knowing how much sediment is produced from such plots helps to answer the question of what is the potential supply of any pollutant (as adsorbed onto this eroded sediment), this does not tell us how much of the eroded sediment is then transported to the watercourse at the catchment scale. In many cases the delivery of sediment to the watercourse represents only a fraction of that eroded and may be up to an order of magnitude less (Novotny & Chesters, 1989). One way of describing this is through delivery ratios; that is the ratio of sediment yield at the catchment outlet to the total potential sediment production within the catchment.

In many cases the delivery ratio is found to be less than unity, implying than sediment and its associated pollutant have been stored within the catchment. This can be explained through a knowledge of the catchments characteristics. This is illustrated in Figure 2, adapted from Novotny & Chesters (1989). The delivery process is represented as a black box which represents the catchments characteristics. Detached sediment enters the box where its movement is influenced by the various factors. The response of the entrained sediment to each of these factors will determine the amount eventually delivered to the water course and this effect on sediment delivery will be discussed below.

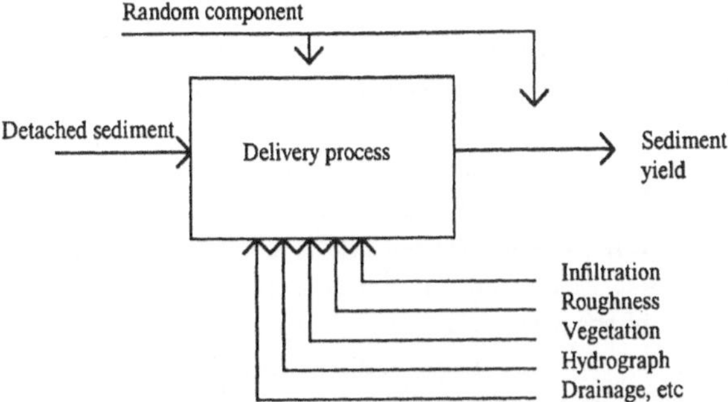

Figure 2. Stochastic representation of the sediment delivery process (from Novotny & Chesters, 1989)

Infiltration rates may vary considerably over a catchment because of changing soil characteristics and land use. There will also be considerable temporal and spatial variation within soil/land use units. These variations can lead to the generation of runoff in one area of the catchment and infiltration in another. Thus pollutants may be transported locally within the catchment but not necessarily out of it.

Surface morphology affects runoff velocities: the rougher the micro-topography (due to soil and plant induced roughness for example), the slower the flow. This will reduce the amount of material transported, and decrease the travel distance of eroded particles. Human-induced roughness, such as tractor wheelings perpendicular to the slope, can also trap runoff allowing the water to infiltrate and the sediment to be deposited locally. However, when these temporary storage sinks fail, because of over-topping or breaching of runoff after heavy rain, concentrations in the runoff can be very large, as shown on the across slope treatments on the Woburn erosion plots (Catt *et al.*, in press).

Vegetation is known to restrict the delivery of sediment to the water course, and is now forming the basis for pollution control measures (Coppin & Richards, 1990). Not only does vegetation increase infiltration rates (Holtan, 1961), reducing runoff and erosion, the use of grass strips has been shown to filter out sediment, and presumably any pollutants within the sediment (Tollner *et al.*, 1982). In experimental work grass strips of 2m width have been shown to have sediment trapping efficiencies of up to 90% (Hayes *et al.*, 1982; Novotny & Chesters, 1989). Whether it is physical filtering or the reduction of runoff velocities which causes deposition is unclear. Also, it must be remembered that it is the fine fractions (clay) which carry the largest concentration of pollutants. These particles may not be trapped by the grass strips as the sand and silt fractions would, but flow through them with the runoff.

The shape of the slope will also influence the transport of sediment and pollutants to the water course. In many catchments slope gradient decreases nearer the valley bottom, unless dramatic rejuvenation of the water course is taking place. These concave slopes generate slower overland flow, making deposition of sediment more likely. This riparian zone has been the focus of many attempts to control sediment and pollution, whereby the amount of deposition has been increased still further by manipulating infiltration rates and roughness by planting vegetative barriers.

The size of the storm and the amount of runoff it generates will also influence sediment delivery. Large storms are more likely to give delivery ratios close to one than small ones, because heavy rain can re-distribute sediment held in storage from previous, smaller events, causing delivery ratios to increase towards 1. Indeed, Heathwaite *et al.* (1990) state that a substantial proportion of non-point pollutants are transported in overland flow associated with storms. However, Young *et al.* (1986) argue that selectivity of the erosion is more pronounced for light frequent rainfall. This is because this is responsible for a disproportionate amount of smaller and more chemically active particles being transported from a site (especially if they are unconsolidated, unaggregated eroded sediments).

Other catchment characteristics have been linked to sediment delivery. Roehl (1962) showed a good correlation between catchment size and sediment delivery ratios, whereas Beer *et al.* (1966) found a poor relation. Numerous other factors may influence sediment delivery, and these are thoroughly reviewed by Novotny & Chesters (1989).

Modelling

Given the problems of relating plot scale measurements to the catchment scale, predictive models may provide a more manageable solution to the assessment of pollutant movement soil erosion. To date the most widely used model for predicting soil erosion has been the Universal Soil Loss Equation (USLE). The USLE is a statistical model developed from the results obtained from a series of standard experimental plots throughout the United States (Wischmeir & Smith, 1978). It requires the user to derive a series of factors reflecting rainfall erosivity, soil erodibility, crop, relief and erosion control practices. The form of the equation has no physical significance and its inability to represent processes spatially and at catchment scales means that it is unsuitable for predicting pollutant movement. It is also hindered by the fact that it should not be applied outside the conditions under which it was developed. This limits it geographically to the United States and to areas of gentle slopes, since the data from which it was developed were from plots of 9% slope. Attempts at calibrating the USLE outside the United States are likewise limited in application to the specific conditions under which any of the input parameters were re-assessed and calibrated (Auerswald, 1988; Schwertmann *et al.*, 1987; Jayakumar *et al.*, 1982). Even the linking of the USLE to a distributed hydrology model, such as CREAMS (Knisel *et al.*, 1980) does not allow the erosion component to be sufficiently reactive to changing conditions.

The weaknesses of the USLE have been recognised for some time (Hudson, 1981) and new approaches based around physical theory are being developed. These "new generation" models such as EUROSEM (Morgan *et al.*, 1992) and WEPP (Lane & Nearing, 1989) model the erosion process using a series of physically-based equations. EUROSEM has at its core the equation for dynamic mass balance (Bennett, 1974; Kirkby, 1980; Woolhiser *et al.*, 1990):

$$\frac{\partial(AC)}{\partial t}+\frac{\partial(QC)}{\partial x}-e(x,t)=q_s(x,t) \tag{1}$$

where C = sediment concentration (dimensionless)
A = cross sectional area of flow (m^2)
Q = discharge (m^3 s^{-1})
q_s= external input or extraction of sediment per unit length of flow (m^3 s^{-1} cm^{-1})
e = net detachment rate or rate of erosion of the bed per unit length of flow (m^3 s^{-1} cm^{-1})
x = horizontal distance (m)
t = time (s).

Equation (1) describes the movement of sediment over the soil surface. It is coupled with equations to describe catchment characteristics, soils and vegetation. These models are being developed even further with the formulation of new models which not only simulate erosion processes but also soil water fluxes, plant growth and the fate of chemicals. An example is the OPUS model (Smith, 1992).

At face value these models can predict pollutant movement throughout the catchment, highlighting source and sink areas, and because of their physical basis, there is the potential of transferring these processes to other situations (Abbott *et al.*, 1986). Despite this, there are constraints to this in reality. Many data are needed to run these models. Even the smaller physically based models such as EUROSEM still require upwards of 30 inputs to run. Many of these inputs are difficult to measure accurately in the field or may exhibit temporal and spatial variability, yet often the model output is sensitive to their assigned input values. For example, in EUROSEM, we have found that the model output is sensitive to saturated hydraulic conductivity, which remains one of the most difficult soil characteristics to measure accurately. At present these models remain research tools, requiring calibration before they can give adequate predictions of runoff and sediment behaviour, as Smith (1978), De Roo & Walling (in press), and Quinton (in press) have illustrated.

Conclusion

We have shown above that we still cannot assess adequately the role played by soil erosion in pollutant transport at a catchment scale. This is crucial, as control strategies might be most effective at this scale. Many papers detail the amount of a certain pollutant measured in runoff and sediment, but though these studies are of regional interest and importance, they do not further our understanding of the processes involved. It is through better understanding of the mechanisms of erosion as a selective agent of sediment detachment, transport and deposition that we shall be able to predict the fate of pollutants within the environment.

References

Abbott, M.B., Bathurst, J.C., Cunge, J.A., O'Connel, P.E. & Rasmussen, J. 1986. An introduction to the European Hydrological System - Système Hydrological European, "SHE", 1. History and philosopy of a physically based distributed modelling system. *Journal of Hydrology*, **87**, 45-59.

Albaladejo, J. & Diaz, E. 1990. *Degradation y regeneracion del suelo en el littoral Mediterraneo Espanol: experiencias en el proyecto LUCDEME.* In Albaladejo, J., Stocking, M.A. & Diaz, E. (eds), Soil degradation and rehabilitation in Mediterranean environmental conditions. Consejo Superior de Investigaciones Cientificas (CSIC), Spain.

Auerswald, K. 1988. Erosion hazard maps for Bavaria. In Morgan, R.P.C. & Rickson, R.J. (eds), *Erosion assessment and modelling.* CEC EUR 10860 EN.

Baker, J.L. & Johnson, H.P. 1983. Evaluating the effectiveness of beet management practices from field studies. pp281-304. In Schaller, F.W. & Baily, G.W. (eds), *Agricultural management and water quality.* Iowa State University Press.

Barrow, C.J. 1991. *Land degradation.* Cambridge University Press.

Barrow, N.J. 1978. The description of phosphate adsorbtion curves. *Journal of Soil Science* **29**, 447 - 462.

Beer, C.E., Farnbam, C.W. & Heinemann, H.G. 1966. Evaluating sediment prediction techniques in western Iowa. *Transactions of the American Society of Agricultural Engineers* **9**, 828 - 831, 833.

Bennett, J.P. 1974. Concepts of mathematical modeling of sediment yield. *Water Resources Reasearch* **10**, 485-492.

Catt, J.A., Quinton, J., Rickson, R.J. & Styles, P.D.R. (in press). Nutrient Losses and crop yields in the Woburn Erosion Reference Experiment. In Rickson, R.J. (Ed.), *Conserving our soil resource. Proceedings of the first international congress of the European society of soil conservation, Silsoe 1992.* Soil Technology, Cremlington, Germany.

Coppin, N.J. & Richards, I.G. 1990. *Use of vegetation in civil engineering.* CIRIA/Butterworths, London.

DePloey, J. & Gabriels, D. 1980. Measuring soil loss and experimental studies. In Kirkby, M.J. and Morgan, R.P.C. (Eds), *Soil Erosion.* Wiley, Chichester, England.

De Roo, A.P.J. & Walling, D.E. (in press). Validating the "ANSWERS" soil erosion model using 137Cs in the Yendacott catchment, Devon, UK. In Rickson, R.J. (Ed.), *Conserving our soil resource. Proceedings of the first international congress of the European society of soil conservation, Silsoe 1992.* Soil Technology. Cremlington. Germany

Fullen, M.A. 1991. A comparison of runoff and erosion rates on bare and grassed loamy soils. *Soil Use and Management* **7,** 136-139.

Hadas, A., Hadas, A. & Quinton, J.N. 1989. Long-term effects of high application rates of NPK fertilizer on tensile strength and water stability of the soil structure. *Geoderma* **47,** 381-392.

Harrod, T.R. (in press). Runoff, soil erosion and pesticide pollution in Cornwall. In Rickson, R.J. (Ed.), *Conserving our soil resource. Proceedings of the first international congress of the European society of soil conservation, Silsoe 1992.* Soil Technology, Cremlington, Germany.

Hayes, J.C., Barfield, B.J. & Barnbisel, R.I. 1982. *The use of grass filters in strip mine drainage. Vol. 3, Empirical verification of procedures using real vegetation.* Technical report IMMR 92/070. Institute for Mining and Mineral Resources, University of Kentucky, Lexington.

Heathwaite, A.L., Burt, T.P. & Trudgill, S.T. 1990. Land use controls on sediment production in a lowland catchment, south west England. In Boardman, J., Foster, I.D.L. & Dearing, J.A. (Eds), *Soil erosion on agricultural land.* pp. 69 -86.Wiley, Chichester.

Holtan, H.N. 1961. *A concept for infiltration estimates in watershed engineering.* USDA-Agricutural Research Service, ARS-41-51.

Hudson, N.W. 1981. *Soil conservation.* Batsford, London.

Jayakumar, M., Padmanabhan, V. & Jeevarathnam, K. 1982. *Hydrology and engineering. 25 years research on soil and water conservation.* Monograph No. 4, Central Soil and Water Conservation Research and Training Institute, Tamil Nadu, India.

Kirkby, M.J. 1980. Modelling water erosion processes. In Kirkby, M.J. & Morgan, R.P.C. (Eds), *Soil Erosion,* pp. 183-216. Wiley, Chichester.

Klaine, S.J., Hinmann, M.L., Winkelman, D.A. & Martin, J.R. 1988. Characterisation of Agricultural non point pollution: nutrient loss and erosion in a West Tennessee catchment. *Environmental Toxicology and Chemistry* **7,** 601-607.

Knisel, W.G. 1980. *CREAMS - A field scale model for chemical, runoff and erosion from agricultural management systems.* Conservation report no. 26. United States Department of Agriculture, Washington D.C.

Kronvang, B. 1990. Sediment associated phosphorous transport from two intensively farmed catchment areas. In Boardman, J., Foster, I.D.L. & Dearing, J.A. (Eds), *Soil erosion on agricultural land.* pp 313 - 330. Wiley, Chichester.

Lane, L.J. & Nearing, M.A. 1989. *USDA - water erosion prediction project: hillslope model documentation.* NSERL Report No. 2. National Soil Erosion Laboratory, USDA - Agricultural Research Service, W. Lafayette, Indiana 47907.

Moldenhauer, W.C. & Foster, G.R. 1981. Empirical studies of soil conservation techniques and design procedures. pp13-29. In Morgan, R.P.C. (Ed.), *Soil Conservation: problems and prospects.* Wiley, Chichester.

Morgan, R.P.C. 1986. *Soil erosion and conservation.* Longman, London.

Morgan, R.P.C., Quinton, J.N. & Rickson, R.J. 1992. *EUROSEM documentation manual.* Silsoe College, Cranfield Institute of Technology.

National Rivers Authority (1992). *The influence of agriculture on the quality of the natural waters in England and Wales.* Water quality series number 6, National Rivers Authority, Bristol.

Novotny, V. & Chesters, G. 1989. Delivery of sediment and pollutants from non-point sources; a water quality perspective. *Journal of Soil and Water Conservation.* **44,** 568 -576.

Quinton, J.N. (in press). Validation of physically based erosion models - with particular reference to EUROSEM. In Rickson, R.J. (Ed.), *Conserving our soil resource. Proceedings of the first international congress of the European society of soil conservation, Silsoe 1992.* Soil Technology, Cremlington, Germany

Roehl, J.E. 1962. Sediment source areas and delivery ratios influencing morphological factors. *International Association of Hydrological Sciences* **59,** 202-213.

Rubio, J.L., Andreou, V. & Cerni, R. 1990. Degratdacion del suelo por erosion hidrica: Diserno experimental y resultados preliminares. pp215-235. In Albaladejo, J., Stocking, M.A. & Diaz, E. (Eds), *Soil degradation and rehabilitation in Mediterranean environmental conditions.* Consejo Superior de Investigaciones Cientificas (CSIC), Spain.

Schwertmann, U., Vogl, W. & Kainz, M. 1987. *Bodenerosion durch Wasser.* Ulmer Verlag, Stuttgart.

Sharpley, A.N. 1985. The selective erosion of plant nutrients in runoff. *Journal of the Soil Science Society of America* **49,** 1527 - 1534.

Sharpley, A.N. & Smith, S.J. 1990. Phosphorous transport in agricultural runoff: the role of soil erosion. In Boardman, J., Foster, I.D.L. & Dearing, J.A. (Eds), *Soil erosion on agricultural land.* pp. 351 - 368. Wiley, Chichester.

Smith, R.E. 1977. Field tests of a distributed watershed erosion / sedimentation model. In *Soil erosion; prediction and control, Proceedings of a National Conference on Soil Erosion. May 24 - 26, 1976. Purdue University, West Lafayette, Indiana.* pp. 201 - 209. Soil Conservation Society of America, Ankeny, Iowa.

Smith, R.E. 1992. *Opus, an integrated model for transport of Nonpoint source pollutants at the field scale: volume 1, Documentation.* United States Department of Agriculture, Agricultural Research Service, ARS-98, 120pp.

Tollner, E.W., Barfield, B.J. & Hayes, J.C. 1982. Sedimentology of erect vegetal filters. *Journal of the Hydraulics Division American Society of Civil Engineers* **108,** 1518-1531.

Wauchope, R.D.1978. The pesticide content of surface water draining from agricultural fields - a review. *Journal of Environmental Quality* **7,** 459-472.

Wischmeir & Smith (1978). *Predicting rainfall erosion losses - a guide to conservation planning.* USDA, Agricultural Handbook No. 537. Washingtion D.C.

Woolhiser, D.A., Smith, R.A. & Goodrich, D.A. 1990. *KINEROS, A kinematic runoff and erosion model: documentation and user manual.* United States Department of Agriculture, Agricultural Research Service, Fort Collins. Colorado. U.S.A.

Young, R.A., Olness, A.E., Mutchler, C.K. & Moldenhauer, W.C. 1986. Chemical and physical enrichments of sediment from cropland. *Transactions of the American Society of Agricultural Engineers* **29,** 165 - 169.

J. N. Quinton & R. J. Rickson, Department of Rural Land Use, Silsoe College, Cranfield University, Silsoe, Bedford MK45 4DT, Great Britain.

Discussion

Question by S. van der Zee
What is the physical-basis for modelling erosion and runoff and how can you scale the model up to the field scale? Please note I am not interested in the USLE approach.

Answer by J.N.Quinton
In the European Soil Erosion Model soil loss is computed as a sediment discharge to give a volume of sediment passing a given point in a given time. This computation is based on the dynamic mass balance equation (equation 1 in this paper).

Question by W. Jury
Are there stochastic models of water, sediment and chemical transport in the field of erosion research, and are residence time distributions and transfer functions well used?

Answer by J.N. Quinton
Yes there are, although I am not acquainted with their details.
The American model WEPP uses transfer functions to obtain many of its inputs, perhaps more than some of us would like to see.

Question by P. Lagas
In the last few months there were some "disasters" as a consequence of river erosion in France and in Italy.
Is it possible to avoid such erosional time-bombs with some kind of soil monitoring? Is it possible to produce erosion vulnerability maps for example?

Answer by J.N. Quinton
Yes it is possible in areas which are vulnerable to erosion. These areas can then be protected using designs based on storms with a desired return period.

Soil Monitoring, Monte Verità, © Birkhäuser Verlag Basel

CONTAMINANT UPTAKE BY PLANTS

R. Frossard

Some of the factors influencing the uptake of heavy metals by plants are presented based on the literature and on original results.
Heavy metal concentrations in plants grown on the same substrate may vary according to plant species, degree of contamination of the substrate and the metal considered. Concentrations in above ground plant parts or in leaves can vary up to 100-fold between species, but mostly between 4- and 20-fold. Heavy metal uptake may also differ with plant varieties, with concentrations usually varying between 1- and 4-fold, though cases of 20-fold and more have been known.
Ambient temperature is shown to affect the uptake of Cd from contaminated soil into shoots and roots of Italian ryegrass, red clover and beans. Shoots of Italian ryegrass contained slightly more and roots considerably more Cd when grown at 11/17 °C (night/day) as compared to 18/24 °C. Red clover and beans contained more Cd in the shoots at warmer temperatures, but clover roots contained less and bean roots rather more Cd at warmer temperatures. The soluble Cd content of the soil (extractant 0.1 M $NaNO_3$) did not differ between the temperature treatments.
Application of fertilizer may enhance the uptake of some heavy metals by plants from contaminated and uncontaminated soil in pot experiments and in the field. The Cd concentration in pot-grown spinach shoots was 40 % lower when no macronutrients in the form of analytical grade reagents were applied. Zn concentration in shoots was not affected, although the soluble Zn content of the soil varied significantly.
Plants influence the bioavailability of heavy metals by changing the pH of the rhizosphere as a result of nutrient uptake, proliferation of root exudates, and respiratory activity, and these characteristics can vary considerably with species. In line with this the soluble contents of Cd and Ni in the soil differed significantly between several plant species grown in pots, the maximum differences being 4-fold.

Introduction

Within some narrow physical, chemical and biological limits, plants will take up almost any compound offered to them including mineral nutrients and heavy metals. The differences between individual compounds are the rates at which they are taken up, the amounts accumulated, and the sites within the plants to which they are translocated.

For environmental risk assessment it is often necessary to know the uptake of contaminants by plants. This is the case with mass balancing of contaminants in agriculture, where the amounts of pollutants taken up by plants and exported from a given area through harvesting have to be known. Data are usually obtained by measuring contaminant concentrations in plants or plant parts and determining the masses of the plants or parts exported (see von Steiger & Baccini, 1990). Inevitably, this method reflects the uptake characteristics of the particular crops studied and the conditions reigning during the investigation.

Attempts have also been made to predict the uptake by plants of heavy metals in polluted soils from concentrations measured in the soil using specific extractants (Andersson, 1976a; Symeonides & McRae, 1977; Häni & Gupta, 1981). Based on such procedures, guide levels for some heavy metals considered critical have been established for Swiss soils in order to protect their fertility in the long term (VSBo, 1986). Two types of guide levels have been chosen: one regarding the total content (2 M HNO_3 extraction) and one regarding the soluble content (0.1 M $NaNO_3$ extraction). The guide level of a heavy metal is considered as exceeded if either the total or the soluble content is exceeded. According to the Swiss Federal Office of Environment, Forests and Landscape (1987), the *soluble content* indicates the amount of metal which can actually be absorbed by plants. It depends on total content as well as on a range of soil properties such as pH and soil type. Large soluble concentrations of metals are detrimental to soil fertility including plant development and to the utilization of plants as food for animals and man. The significance of the *total content* as a guide level is based mainly on long-term considerations (Swiss Federal Office of Environment, Forests and Landscape, 1987). It helps in monitoring pollution, and it serves as an indicator of potential future risk.

Both forms of assessing environmental risk use contaminant uptake by plants as a characteristic, either directly or indirectly. Here I draw attention to some manifestations of plant life that might necessitate a more complex approach to future investigations of this kind. I shall show that several factors might modify contaminant uptake by plants. Some of these cannot be explained by soil properties, or at least not by soil properties alone.

Effects of plant species and varieties

Species

Different plant species can exhibit different uptake patterns for heavy metals, as has been shown for cadmium added in various amounts to plants growing either in nutrient solution (Page *et al.*, 1972) or in soil (John, 1973; Bingham *et al.*, 1975, 1976). The Cd concentrations in above ground plant parts or in leaves had ratios (largest Cd concentration measured in the dry matter of any one plant species to smallest concentration measured in another species subject to the same treatment) between 1.6 and 39 for the species tested and the treatments used. Nearly three quarters of the ratios were less than 7.5, however. Rice, as tested by Bingham *et al.* (1975), was exceptional. It took up very little Cd even at large concentrations in the soil, so that a maximum ratio of 580 would have resulted.

Davis & Carlton-Smith (1980) grew up to 39 commercially available cultivars of crops representing 28 plant species in two soils A and B, which were regarded by the authors as "heavily contaminated following a history of sludge deposition on a "sacrificial' basis". Heavy metal concentration ratios as specified above were approximately 90 and 79 for Cd on soil A and B, respectively. However, the ratio on soil A rises to 231 if the heavily accumulating species tobacco, which was sown only on this soil, is included. For Pb the corresponding concentration ratios (not including tobacco) were 104 and 49.5, for Zn 36.4 and 17.6, for Cu 6.3 and 3.5, and for Ni 21.2 and >33. Since organic matter content, cation exchange capacity and degree of contamination of the soils varied considerably, differences in plant uptake of heavy metals were to be expected. Interspecific variation, nevertheless, remained large on both soils. This was also true for the heavy metal concentrations found in the edible parts of the crops.

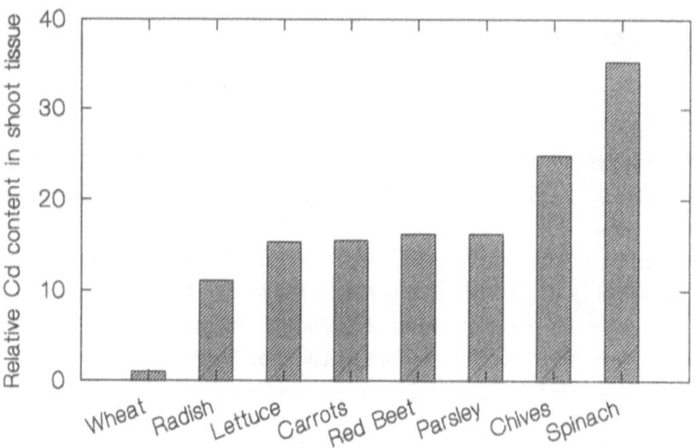

Figure 1. *Relative Cd content (wheat = 1) in shoot dry matter of eight plant species grown in pots on the same soil (sandy loam, pH (H_2O) ca. 6.0, organic C 1.7 %, CEC 31 mmol Me^{2+}/100 g, total Cd (2M HNO_3) 0.18 mg/kg; all values per dry soil). Plants were harvested when deemed ripe for commerce.*

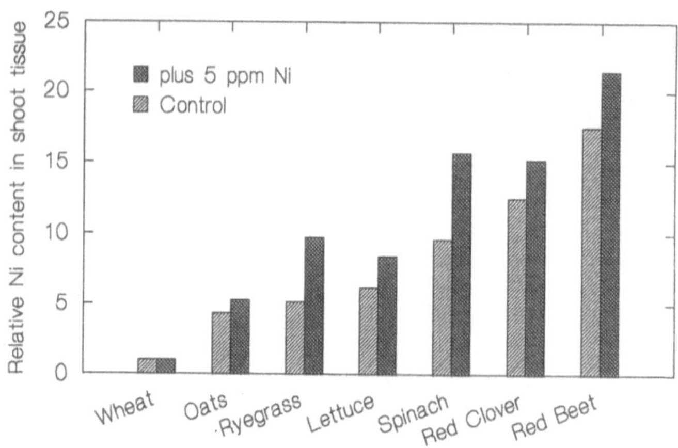

Figure 2. *Relative Ni content (wheat = 1) in the shoot dry matter of seven plant species grown in pots on the same soil. Soil data as in Figure 1; total Ni 19 mg/kg (control), additional Ni added as sulphate. Plants were harvested when deemed ripe for commerce.*

The composition of different species can vary under less dramatic circumstances as well. Seventeen crop species were grown side by side at nine field sites in Finland, and the two-year averages of the micronutrient contents in the dry matter varied appreciably between plant species and plant parts (Yläranta & Sillanpää, 1984). These authors included the data for individual plant parts in their comparison and found that the largest mean content of Cu was 4-times that of the smallest. The corresponding values for Fe and Zn were 7, for B 21, for Mn 35 and for Mo 46.
Similarly, differences in uptake of Cd or Ni can be observed between species grown in pots in a Swiss agricultural soil not considered as contaminated (Figures 1 and 2). Interestingly, or, for those engaged in mass balancing activities perhaps rather annoyingly, the ranking of species changes when the soil is contaminated (Figure 2). This is also evident from the data of other workers mentioned (e.g. Page *et al.*, 1972; Bingham *et al.*, 1975).

Varieties
While most researchers recognize interspecific variation, those tackling environmental problems from the viewpoint of soil science might be less aware of the fact that varietal differences within plant species can sometimes be as great or even greater than interspecific variation (see references in Yuran & Harrison, 1986). Impressive are the differences in Cd uptake between nine lettuce cultivars grown in solution culture containing Cd at 0.1 µg/ml, as reported by John & van Laerhoven (1976), where the largest to smallest Cd concentration ratio in tops was 21 after five weeks of exposure. Although reports exist of similar or even of larger Cd concentration ratios for varieties of other crops grown in soil (Marschner, 1983; Yuran & Harrison, 1986), most observed varietal differences in heavy metal uptake are much smaller. Typical values stemming from investigations on Cd, Ni, Pb, Cu or Zn lie between 1 and 3.7 (Davis & Carlton-Smith, 1980; Crews & Davies, 1985; Yuran & Harrison 1986; Florijn *et al.*, 1991). Also, the Cu and Zn contents in the leaves of five varieties of spinach grown in the field differed over a two-year period, but Cd, Mn and Pb contents did not, whereas in the stalks only Cd and Mn varied (Willaert *et al.*, 1990).
The problem for the researcher is that without preliminary trials he cannot know whether the variety of his test crop will accumulate much of a metal or not. Furthermore, data gained about interspecific variation of contaminant uptake are always subject to varietal variation as well, depending on the cultivar chosen to represent the species.
Florijn *et al.* (1991) have recently questioned the results of John & van Laerhoven (1976). They repeated John and van Laerhoven's experiment in detail with eight lettuce varieties, seven of which were the same as those used in the original experiment, and they obtained much less variation. The maximum shoot Cd concentration ratio was 2.1 at the 0.1 µg/ml treatment. One of their conclusions is that the data of John & van Laerhoven (1976) and their inclusion in textbook material (Marschner 1983) should at least be re-evaluated.
This reasoning, however, takes no account of the fact that the uptake characteristics of varieties do not necessarily remain genetically stable over the years. Moreover, Florijn and coworkers used high pressure lamps (HPL) as a light source, giving a light intensity of 125 W/m^2, the equivalent of which was probably difficult to achieve with the mixture of cool fluorescent and incandescent

lamps used by John and van Laerhoven. Apart from that, the spectral qualities of the two light sources differ considerably (McFarlane, 1978), and this is known to affect plant metabolism. Further differences are the times before plants were exposed to Cd (four weeks against three), the relative humidities reigning (82 % against 70%) and the number of plants grown per container (1 against 8). Such seemingly trivial environmental detail can affect the uptake of contaminants by plants and hence the outcomes of experiments. Some of these will be discussed later.

Effects of temperature

A common observation in field experiments is that heavy metal contents of plants vary from year to year (Yuran & Harrison, 1986; Diez & Krauss, 1990; Willaert *et al.*, 1990). This is often attributed to diverse climatic conditions such as temperature or precipitation or both (Diez & Krauss, 1990; Willaert *et al.*, 1990). The effects of temperature on heavy metal uptake by plants have been studied mainly in connection with soil temperatures (see Sheaffer *et al.*, 1979 and references therein). In many experiments warm soil increased the uptake of Cd and Zn and sometimes of other metals, but there have also been cases where little or no effect was observed, or where some metal contents decreased when the soil was warmer. Results also differed with plant species.

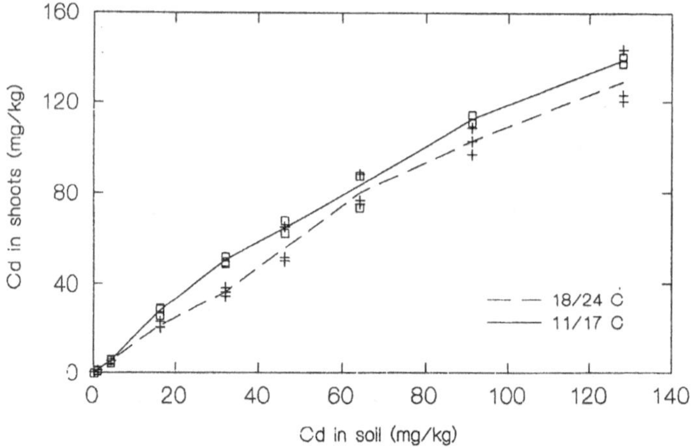

Figure 3. *Cd content in shoots of Italian ryegrass grown in pots at 11/17 °C (night/day) or 18/24 °C ambient temperature in a growth cabinet. Cd added as sulphate. Concentrations refer to dry matter. Symbols (squares = 11/17 °C, crosses = 18/24 °C) indicate replicate values (three per treatment and temperature regime). Lines connect the respective averages. For further details see text.*

In contrast, Tanner (1991) studied the uptake of Cd by plants grown at different ambient temperatures in a growth cabinet. Italian ryegrass, red clover and beans were grown in soil at 11/17 °C (night/day) and at 18/24 °C with increasing concentrations of Cd supplied as sulphate. Watering was measured, and relative humidity was such that approximately the same atmospheric vapour pressure deficit reigned during the light periods (14 h). Plants (including roots) were harvested at the same stage of development in both temperature regimes.

The shoots of Italian ryegrass contained slightly more Cd under the cool regime (Figure 3), and roots contained considerably more Cd at these temperatures (Figure 4). Red clover and beans contained more Cd in their shoots when kept at warmer temperatures, whereas the roots of clover contained less and those of beans rather more Cd with this treatment (Figures 5, 6, 7, 8). Increasing Cd concentrations inhibited the production of total plant biomass. With red clover and beans, this effect was more pronounced at warm temperatures than at cool ones, while total biomass of Italian ryegrass was more affected by Cd at low temperatures (Tanner 1991).

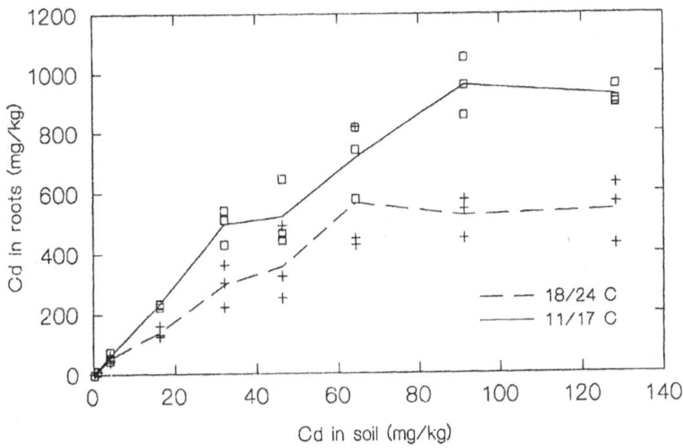

Figure 4. Cd content in roots of Italian ryegrass grown at different temperatures (see legend of Figure 3).

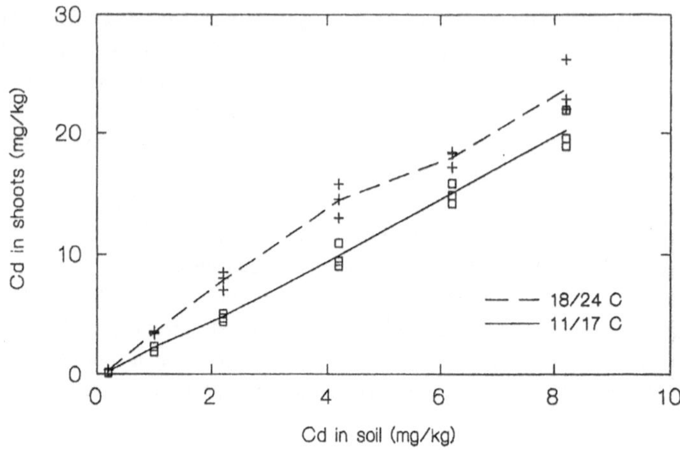

Figure 5. *Cd content in shoots of red clover grown at different temperatures (see legend of Figure 3).*

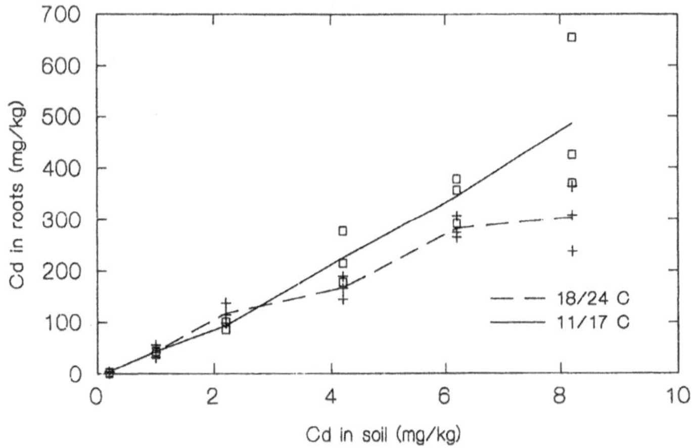

Figure 6. *Cd content in roots of red clover grown at different temperatures (see legend of Figure 3).*

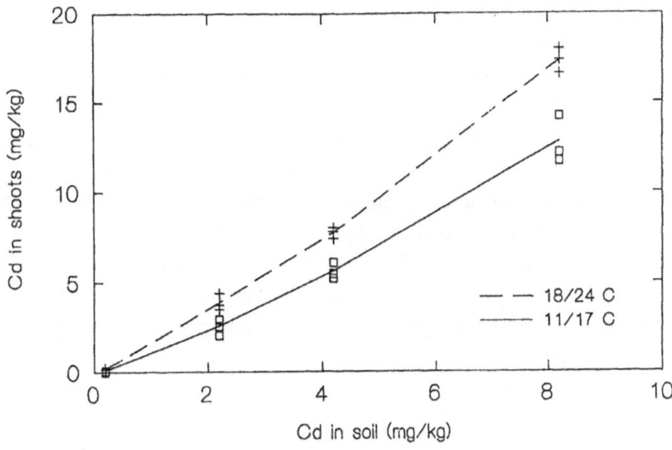

Figure 7. *Cd content in shoots of beans grown at different temperatures (see legend of Figure 3).*

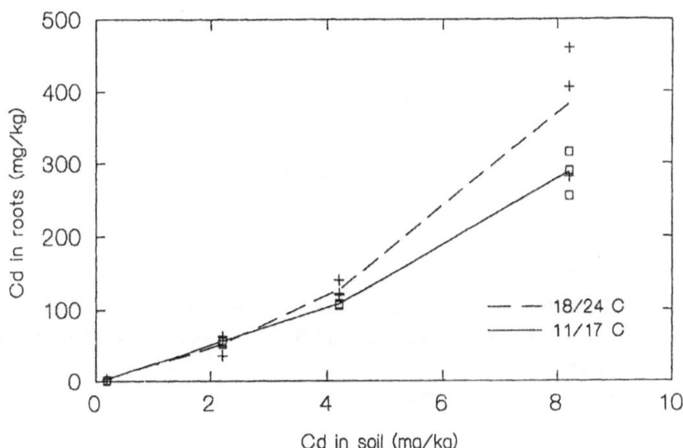

Figure 8. *Cd content in roots of beans grown at different temperatures (see legend of Figure 3).*

At the end of the experiments and immediately after harvesting the shoots, soil samples were taken and the soluble contents of Cd were determined according to the procedures recommended for the Swiss Ordinance Relating to Pollutants in Soil (VSBo, 1986). Extraction with 0.1 M $NaNO_3$ occurred at 21 °C. No differences in soluble Cd were observed between the two temperature treatments (Figures 9, 10 and 11). Yet Cd accumulation into shoots or roots or both varied with temperature. Since the extent of yield reduction due to Cd was also influenced by temperature (Tanner, 1991) this means that climatic conditions are important when establishing guide levels for Cd (and probably for other contaminants) by way of plant experiments in phytotrons and greenhouses. Enhanced uptake of Cd, Zn, Mn and (partly of) Fe by lettuce with increasing ambient temperatures has previously been reported by Siriratpiriya *et al.* (1985). Should the findings presented above turn out to apply to the field too, then guide-levels for heavy metals in soil might have to be reassessed.

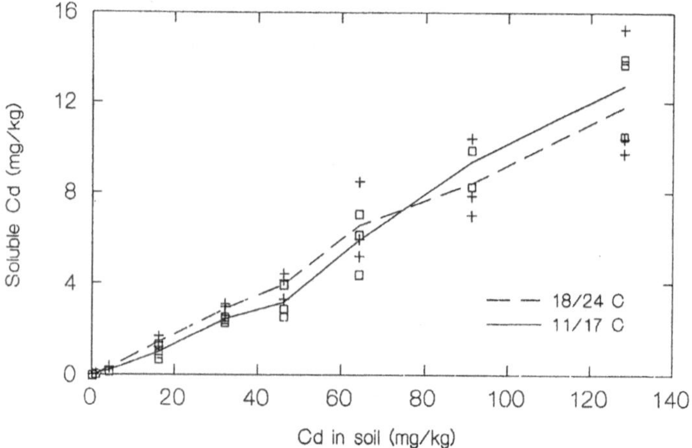

Figure 9. Relationship between total and soluble Cd content in soil (dry matter) grown with Italian ryegrass at 11/17 °C (night/day) or 18/24 °C (see Figure 3 and 4).

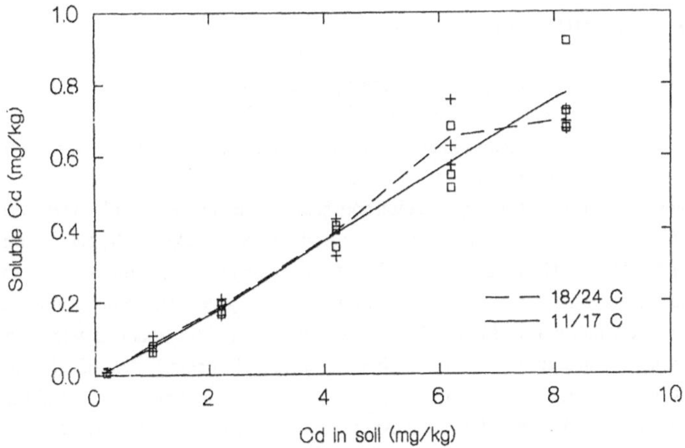

Figure 10. Relationship between total and soluble Cd content in soil (dry matter) grown with red clover at 11/17 °C (night/day) or 18/24 °C (see Figure 5 and 6).

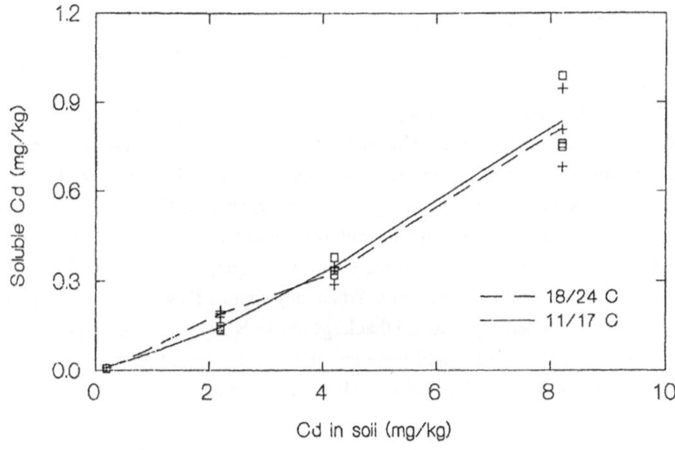

Figure 11. Relationship between total and soluble Cd content in soil (dry matter) grown with beans at 11/17 °C (night/day) or 18/24 °C (see Figure 7 and 8).

Effects of fertilization

The problem of heavy metal input into agrosystems through application of mineral fertilizers has long been known and is widely recognized (Williams & David, 1976; von Steiger & Baccini, 1990). Much less attention seems to have been paid to the fact that adding fertilizer can lead to enhanced uptake of elements other than those contained in the fertilizer itself.

Németh & Grimme (1974) studied the cation concentrations in the solution of several soils used in pot experiments. Fertilizing with K not only increased the concentration of K, but also the concentrations of Na, Mg and Ca. Fertilizing with Mg also increased the concentrations of K and Mn. In a laboratory experiment without plants, addition of K or Mg to a calcareous soil of pH 7.4 increased the concentrations of Mn and Zn to values exceeding those obtained when applying the usual amounts of Mn and Zn fertilizers in pot experiments. The enhancement of soil solution concentrations was reflected in greater plant cation content, with the exception of Na and Mg in the K fertilizer treatments. (As is now known, these exceptions were probably due to antagonistic effects of the increasing K concentrations on the uptake and translocation by plants of Mg and Na).

The effects of fertilization were by no means short-lived. Increased concentrations of Ca, Mg, Mn and, in some cases, of Fe and Zn still prevailed at the end of the vegetation period in soils fertilized with K. Németh and Grimme explained their results by a shift in the equilibrium between adsorbed and dissolved cations in the soil after application of mineral fertilizers in accordance with the laws of ion exchange. They concluded that the effect was of special importance for pot experiments, in which large concentrations of salt occur, and in which there is no leaching, and that the effect would probably be very much smaller in the field.

In contrast, Andersson (1976b) found that increases in the Cd content of wheat grain due to fertilizer application also occurred in a field experiment. He, too, suggested ion exchange as an explanation. Williams & David (1976) investigated the effect of phosphate fertilizers containing Cd on plant Cd content and obtained a linear relationship between the Cd content of wheat grain and the amount of superphosphate applied during a five-year field trial. To their surprise, addition of ammonium nitrate fertilizer, itself containing very little Cd, increased by up to 50 % the uptake of soil Cd from plots that had received no superphosphate as well as increasing the uptake of Cd from plots that had been treated with various amounts of superphosphate. The Zn content of the grain was not related to fertilizer treatments. When they applied P as monocalcium phosphate to oat plants in a pot experiment, they found that large amounts increased the Cd content of shoots, though again the Zn content and yield were not affected. It seems therefore probable that the fertilizer effect reported by Németh & Grimme (1974) will also be important in the field.

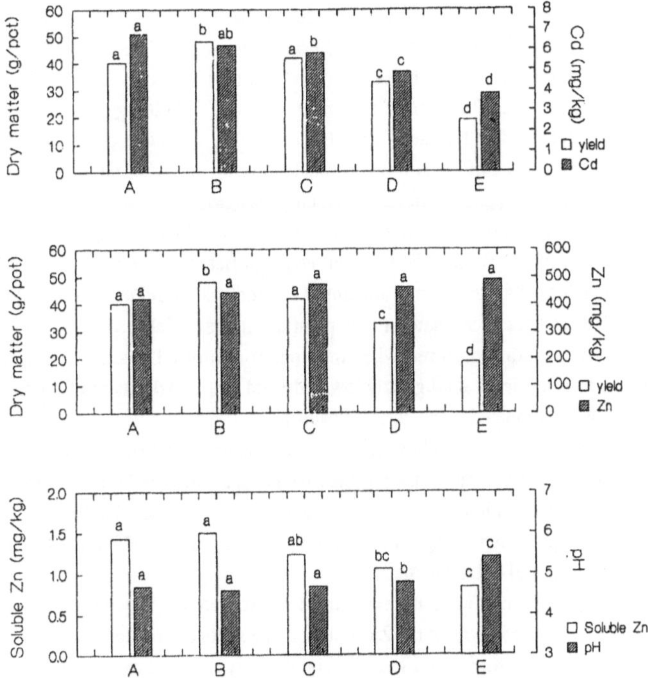

Figure 12 - 14. Effect of fertilization on yield, Cd and Zn content of spinach shoots (Figures 12 and 13) and on the soluble Zn content and pH of the soil after harvest (Figure 14) in a pot experiment. Soil data as in Figure 1; total Zn 52 mg/kg. Two successive crops were grown and their combined yield is shown in Figure 12 as well as in Figure 13. Cd and Zn concentrations refer to dry matter and in the case of shoot contents represent weighted averages of the two crops. Fertilizer treatments were (mg/pot):

	A	B	C	D	E
N	200	100/100*	100	50	0
P	87	43/43*	43	22	0
K	300	150/150*	150	75	0
Mg	48	24/24*	24	12	0*

* split dressing at 0 and 4 weeks (first crop) and at first harvest and 12 days thereafter (second crop).
Identical bars not marked with the same letter are significantly different (P<0.05, Duncan's Multiple Range Test).
For further details see text.

Figure 12 demonstrates what happens when the fertilizing with macronutrients traditionally used for pot experiments is reduced. Pots containing 7.5 kg dry weight of a Swiss agricultural soil were sown with spinach (cultivar Carambole). Nutrients had been added as analytical grade reagents two weeks before sowing. The 'normal' treatment consisted of 200 mg N as NH_4NO_3, 87 mg P as $Ca(H_2PO_4)_2$, 300 mg K as KCl and 48 mg Mg as $MgSO_4$, all values per kg of dry soil. All the pots received 1.13 mg B/kg of dry soil two weeks before sowing. After the first harvest fertilizing with macronutrients was repeated, and a second crop was grown.

Splitting the normal doses of nutrients in two halves, the second half being applied four weeks after sowing and twelve days after the first harvest, significantly increased the yield of spinach shoots. Supplying only half the normal amount of macronutrients did not affect yield, but significantly decreased the Cd concentration in the plant dry matter. Yield reduction and a further decrease in Cd concentration occurred when only one fourth of all macronutrients was applied. Supplying no macronutrients at all greatly lowered yield, but also decreased Cd concentration by approximately 40 % compared to the normal treatment.

Zn content of shoots increased slightly when no nutrients were applied, but the changes were not significant (Figure 13). By contrast, the soluble Zn content as measured after harvesting decreased significantly when fertilization was reduced, while the soil pH increased at the same time (Figure 14). Here the concept of increasing Zn concentration in plant shoots (Figure 13) with increasing solubility of Zn in the soil did not apply.

Expectations were better met when the experiment was repeated with perennial ryegrass (cultivar Bastion). The weighted averages of the Zn content in the shoot dry matter of two successive crops varied between 58 and 68 mg/kg and were not significantly different between treatments, thus reflecting the trend of the solubility curve (Figure 15). However, this time Zn solubility did not change, although pH changed significantly with treatments. The pH values were always larger than with spinach. Unfortunately, it was not possible to measure soluble Cd in the soil in these experiments due to analytical limitations. Values for shoot Cd content of perennial ryegrass were too small (<0.15 mg/kg in dry matter) to give reliable analytical results.

On orchard soils contaminated by sprays Merry *et al.* (1986) found small effects of P and S fertilization with analytical grade reagents on the concentrations of Cu, Pb and As in silver beet, a leaf vegetable. Contaminant contents mostly decreased with increasing P or S treatments, while yields remained practically unaffected. Fertilizing with NH_4NO_3 increased the yield of shoots, had no effect on Cu and Pb concentrations, but greatly reduced As concentration in shoots, probably by decreasing As translocation from roots to shoots.

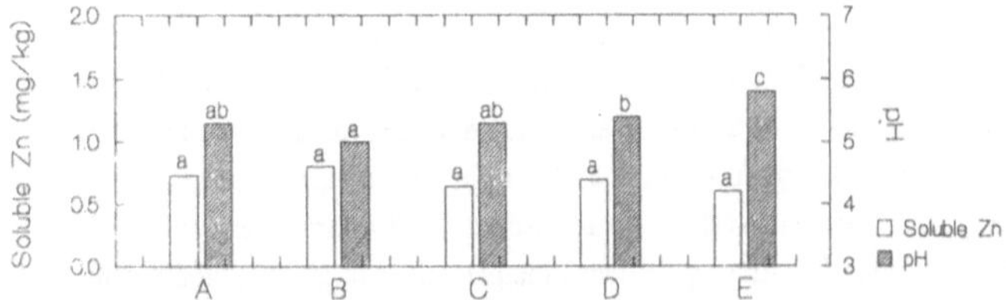

Figure 15. Effect of fertilization on the soluble Zn content and pH of the soil grown with perennial ryegrass after harvest in a pot experiment. Soil data as in Figure 1; total Zn 52 mg/kg. Two successive crops were grown. Soluble Zn concentrations refer to dry matter. Fertilizer treatments were (mg/pot):

	A	B	C	D	E
N	*200*	*100/100**	*100*	*50*	*0*
P	*87*	*43/43**	*43*	*22*	*0*
K	*300*	*150/150**	*150*	*75*	*0*
Mg	*48*	*24/24**	*24*	*12*	*0*

Plant effects on bioavailability of heavy metals

The results that I have reported strongly suggest that contaminant uptake by plants cannot be understood by looking at soil processes alone. Decreasing the pH of a soil, for instance, increases the solubility of most heavy metals, but also influences the ion uptake of plant roots. In fact, Hatch *et al.* (1988) showed that total Cd uptake from a flowing solution culture by four plant species actually *increased* with increasing pH of the solution. At pH 7.0, as compared to pH 5.0, the uptake was enhanced four fold in cocksfoot (*Dactylis glomerata* L.), eight fold in perennial ryegrass, and ten fold in lettuce and in watercress (*Rorippa nasturtium-aquaticum* (L.) Hayek).
Many plants show optimal growth in soil of pH 6 to 7.5. This means they also take up large amounts of nutrients. Since plants cannot differentiate between absorbing beneficial and detrimental substances, contaminants will probably be subject to the same uptake mechanisms as are nutritional elements. In other words, well-fertilized and well-watered fast-growing plants having a sound, efficient root system are also very likely to absorb more contaminants (cf. Figure 12).

On the other hand, plants modify the root environment by the secretion of protons, and this is a most important way of taking up nutrients (Schubert & Mengel, 1989). This secretion, probably interacting with imbalances in the cation and anion uptake, root respiration and microbial activities, can easily result in pH reductions of 1 to 1.5 units in the rhizosphere, and it is influenced by amount and type of fertilization (Schaller & Fischer, 1985; Marschner *et al.*, 1986). Depending on plant species and type of fertilizing nutrients, the pH values in the rhizophere can also increase as a result of shifts in the cation-anion uptake ratio. Plants growing together on the same soil may therefore have different pH values in their own rhizospheres, and it is possible that these values will not be expressed in the pH values measured in the bulk soil (Marschner *et al.*, 1986). Thus different metal solubilities in the immediate vicinity of the root might result, leading to varying uptake patterns among plant species (cf. Figures 1 and 2).

Recently, Mench & Martin (1991) demonstrated that root exudates obtained from two species of tobacco and from maize affect the solubilities of several heavy metals in soil samples. The amounts of Mn, Cu, Cd and Fe extracted from soils by root exudates of identical carbon content differed with plant species. Tobacco root exudates enhanced Cd concentrations but diminished Fe concentrations in the extracts, compared to double-distilled water, whereas maize exudates had no effect upon the two concentrations. Mench & Martin conclude that an increase in Cd solubility in the rhizosphere of apical root zones caused by root exudates may be the cause of large Cd accumulation in tobacco. While Cd and Mn content of the three plant species reflected approximately the amount of metal extracted by the corresponding root exudates, this was not true for Cu. Total root absorption of Cd as well as Cd translocation into shoots of beans has been shown to be influenced by transpiration (Hardimann & Jacoby, 1984). Rapidly transpiring plants took up more Cd from solution and translocated a larger percentage into tops. Transpiration might therefore have contributed to Cd transport into the tobacco leaves studied by Mench & Martin (1991).

Considering all aspects discussed above, the results from two pot experiments presented here for the first time in Figures 16 to 19 will hardly be surprising. By analyzing the soluble content of Cd or Ni in the soil according to VSBo (1986) after harvest it was found that significant differences existed between plant species (Figure 16 and 18). In addition, the ranking did not remain constant when the heavy metal concentrations were increased (Figure 17 and 19). Plants obviously influence the soluble metal contents in soils. If these results are confirmed to the same extent in the field then guide-levels for heavy metals in soil must be reassessed.

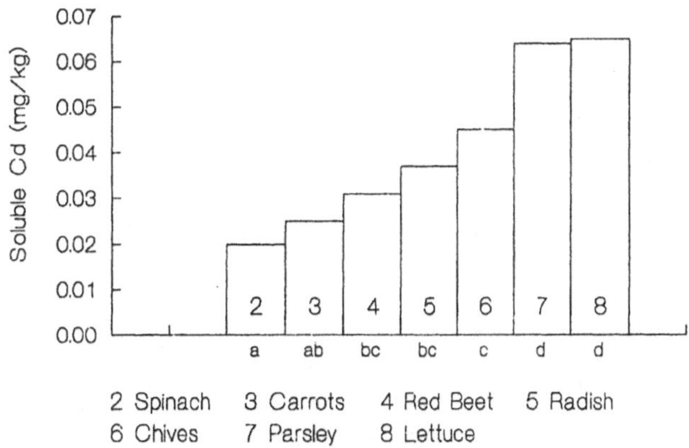

Figure 16. Effect of plant species on the soluble Cd content in soil (dry matter) contaminated with 0.4 mg/kg Cd as sulphate. Soil data as in Figure 1. Bars not marked with the same letter are significantly different (P<0.05, Duncan's Multiple Range Test). Plants had been harvested when deemed ripe for commerce.

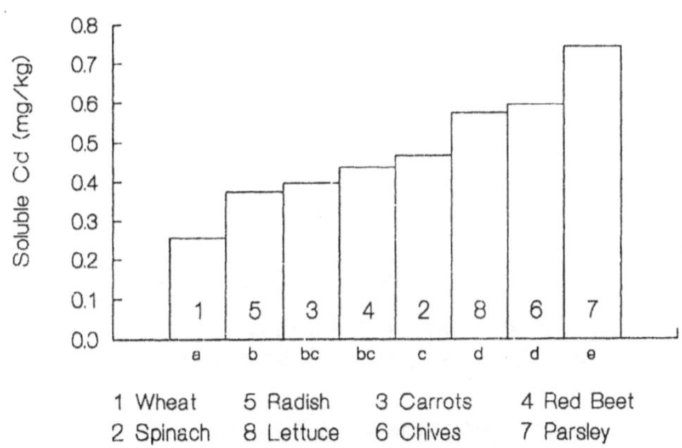

Figure 17. Effect of plant species on the soluble Cd content in soil (dry matter) contaminated with 4 mg/kg Cd (see legend of Figure 16).

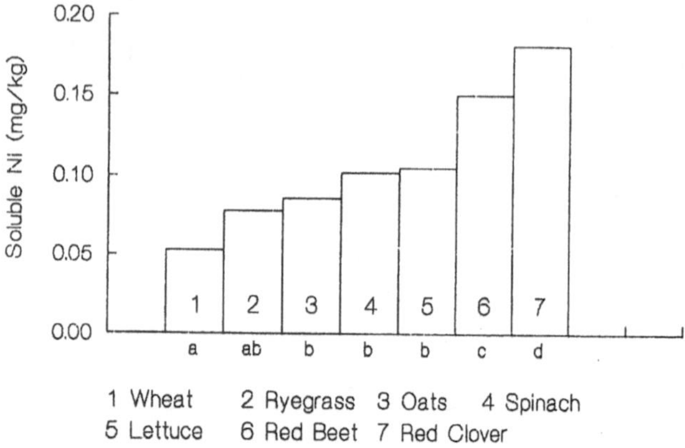

Figure 18. Effect of plant species on the soluble Ni content in soil (dry matter). Total Ni 19 mg/kg, no Ni added. For soil data and statistics see legend of Figure 16.

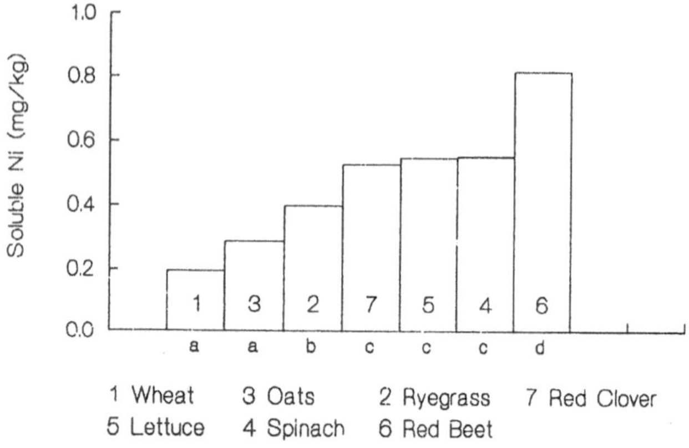

Figure 19. Effect of plant species on the soluble Ni content in soil (dry matter) contaminated with 5 mg/kg Ni as sulphate (see legend of Figure 18).

Acknowledgement

I thank F. X. Stadelmann for supporting the experiments reported here and M. Aegerter, E. Kratochvil, J. Niederhauser, V. Tanner and O. Wyss for their technical assistance. P. Lischer kindly helped me to prepare the graphs.

References

Andersson, A. 1976a. On the determination of ecologically significant fractions of some heavy metals in soils. *Swedish Journal of agricultural Research* **6**, 19-25.

Andersson, A. 1976b. On the influence of manure and fertilizers on the distribution and amounts of plant-available Cd in soils. *Swedish Journal of agricultural Research* **6**, 27-36.

Bingham, F.T., Page, A.L., Mahler, R.J. & Ganje, T.J. 1975. Growth and cadmium accumulation of plants grown on a soil treated with a cadmium-enriched sewage sludge. *Journal of Environmental Quality* **4**, 207-211.

Bingham, F.T., Page, A.L., Mahler, R.J. & Ganje, T.J. 1976. Yield and cadmium accumulation of forage species in relation to cadmium content of sludge-amended soil. *Journal of Environmental Quality* **5**, 57-60.

Crews, H.M. & Davies, B.E. 1985. Heavy metal uptake from contaminated soils by six varieties of lettuce (*Lactuca sativa* L.). *Journal of agricultural Science Cambridge* **105**, 591-595.

Davis, R.D. & Carlton-Smith, C. 1980. Crops as indicators of the significance of contamination of soil by heavy metals. *Technical Report* **140**. Stevenage Laboratory, Stevenage, Great Britain.

Diez, Th. & Krauss, M. 1990. Schwermetallaufnahme durch Gemüsepflanzen bei extremer Bodenbelastung. *Landwirtschaftliches Jahrbuch* **67**, 549-559.

Florijn, P.J., Nelemans, J.A. & Van Beusichem, M.L. 1991. Cadmium uptake by lettuce varieties. *Netherlands Journal of Agricultural Science* **39**, 103-114.

Häni, H. & Gupta, S. 1981. Ein Vergleich verschiedener methodischer Ansätze zur Bestimmung mobiler Schwermetallfraktionen im Boden. *Landwirtschaftliche Forschung* **37**, 267-274.

Hardimann, R.T. & Jacoby, B. 1984. Absorption and translocation of Cd in bush beans (*Phaseolus vulgaris*). *Physiologia Plantarum* **61**, 670-674.

Hatch, D.J., Jones, L.H.P. & Burau, R.G. 1988. The effect of pH on the uptake of cadmium by four plant species grown in flowing solution culture. *Plant and Soil* **105**, 121-126.

John, M.K. 1973. Cadmium uptake by eight food crops as influenced by various soil levels of cadmium. *Environmental Pollution* **4**, 7-15.

John, M.K. & Van Laerhoven, C.J. 1976. Differential effects of cadmium on lettuce varieties. *Environmental Pollution* **10**, 163-173.

Marschner, H. 1983. General introduction to the mineral nutrition of plants. In: Läuchli, A., Bieleski, R.L. (eds): Inorganic plant nutrition, *Encyclopedia of plant physiology N.S.* **15A**, 5-60, Springer, Berlin.

Marschner, H., Römheld, V., Horst, W.J. & Martin, P. 1986. Root-induced changes in the rhizosphere: Importance for the mineral nutrition of plants. *Zeitschrift für Pflanzenernährung und Bodenkunde* **149**, 441-456.

McFarlane, J.C. 1978. Light. In: R.W. Langhans (ed.): *A growth chamber manual,* 15-44, Comstock Publishing Associates, Cornell University, Ithaca USA.

Mench, M. & Martin, E. 1991. Mobilization of cadmium and other metals from two soils by root exudates of *Zea mays* L., *Nicotiana tabacum* L. and *Nicotiana rustica* L. *Plant and Soil* **132**, 187-196.

Merry, R.H., Tiller, K.G. & Alston, A.M. 1986. The effects of contamination of soil with copper, lead and arsenic on the growth and composition of plants. I. Effects of season, genotype, soil temperature and fertilizers. *Plant and Soil* **91**, 115-128.

Németh, K. & Grimme, H. 1974. Einfluss einer Düngung auf die Aufnahme nicht gedüngter Nährstoffe im Gefässversuch. *Zeitschrift für Pflanzenernährung und Bodenkunde* **137**, 203-213.

Page, A.L., Bingham, F.T. & Nelson, C. 1972. Cadmium absorption and growth of various plant species as influenced by solution cadmium concentration. *Journal of Environmental Quality* **1**, 288-291.

Schaller, G. & Fischer, W.R. 1985. Der Einfluss unterschiedlicher Kalium- und Nitraternährung auf pH-Aenderungen in der Rhizosphäre. *Zeitschrift für Pflanzenernährung und Bodenkunde* **148**, 349-355.

Schubert, S. & Mengel, K. 1989. Important factors in nutrient availability: root morphology and physiology. *Zeitschrift für Pflanzenernährung und Bodenkunde* **152**, 169-174.

Sheaffer, C.C., Decker, A.M., Chaney, R.L. & Douglass, L.W. 1979. Soil temperature and sewage sludge effects on metals in crop tissue and soils. *Journal of Environmental Quality* **8**, 455-459.

Siriratpiriya, O., Vigerust, E. & Selmer-Olsen, A.R. 1985. Effect of temperature and heavy metal application on metal content in lettuce. *Scientific Reports of the Agricultural University of Norway* **64**. Ås-NLH, Norway.

Swiss Federal Office of Environment, Forests and Landscape. 1987. Commentary on the ordinance relating to pollutants in soil (VSBo), Bern.

Symeonides, C. & McRae, S.G. 1977. The assessment of plant-available cadmium in soils. *Journal of Environmental Quality* **6**, 120-123.

Tanner, V. 1991. Auswirkungen von Cadmium auf das Wachstum, den Wasser- und den Nährstoffhaushalt unterschiedlich Cd-resistenter Kulturpflanzen bei verschiedenen Umgebungsbedingungen. *Inauguraldissertation.* Philosophisch-Naturwissenschaftliche Fakultät der Universität Bern, Bern.

VSBo 1986. *Ordinance Relating to Pollutants in Soil of June 9, 1986. SR 814.12*, Government of Switzerland, Bern.

Von Steiger, B. & Baccini, P. 1990. Regionale Stoffbilanzierung von landwirtschaftlichen Böden mit messbarem Ein- und Austrag. *Bericht des Nationalen Forschungsprogramms "Boden"* **38**. Liebefeld-Bern.

Willaert, G., Verloo, M. & Sarrazyn, R. 1990. Invloed van de cultivar op de gehalten van minerale elementen in spinazie. *Mededelingen Faculteit van Landbouwwetenschappen der Rijksuniversiteit Gent* **55**, 83-92.

Williams, C.H. & David, D.J. 1976. The accumulation in soil of cadmium residues from phosphate fertilizers and their effect on the cadmium content of plants. *Soil Science* **121**, 86-93.

Yläranta, T. & Sillanpää, M. 1984. Micronutrient contents of different plant species grown side by side. *Annales Agriculturae Fenniae* **23**, 158-170.

Yuran, G.T. & Harrison, H.C. 1986. Effects of genotype and sewage sludge on cadmium concentration in lettuce leaf tissue. *Journal of the American Society for Horticultural Science* **111**, 491-494.

R. Frossard
Swiss Federal Research Station for Agricultural Chemistry and Hygiene of Environment
CH-3097 Liebefeld-Bern
Switzerland

Discussion

Question by S. van der Zee

Did you assess dissolved heavy metal in $NaNO_3$ at the same temperatures as used in the plant uptake experiments, because if you did not you have eliminated the temperature effect and showed only that e.g. Cd-sorption was reversible in your plots of dissolved and total soil heavy metals. (I do not imply that the temperature effect in this case was significant).

Answer by R. Frossard

Extraction was carried out as recommended by the Swiss Ordinance Relating to Pollutants in Soil (VSBo, 1986). The Ordinance does not specify at which temperature extraction should take place. We chose 21 °C because it corresponded to the room temperature, and because it was average between 17 and 24 °C. During the plant experiments, night temperatures were lower in any case.
(In fact we have extracted six samples at 17, 21 and 24 °C since this discussion was held. No significant differences in the soluble Cd content were found between the extraction treatments).

Question by P. Brookes

What are the mechanisms whereby plants increased the solubility of soil Cd in your experiments?

Answer by R. Frossard

We have not investigated this yet. Possible mechanisms include changes in pH in the rhizosphere as a result of a) proton extrusion following nutrient uptake by roots, b) microbial decomposition of root exudates, c) increased concentration of respiratory CO_2 due to high root densities.

ASSESSMENT OF CONTAMINANT FLUXES THROUGH THE SOIL SURFACE

G. Furrer

Most contaminants enter the soil through its surface, from the air, for example in rain and snow, or by direct application, e.g. in fertilizers, sewage sludge, etc. Some may also rise into the soil from below, for example from polluted ground water, from buried waste, or from leaking installations. Further, there are naturally occuring toxins, such as heavy metals released by weathering and some organic compounds produced microbially, and these too contribute to the burdens in the soil.

The kinetics of transport through the topsoil depends on the physical and chemical properties of the contaminants which themselves can be affected by microbial activities (transformations). The processes include dry, occult and wet atmospheric deposition, snowmelt infiltration, additions of agrochemicals, outgassing of volatile chemicals, and erosion of top soil material by rain and wind.

The methods for measuring atmospheric deposition rates of the pollutants seem not to be sufficiently accurate. Especially for the purpose of soil pollution monitoring on small scales presently available methods cannot be considered as sufficiently accurate. No single method can account for all atmospheric deposition processes and all kinds of contaminant (in particular the broad range of organic pollutants present in rain are difficult to sample and to analyse). It is therefore crucial to combine the best available sampling and measuring devices in order to record all the important deposition processes as accurately as possible.

In forests bulk precipitation sampling below the canopy (including throughfall, stemflow and litter fall) is currently the only direct method to determine the rates of total atmospheric deposition, although internal cycling via soil, root and canopy may lead to significant overestimates for some chemicals. For vegetation with low canopies, where it is not possible to sample precipitation below the canopy, micrometeorological techniques appear to be better suited. These methods, however, still need further development for this kind of application in soil monitoring. Because of the high labour requirements application of these methods will remain restricted to a few selected sites if included in a long-term monitoring programme.

The study of atmospheric deposition should not be restricted to the monitoring of deposition rates. Additionally, the mechanisms of deposition still need to be investigated in order to improve the understanding of the three pathways of atmospheric deposition (dry, occult and wet) with respect to different kinds of vegetation and soil surfaces. Only so shall we be able (i) to predict the behaviour of different pollutants as they cross the boundary between soil and atmosphere and (ii) to extrapolate results from a specific observation site to other sites.

Much can be learnt about deposition mechanisms and pathways by the study of individual rain and storm events. Such investigations are of particular relevance for episodic events (such as snowmelt and rain after long periods of draught) when the input of atmospheric pollutants into soil can be extraordinarily large.

In conclusion, measuring atmospheric input to soil is costly if it is done properly. Although it is of crucial importance to have some long-term observations in national soil monitoring programmes, measuring atmospheric deposition will always be restricted to a limited number of sites.

Dr. G. Furrer
Institut für terrestrische Ökologie (ITÖ)
ETH Zürich
Grabenstrasse 3
CH-8952 Schlieren
Switzerland

ASSESSMENT OF CONTAMINANT FLUXES ACROSS THE SOIL PLANT INTERFACE AND BELOW THE ROOTING ZONE

E. Matzner

The major objective of the mass balance approach in routine monitoring of soil pollution should be to *assess changes* in soil storage and to *identify relevant pathways* of pollutants by determining and comparing the input and output fluxes to a defined compartment of soil. This approach also provides a useful basis for research programmes, supplementing the monitoring, which are designed to find *explanantions* for the observed changes and the processes involved.

The following discussion deals with two of the major fluxes to be considered in assessing soil contamination in a mass balance approach. These are the uptake of pollutants by plants and the movement of pollutants by seepage below the rooting zone. Leaching with the seepage water is equivalent to a loss of pollutants from the soil, whereas material flowing from the soil into the plant compartment will partially return into the soil compartment at a site.

The mass balance approach can be used to establish balances over different *scales* (e.g. global, regional, plot, soil horizon). Here focus is on the problems related to mass balances on a plot scale and especially on the methods to measure fluxes and the associated uncertainties.

Contaminant uptake by plants

The *uptake of pollutants by plants* and the subsequent removal by harvesting can play a major role in the mass balance of some elements. The determination of this flux within an acceptable range of uncertainty seems to be possible for most pollutants by analysing the plant material harvested.

Plant uptake and leaching of pollutants from canopy surfaces will affect the determination of actual rates of atmospheric deposition by the 'throughfall method'. Although this leaching effect appears to be of small import for most of the pollutants so far studied more research should be done to prove it.

Contaminant displacement by deep seepage

When the merits and disadvantages of various methods for assessing pollutant fluxes with *seepage water* were evaluated in the workshop it was concluded that observation sites in which the seepage water is channelled on and collected at an impermeable sublayer represent an ideal situation. The use of such 'natural lysimeters' has the following advantages.

- The total amount of seepage water can be measured more easily and accurately than by other methods (e.g. numerical simulation based on Darcy flux).
- The flux is integrated over a fairly large area, and transport of pollutants by preferential flow, e.g. in macropores, is included.
- The soil remains largely undisturbed.
- There is no need to extract soil solution by suction lysimeters, free draining lysimeters or destructive methods, such as centrifuging. Suction lysimeters have drawbacks such as potential retardation of pollutants on the cup material, disturbance of the soil during installation and a high risk of missing preferential flow in macropores.

Disadvantages of the natural lysimeters might be:

- lack of spatial resolution of sources or sinks in the soil profile,
- uncertainity about the impermeability of the sublayer,
- effect of water saturation above the impervious layer on the chemical composition of water samples,
- the fact that these conditions might be difficult to find.

The extrapolation of flux measurements at the plot scale to larger areas appears in principle to be feasible. It should be based on a stratification of the domain according to the major factors that influence fluxes such as vegetation, management, soil properties, pollution climate and physical climate.

Assessment of vertical gradients of pollutant fluxes within the rooting zone, as well as of episodic events, should be part of a monitoring programme because both provide valuable information about local sources and sinks and about the processes involved in the transport of pollutants. This information can be very helpful for understanding the causes of change and provides a bridge to research related with soil monitoring. Such additional investigations can be more cost-effective if they are limited to a small selection of monitoring sites in a network, rather than being performed everywhere.

Uncertainties in measuring mass fluxes

In conclusion, the mass balance approach should be seen as an essential part of a soil monitoring programme because it might provide early forecasts of changes in the soil. However, one must be aware of the potential errors in establishing a flux balance. Many fluxes are hard to quantify at present. This is especially true for organic substances, while for some other pollutants also estimation uncertainties much smaller than the estimated rates have been obtained in mass balance studies. Future scientific work should focus on the quantification of errors as well as on methods to minimize them.

Uncertainties in flux balances depend not only on the type of pollutant, but also on site conditions. For instance in forest ecosystems the input of metals to the soil occurs fairly steadily with time, while in agricultural ecosystems inputs, especially of pesticides and fertilizers, are episodic. In the latter fluctuations of the concentrations of pollutants at the upper boundary of the soil are large. The influence of preferential flow and the uncertainty in establishing seepage water fluxes may thus be much greater in the topsoil of arable land than in a forest.

Conclusions

While the flux balance method has its shortcomings, it seems to be the only one suitable for investigating the balances of pollutants that do not accumulate, such as protons, nitrogen, and some organic compounds. Periodic soil inventories also have their shortcomings in determining changes within a reasonable time, and so both methods, the mass balance as well as the soil inventory have their proper roles. They might be used as alternatives or together depending on circumstances. The two methods need to be compared and further developed to maximize their reliability for determining long term trends.

Finally, mass balances on the small scale of individual soil monitoring sites might be useful for validating mass balances that are established for larger scales and *vice versa*.

E. Matzner
Lehrstuhl für Bodenökologie
Universität Bayreuth
Postfach 101251
D-8580 Bayreuth
Germany

SECTION 3

Assessing Soil Contamination

Soil Monitoring, Monte Verità, © Birkhäuser Verlag Basel

SAMPLING EFFICIENCY FOR MASS RECOVERY CALCULATIONS

Markus Flury

In the study of solute transport in field soils, the amount of chemicals recovered by a sampling campaign largely determines the quality of the experiment as well as the verification of any model. However, generally only a small part of the experimental domain can be sampled, and the mass recovery has to be estimated from these samples. The mass recovery calculation is influenced by the instruments used to sample the chemicals. This can be shown by representing a physical instrument, for instance a soil corer or a suction cup, by a mathematical weighting function. A measurement is considered as an average of a microscopic property over space and time. Volumetric sampling by soil cores is compared with non-volumetric sampling by suction cups. Given certain information on the spatial structure of the chemical distribution, the quality of the mass recovery calculation can be assessed.

Introduction

Tracer experiments play a fundamental role in investigating the mass flux of solute in the field scale. Such an experiment consists of an application of a tracer on to the soil surface and a sampling some time later. The amount of tracer in the sample is then measured, for example by chemical analysis. One important quantity we want to know is the amount of solutes that could be recovered. This is different from the amount actually recovered by the sampling, and we shall call the latter the *effective* mass recovery. Usually, we are interested in estimating the mass of chemicals recovered in relation to the total mass of chemicals applied, what we shall call the *estimated* mass recovery, or simply the mass balance.

The fundamental problem in calculating this value is that, in general, we can sample only a small subset of the experimental domain. The theoretical value of the mass of a chemical in a domain V has to be estimated from the samples of domain v. This is a statistical problem, and more particularly a geostatistical problem. To apply geostatistics we assume that the property of interest, here the mass density of a chemical, is a random variable in space, i.e., a regionalized variable, which satisfies the intrinsic hypothesis (Matheron, 1965). The intrinsic hypothesis implies that the mean of the difference between two points is constant or depends only on the separation in distance and direction of the two points and that the variance of the difference depends only on the separation in distance and direction. By block kriging we obtain an unbiased minimum variance estimate of the mean mass density in V and the estimation variance itself, often denoted by σ_E^2. The main disadvantage of this approach is that the spatial structure of the regionalized variable has to be determined beforehand, and this often requires more data than are generally available or can be afforded.

I have chosen a somewhat different approach, the kernel of which is the concept of measurement as introduced by Marle (1967) using ideas from Matheron (1965). Its basis is that a measurement can be expressed mathematically by a convolution of a microscopic property with a weighting function. The result is a macroscopic property. This is a fundamental concept in the theory of flow of fluids in porous media (Bear, 1972). It was further developed by Ene (1981), Marle (1982), Cushman (1983, 1984, 1986), and Baveye and Sposito (1984, 1985). Cushman (1984, 1986) introduced the concept of scale of an instrument. Baveye and Sposito (1985) showed that a mathematical representation of an instrument can vary in space and time and that the effective volume of measurement is influenced by instrumental and soil variability. For the interpretation of measurements it is therefore important to consider the limitations of the instruments.

I apply the concept of measurement to calculate mass recovery or mass balances in field experiments. I shall show that mass balance calculations depend on the instruments used to sample the given domain, and discuss the elements of mass recovery calculations in field experiments. Next, mass recovery calculations based on volumetric samples (soil cores) and non-volumetric samples (suction cups) are discussed. To quantify the errors in the estimation of the mass recovery, it is necessary to assume a spatial structure for the mass density of a chemical. This spatial structure is given in the form of the existence of a representative elementary volume (Bear, 1972). In the last section the effect of scale and orientation of the instrument is investigated.

Mass recovery calculation

The microscopic mass density $c(\mathbf{x})$ $[ML^{-3}]$ of molecules, visualized as mathematical points, at a point $\mathbf{x}$ in space, where $\mathbf{x}$ denotes the spatial coordinates, can be represented by (Sposito, 1978)

$$c(\mathbf{x}) = \sum_{k=1}^{N} m\, \delta(\mathbf{X}_k - \mathbf{x}) \,, \qquad \mathbf{x} \in \mathbb{R}^3 \,, \tag{1}$$

where m is the mass of any one of the N molecules in space, $\delta(\mathbf{x})$ is the Dirac delta distribution (e.g., Arfken, 1985), $\mathbf{X}_k$ is the position of the kth molecule, and $\mathbb{R}^3$ is the three dimensional space of the real numbers. We can now define the total mass of solutes in a set B by integration of the mass density

$$\int_B c(\mathbf{x})\, d\mathbf{x} \,, \tag{2}$$

where the set B corresponds to a spatial domain of soil, $c(\mathbf{x})$ $[ML^{-3}]$ is the microscopic solute mass density, and $\mathbf{x}$ denotes the spatial coordinates. Note that the domain B does not have to be connected (forming one connected domain).

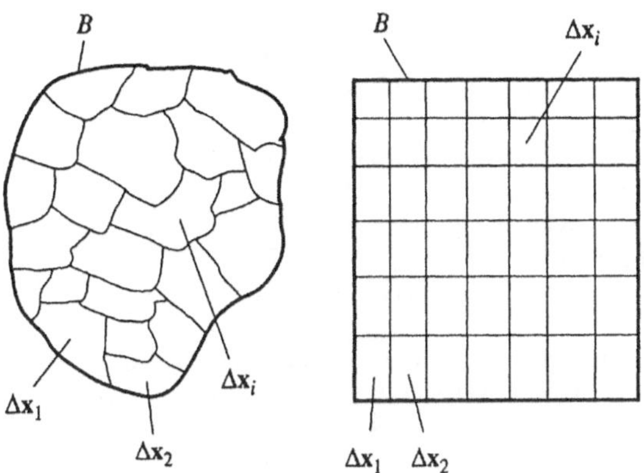

Figure 1 Two arbitrary representations of the spatial discretization of a domain B by subdomains $\Delta \mathbf{x}_i$.

As we can measure only macroscopic field variables with our instruments, (2) can in general not be determined exactly. The microscopic solute mass density $c(\mathbf{x})$ therefore has to be replaced by the macroscopic mass density $\bar{c}(\mathbf{x})$ $[ML^{-3}]$, defined as the mass of a chemical in a non-vanishing volume divided by this volume. The integral (2) has to be approximated by

$$\sum_i \bar{c}(\mathbf{x}_i)\,\mu(\Delta\mathbf{x}_i)\,, \tag{3}$$

where i are the different samples taken, $\bar{c}(\mathbf{x}_i)$ is the *local macroscopic* mass density at the point $\mathbf{x}_i$, $\Delta\mathbf{x}_i$ is the spatial discretization of B such that

$$\bigcup_i \Delta\mathbf{x}_i = B\,, \quad \Delta\mathbf{x}_i \cap \Delta\mathbf{x}_j = \emptyset\,, \qquad \forall i \neq j\,, \tag{4}$$

and

$$\sum_i \mu(\Delta\mathbf{x}_i) = \mu(B)\,. \tag{5}$$

The symbol $\emptyset$ denotes the empty set and $\mu(\mathcal{F})$ is volume of the argument $\mathcal{F}$. This discretization is schematically represented in Figure 1. The sum in (3) is the *estimated* mass recovery or the mass balance. The relation between microscopic and macroscopic concentrations can be given by definition as (Marle, 1967)

$$\bar{c}(\mathbf{x}_i) = \int c(\mathbf{x}_i + \mathbf{x}')m_i(\mathbf{x}')\,d\mathbf{x}'\,, \tag{6}$$

where

$$\int m_i(\mathbf{x})\,d\mathbf{x} = 1\,. \tag{7}$$

The weighting function $m_i(\mathbf{x})$ is supposed to be a regular function, and the subscript i indicates that the function can vary in space. Physically, equation (6) and (7) express that the macroscopic concentration results from averaging the microscopic concentrations over a certain volume. The exact averaging procedure is given by the function $m(\mathbf{x})$. In a more general case, the averaging can also be dependent on time. Note that (6) makes sense only if the integration gives a physically reasonable property, which means that the variable $c(\mathbf{x})$ multiplied by the infinitesimal element of integration has to be additive (Marle, 1967; Hassanizadeh and Gray, 1979). This is clearly the case for mass densities integrated over a volume. I shall call a function $m(\mathbf{x})$, which satisfies (7), an instrument (Cushman, 1984). I can now define the deviation b between experimental and theoretical mass recovery as the difference of (3) and (2)

$$b := \sum_i \bar{c}(\mathbf{x}_i)\, \mu(\Delta\mathbf{x}_i) - \int_B c(\mathbf{x})\, d\mathbf{x} \,. \tag{8}$$

The next sections discuss equation (8) for two different physical instruments commonly used in field experiments to sample chemicals, a core sampler and a suction cup.

Volumetric sampling

When taking volumetric soil samples the weighting function can be expressed by

$$m_i(\mathbf{x}) = \chi_{D_i}(\mathbf{x}_i + \mathbf{x}) \frac{1}{\mu(D_i)} \,, \tag{9}$$

where D_i is the sampling domain centered at point $\mathbf{x}_i$, $\chi_{D_i}(\mathbf{x})$ is the characteristic function of D_i, i.e., $\chi_{D_i}(\mathbf{x}) = 1$ if $\mathbf{x} \in D_i$ and $\chi_{D_i}(\mathbf{x}) = 0$ otherwise, and $\mu(D_i)$ is the sampling volume. For instance, if the samples are taken by a rigid cylindrical core of radius R and length h, then

$$\mu(D_i) = \pi R^2 h \,. \tag{10}$$

Equations (9) and (10) were given by Marle (1967) for the sphere (then $\mu(D_i) = \frac{4}{3}\pi R^3$) and by Baveye and Sposito (1984) for the cylinder. Inserting (9) in (6) yields

$$\bar{c}(\mathbf{x}_i) = \int c(\mathbf{x}_i + \mathbf{x}') \chi_{D_i}(\mathbf{x}_i + \mathbf{x}') \frac{1}{\mu(D_i)}\, d\mathbf{x}' \,, \tag{11}$$

using the substitution $\mathbf{y} = \mathbf{x}_i + \mathbf{x}'$, then

$$\begin{aligned} \bar{c}(\mathbf{x}_i) &= \frac{1}{\mu(D_i)} \int c(\mathbf{y}) \chi_{D_i}(\mathbf{y})\, d\mathbf{y} \\ &= \frac{1}{\mu(D_i)} \int_{D_i} c(\mathbf{x})\, d\mathbf{x} \,. \end{aligned} \tag{12}$$

This is an important result, as the integral in (12) corresponds to the total mass of solutes in the domain D_i. This value can be determined experimentally for instance by extracting the soil sample by an appropriate procedure and quantifying the solutes by an analytical method. Hence, when $\bar{c}(\mathbf{x}_i)$ is known, expression (3) could be calculated. However, the approximation (3) might not be reliable, because one cannot sample the entire volume of the experimental plot. Only a fraction of the entire volume can be sampled. The deviation b of the experimental recovery from the actual recovery (8) is given by inserting (12) in (8):

$$b = \sum_i \frac{1}{\mu(D_i)} \int_{D_i} c(\mathbf{x})\, d\mathbf{x}\, \mu(\Delta\mathbf{x}_i) - \int_B c(\mathbf{x})\, d\mathbf{x} \,. \tag{13}$$

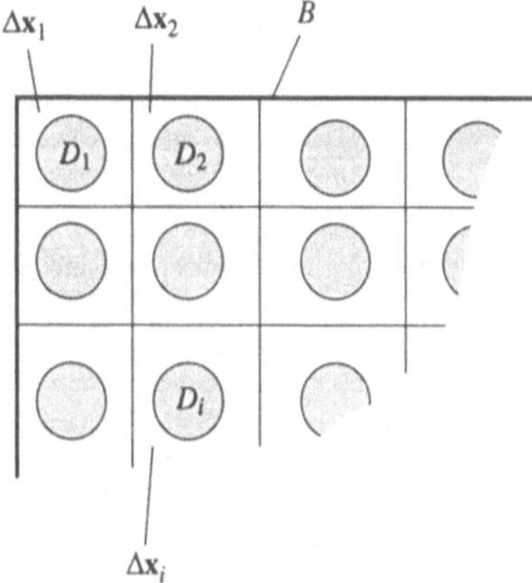

Figure 2 Relation between B, the discretization domain $\Delta \mathbf{x}_i$, and the sampling domain D_i. For instance, the $\Delta \mathbf{x}_i$s might form a grid and the D_is might represent cylindric core samplers.

Let us assume for simplicity that the ratio $\kappa := \mu(\Delta \mathbf{x}_i)/\mu(D_i)$ is constant and $\kappa \geq 1$ (the latter simply means that the volume of the space element $\Delta \mathbf{x}_i$ is larger than that of the core sampler); then (13) can be written as

$$b = \kappa \int_{\bigcup_i D_i} c(\mathbf{x})\, d\mathbf{x} - \int_B c(\mathbf{x})\, d\mathbf{x} \,, \tag{14}$$

where it has been assumed that the D_is are disjoint. Figure 2 shows the relation between the domains B, $\Delta \mathbf{x}_i$, and D_i. We define the union of the sampling volume as $A := \bigcup_i D_i$. Then the first integral in (14) can be split:

$$b = \kappa \int_B c(\mathbf{x})\, d\mathbf{x} - \kappa \int_{B \setminus A} c(\mathbf{x})\, d\mathbf{x} - \int_B c(\mathbf{x})\, d\mathbf{x} \,, \tag{15}$$

where $B \setminus A$ is the complement of A in B. It follows that as $B \setminus A \to \emptyset$ for $\kappa \to 1$ then all mass is recovered by the chosen sampling procedure, or equivalently

$$\lim_{\kappa \to 1} b = 0 \,. \tag{16}$$

This result shows that the experimental mass recovery improves as the ratio of the sampling volume $\mu(D_i)$ relative to the discretization volume $\mu(\Delta \mathbf{x}_i)$ increases. This is intuitively clear and is an expression of the linearity of integration and summation. Note that (16) has an analogue in geostatistics. There, the dispersion variance σ_E^2 of the mean value in V estimated by the samples of size v goes to zero as $v \to V$.

In practical situations the limit $\kappa \to 1$ in (16) can hardly ever be achieved, as one cannot sample the entire soil volume, and the bias (15) will not cancel. In special situations, as for instance when $c(\mathbf{x})$ is approximately constant or when κ is close to unity, then the deviation b will cancel. For discussing (15) when $\kappa \not\to 1$, I introduce the concept of the representative elementary volume (REV). For our purposes it suffices to define the REV in the sense of Hubbert (1956) and Bear (1972) as the volume whose characteristic length d is given by the relation

$$\ell \ll d \ll L\,, \tag{17}$$

where ℓ is the characteristic length of the microscopic scale and L is the scale of the gross inhomogeneities (Hassanizadeh and Gray, 1979). Here, the lengths ℓ and L are scales of the mass density. Other properties of a porous medium may have different scales. Cushman (1983) discusses relation (17) in comparison to a more restrictive relation. The selection of d in (17) is based on the requirement that a *measured* quantity, for instance the mass density, is independent of the size of the averaging volume W, i.e.,

$$\left(\frac{\partial(\text{mass density})}{\partial W}\right)_{W=\text{REV}} = 0\,, \tag{18}$$

and d is the characteristic length of the REV (Baveye and Sposito, 1984; Bear and Bachmat, 1984). The REV is usually the minimal volume, V, for which (18) holds. Here, I will call all volumes for which (18) is valid an REV. Hence, the REV is not constant, but I require at least that the REVs form a connected space.

Let me now apply this concept of the REV to (15) and consider two cases:

$$\text{(i)} \quad \mu(D_i) \approx \text{REV} \quad \text{and} \quad \mu(\Delta\mathbf{x}_i) \approx \text{REV} \tag{19}$$

$$\text{(ii)} \quad \mu(D_i) \ll \text{REV}\,. \tag{20}$$

If $\kappa = 1$ then both cases are equivalent in terms of equation (15), because no assumption about existence or size of an REV was made in deriving it.

Next consider $\kappa \neq 1$. Then in case (i) the solute mass density in the sample space D_i equals approximately the solute mass density in the grid space $\Delta\mathbf{x}_i$, because both volumes are within the scale of the REV. The terms in equation (13) cancel, and $b \approx 0$. Hence, mass balance calculations yield accurate results. In case (ii) no simplification of (15) can be made, and the mass recovery estimation can be biased. However, quantification of this bias requires that the function $c(\mathbf{x})$ be specified, which is in general impossible.

Non-volumetric sampling

When *measuring* the macroscopic field variables by soil cores, the weighting function m is given by equation (9). If the physical instrument does not provide volumetric samples then the weighting function is much more difficult to specify. Non-volumetric sampling devices are, for example, suction cups (Barbee and Brown, 1986) and capillary-wick samplers (Holder *et al.*, 1991). Mass recoveries may also be calculated by measuring the mass flux through a given plane. The mass flux in soil can be measured using for instance a lysimeter. I do not discuss non-volumetric sampling in general, I restrict here to the case of suction cups. So let us try to derive the weighting function for a suction cup. Suppose S_i denotes the surface of the suction cup, centered at point $\mathbf{x}_i$, and $\mathbf{n}$ is the unit vector normal to the infinitesimal surface element. The total mass of chemical sampled by a suction cup positioned at $\mathbf{x}_i$ during a time period t is given by

$$\int_0^t \int \mathbf{j}(\mathbf{x}_i + \mathbf{x}', t') \cdot \mathbf{n}_{S_i}\, \chi_{S_i}(\mathbf{x}_i + \mathbf{x}')\, d\mathbf{x}'\, dt' , \tag{21}$$

where $\mathbf{j}(\mathbf{x}, t)$ $[ML^{-2}T^{-1}]$ is the microscopic solute flux at point $\mathbf{x}$ and time t, $\chi_{S_i}(\mathbf{x})$ is the characteristic function of S_i, and $\cdot$ denotes the scalar product. The integral over the space domain corresponds to an area average operator (Hassanizadeh and Gray, 1979). The flux can be written as (Sposito, 1978)

$$\mathbf{j}(\mathbf{x}, t) = \sum_{k=1}^{N} m\, \delta(\mathbf{X}_k(t) - \mathbf{x})\, \mathbf{v}_k , \tag{22}$$

where m is the mass of a single molecule, $\mathbf{X}_k(t)$ is the position of the kth molecule in space at time t, and $\mathbf{v}_k$ $[LT^{-1}]$ is the velocity of the kth molecule (see also equation (1)). By inserting the right-hand side of equation (22) into (21), the total mass of chemical sampled by a suction cup can be written as

$$\int_0^t \int \left[\sum_{k=1}^{N} m\, \delta(\mathbf{X}_k(t) - \mathbf{x}) \mathbf{v}_k \right] \cdot \mathbf{n}_{S_i}\, \chi_{S_i}(\mathbf{x}_i + \mathbf{x}')\, d\mathbf{x}'\, dt' . \tag{23}$$

The macroscopic mass density $\bar{c}^*(\mathbf{x}, t)$ collected by the suction cup is then given by dividing the total mass (equation (23)) by the total volume V_{tot} of liquid collected by the porous cup. The superscript * indicates that this macroscopic mass density does not correspond to a mass per volume of soil. To formulate the weighting function $m(\mathbf{x})$, equation (23) must be first rearranged in a form like (6), and an explicit formulation of $m(\mathbf{x})$ has to be found. To proceed in this direction two fundamental problems arise. First, the flux $\mathbf{j}$ can vary in space and second, it can vary in time. Hence, it is difficult or even impossible to specify (23).

We can give an intuitive explanation for this difficulty by looking at the measurement process on a macroscopic scale. For collecting soil solution a suction has to be applied to the porous cup. A potential gradient builds up in the soil, and water flows toward the porous cup. The soil solution drawn into the suction cup is a mixture of liquid originating from pores of different diameters and lengths. Pores with a water potential less than that maintained in the suction cup will not even contribute any solution. Other pores with a large water potential will preferentially contribute solution. Furthermore, the sampling depends on time. For example, some pores can empty during the sampling period and will not contribute thereafter to the flux. The sampling volume and its shape can change during the measurement. Figure 3 schematically illustrates the different measurement volumes of a suction cup for two different times. Whereas at time 1 the whole surface of the suction cup collects solution, at the later time 2 solution enters only through a part of the cup's surface.

This leads to another problem, which is associated with the sphere of influence of the instrument. This is discussed in the next section.

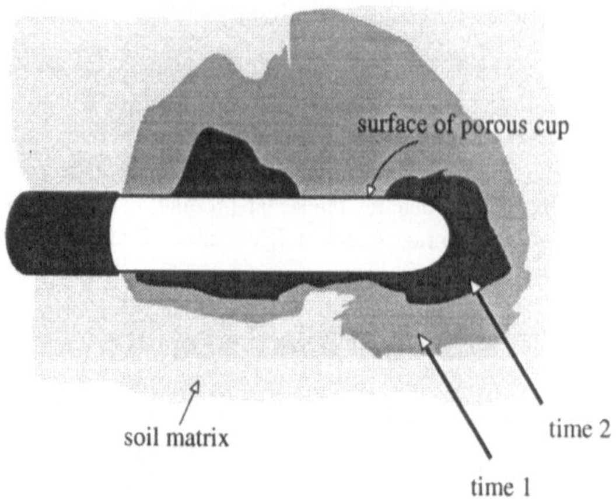

Figure 3 Sampling soil solution by a suction cup. The two dark shaded areas are the domains from where soil solution is collected by the suction cup. At time 1, the whole surface of the cup collects solution. At the later time 2, pores drained and only a part of the cup surface collects solution. Both times are within the measurement period of the sampling.

Instrument Scale

The mathematical representation of an instrument depends on its spatial orientation. This same physical instrument has different mathematical forms for different orientations (Cushman, 1986). A soil corer, which length is usually longer than its radius, can be installed vertically or horizontally in soil. The instrument is still the same, but the measurement is different. To take that point into account I introduce the concept of scale of an instrument according to Cushman (1984). He defines the scale λ of an instrument as

$$\widehat{m}(\omega) \downarrow\downarrow 0 \quad \text{for} \quad |\omega| \gg \lambda^{-1} . \tag{24}$$

The symbol $\downarrow\downarrow$ means a rapid decrease—Cushman does not specify the order of the decrease, but I assume here an order of x^{-1}—and the $\widehat{}$ is the spatial Fourier transform

$$\widehat{m}(\omega) = \int m(\mathbf{x}) \exp(-i\omega \cdot \mathbf{x})\, d\mathbf{x} , \tag{25}$$

where $\widehat{m}(\omega)$ is the weighting function at frequency ω and $i = \sqrt{-1}$. It is very convenient to use Fourier transforms for representing a measurement (Marle, 1967), as the convolution integral is replaced by a multiplication. Note that equation (12) is a convolution when the spatial domain D_i is moved to the origin.

I apply now the concept of scale to the soil core device. Then, the Fourier transform of (9) and (10) in cylindrical coordinates is

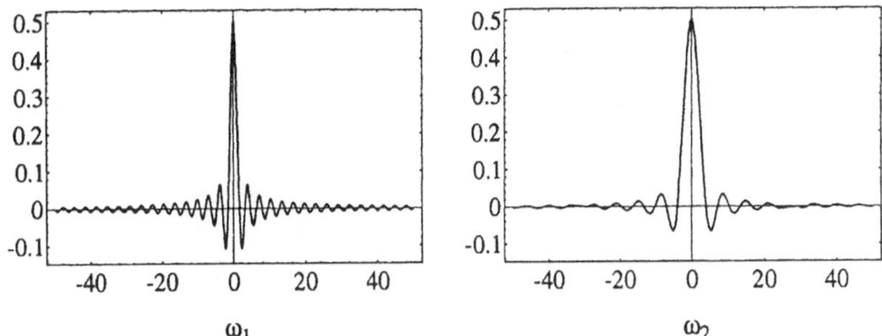

Figure 4 *Fourier transformed weighting function $\widehat{m}(\omega)$ for a cylindric core sampler, Equation (26), with length $h = 4$ and radius $R = 1$. The coordinates are dimensionless and the abscissa is the frequency ω. In the left plot, ω_2 is kept constant near the zero point, in the right plot this is done with ω_1. Note that the function $\widehat{m}(\omega)$ has removable singularities at $\omega_1 = 0$ and $\omega_2 = 0$.*

$$\widehat{m}(\boldsymbol{\omega}) = \frac{8\sin(\frac{h}{2}\omega_1)}{h^2 R\omega_1}\frac{1}{\omega_2}J_1(R\omega_2)\,, \tag{26}$$

where $\boldsymbol{\omega} = (\omega_1, \omega_2)$ and ω_1 corresponds to the direction of length of the core (length h), ω_2 to the plane perpendicular to the direction of h (radius R), and J_1 is the Bessel function of first kind of integral order 1 (Arfken, 1985). Figure 4 shows two plots of the Fourier transformed weighting function (26) for constant ω_1 and ω_2, respectively.

The function $\widehat{m}(\boldsymbol{\omega})$ decreases (with order ω_1^{-1} and $\omega_2^{-3/2}$) for increasing ω, and I define the scales according to (24) and physical reasoning as

$$\lambda_1 = h,\ \lambda_2 = R\,. \tag{27}$$

With the core sampler, the method described of taking the Fourier transform of the measurement function is not satisfactory. The Fourier transform decreases only slowly and the scale of the instrument based on equation (24) is not well defined. The physical scales, however, are obvious. Nevertheless, this example of the core sampler illustrates the principle of the method.

For an example of the application of the instrument scale, I again consider an REV with characteristic length scales (d_x, d_y, d_z) in the three Cartesian directions. I assume that the instrument has scales $\lambda_1 \gg \lambda_2$. In the isotropic case $(d_x \approx d_y \approx d_z)$ the orientation of the instrument does not matter. If one of the characteristic lengths is much larger than the others $(d_x \approx d_y \ll d_z)$, then it is advisable to choose the larger instrument scale (λ_1) in direction of the larger characteristic length (d_z) of the REV.

For example, consider a field experiment where water containing a mobile dye tracer is infiltrated into a soil and suppose that preferential flow in the form of macropore flow or fingering occurred. I further make the plausible assumption that the vertical scale of the flow patterns of the dye tracer is much larger than the horizontal scale. Such a situation is

Figure 5 Vertical view of a soil profile after infiltration of a dye tracer. The dye tracer was applied onto the soil surface, and moved along preferential flow paths through the soil. The shaded area shows where the dye reached.

represented in Figure 5. Finally let us assume that an REV for the mass density of the tracer exists and that its horizontal scale is much larger than the widths of the fingers. This means that in horizontal direction the REV contains many fingers. Then, the characteristic length of the REV is much larger in the horizontal than in the vertical direction ($d_x \approx d_y \gg d_z$). As a consequence, an ordinary core of soil, whose cylindrical axis is much larger than its radius, should preferably be taken horizontally, a procedure which is certainly not an obvious choice from a purely technical and practical point of view.

For suction cups, the scale of the instrument cannot be determined, as the explicit form of the weighting function is not known. In other words, we do not exactly know the spatial volume influenced by the suction cup. Indeed, the geometry of the sphere can be very complicated (cf. Figure 3).

Conclusions

For mass balance calculations the measuring instrument and its scale have to be taken into account. Mass balances calculated from suction cup measurements can be unreliable, because the domain of the pore space from which samples are taken is unknown. This must be considered when interpreting the experimental results. For example, consider a hypothetical experiment where suction cups are installed in the top first meter of soil. The mass balance according to the measurements gives a fraction α of total solute recovery. We may not deduce that the missing fraction $1-\alpha$ of the solute moved deeper than 1 m. Indeed, the whole mass of solutes can still be in the uppermost meter of the soil, but in the portion of the volume not measured by the experimental device.

This does not mean that suction cups are *a priori* unreliable instruments for assessing the mass recovery of a chemical. If the soil solution is in equilibrium, then the results obtained with suction cups can be reliable, provided the spatial sampling is intense enough.

Mass balance calculations based on sampling by soil cores might also fail. However, when the existence of a representative elementary volume is assumed and its size is within the range of the sampling volume, the mass balances will be accurate. If no such assumption about an REV is made, then the ratio κ, that is the discretization volume divided by the sampling volume, at least gives a qualitative criterion for the effectiveness of the sampling.

Representing measurement by a convolution is a tool for estimating mass recoveries in field experiments. It has the advantage that the influence of the instrument on the results can be assessed. To quantify the estimation error the spatial structure of the investigated variable has to be determined or assumed. In geostatistics this is done by means of the variogram or covariance function, whereas the measurement approach uses the concept of the representative elementary volume.

Acknowledgment

I thank H. Flühler, A. Papritz, and K. Roth for helpful comments and discussions. This study was supported by Ciba–Geigy Ltd., Basel.

References

Arfken, G. 1985. *Mathematical methods for physicists.* Third edition. Academic Press, Orlando.

Barbee, G.C. and K.W. Brown. 1986. Comparison between suction and free-drainage soil solution samplers. *Soil Science* **141**, 149–154.

Baveye, P. and G. Sposito. 1984. The operational significance of the continuum hypothesis in the theory of water movement through soils and aquifers. *Water Resources Research* **20**, 521–530.

Baveye, P. and G. Sposito. 1985. Macroscopic balance equations in soils and aquifers: The case of space- and time-dependent instrumental response. *Water Resources Research* **21**, 1116–1120.

Bear, J. 1972. *Dynamics of fluids in porous media.* Dover Publications, New York.

Bear, J. and Y. Bachmat. 1984. Transport phenomena in porous media – basic equations. 3–61. *In:* J. Bear and M.Y. Corapcioglu (eds). Fundamentals of transport in porous media. NATO ASI Series E: Applied Sciences, No. 82. Martinus Nijhoff Publishers, Dordrecht.

Cushman, J. 1983. Multiphase transport equations: I. General equation for macroscopic statistical, local, space-time homogeneity. *Transport Theory and Statistical Physics* **12**, 35–71.

Cushman, J. 1984. On unifying the concepts of scale, instrumentation, and stochastics in the development of multiphase transport theory. *Water Resources Research* **20**, 1668–1676.

Cushman, J. 1986. On measurement, scale, and scaling. *Water Resources Research* **22**, 129–134.

Ene, H.J. 1981. On the thermodynamic theory of mixtures. *International Journal of Engineering Science* **19**, 905–914.

Hassanizadeh, M. and W.G. Gray. 1979. General conservation equations for multi-phase systems: 1. Averaging procedure. *Advances in Water Resources* **2**, 131–144.

Holder, M., K.W. Brown, J.C. Thomas, D. Zabcik, and H.E. Murray. 1991. Capillary-wick unsaturated zone soil pore water sampler. *Soil Science Society America Journal* **55**, 1195–1202.

Hubbert, M.K. 1956. Darcy's law and the field equations of the flow of underground fluids. *A.I.M.E. Petroleum Transactions* **207**, 222–239.

Marle, C. 1967. Écoulements monophasiques en milieu poreux. *Revue de l'institut français du pétrole* **12**, 1471–1509.

Marle, C. 1982. On macroscopic equations governing multiphase flow with diffusion and chemical reactions in porous media. *International Journal of Engineering Science* **20**, 643–662.

Matheron, G. 1965. *Les variables régionalisées et leurs estimation.* Masson, Paris.

Sposito, G. 1978. The statistical mechanical theory of water transport through unsaturated soil. 1. The conservation laws. *Water Resources Research* **14**, 474–478.

Markus Flury, Soil Physics, Institute of Terrestrial Ecology, ETH Zürich, Grabenstrasse 3, CH–8952 Schlieren, Switzerland.

Discussion

Question by R. Schulin

Your distinction between suction cups and coring devices is correct, but only in principle, I think. In reality, the volume sampled by a coring device depends also on the soil properties, and in particular on stones and woody roots which are pushed aside by the core wall when it is driven into the soil. Also swelling of soil can influence the sampled volume. Further, would you think that suction cups could be calibrated experimentally, for a given soil?

Answer by M. Flury

I think calibration of suction cups would not be practicable because soil is too heterogeneous and I do not see right now, how to do it. Core samples would work well at least in soils without coarse structures and in non-swelling soils.

Question by R. Webster

I think that there is no conflict between the geostatistical and the measurement approaches: the latter is embraced within geostatistics by, for example, Krige's relation. The concept of the Representative Elementary Volume, however, assumes that there is some scale at which the instrument embraces all variation. Has this ever been demonstrated?

Answer by M. Flury

Obtaining sufficient data has proved very difficult, and I am not aware of any study that has experimentally demonstrated the existence of the Representative Elementary Volume for any property or chemical in soil.

Question by A. Papritz

Usually the volume of a single soil sample and the block volume the mean content of which we have to estimate differ by several orders of magnitude. Thus, the main uncertainty stems from the extrapolation from the core to the block. The uncertainty caused by lack of precise knowledge about the measurement function seems then irrelevant. Could you comment on that?

Answer by M. Flury

I completely agree with your remark. In the case you mentioned, the measurement approach does not help to estimate the mass recovery. However, in more specific situations it is applicable. I give you three examples:

- The measurement approach helps in the design of the experiment. Assume that your principal goal is the estimation of the mass recovery and you have a given number of soil samples you can take. Then it is better to choose a small field plot instead of a large one, provided this is not in conflict with other experimental goals.
- The measurement approach helps to interpret your measurements. To draw certain conclusions from experimental measurements requires that the weighting function of the instruments used is known, at least approximately.
- The measurement approach forces you to consider the physical functions of the instrument.

Soil Monitoring, Monte Verità, © Birkhäuser Verlag Basel

ANALYTICAL PROBLEMS IN THE DETERMINATION OF INORGANIC SOIL CONTAMINANTS

Peter Blaser and Stefan Zimmermann.

Most of the inorganic materials designated as pollutants in soil are trace elements that occur naturally in *all* soils in a wide range of concentrations. They are considered harmful only where their concentrations exceed certain threshold values. In most cases there is no method that specifically characterizes these elements according to their origin. An assessment of inorganic pollution, based on total element concentrations of top-soil samples only, is therefore almost impossible. More promising results can be obtained using radio isotope techniques or from chemical analysis of entire soil profiles. The latter approach is demonstrated in a case study with lead, extracted with nitric acid and NH_4-acetate-EDTA from soil samples collected by horizons in 100 profiles from Switzerland. The total (extracted by nitric acid) and the extractable (extracted by NH_4-acetate-EDTA) amounts of lead varied significantly among the different parent materials.

Using at least two different extraction procedures, varying in their ability to affect the lithogenic fraction of the soil, will strongly improve the assessment of soil pollution.

Introduction

The ordinance relating to pollutants in soil (VSBo), published by the Swiss Federal Office of Environment, Forest and Landscape, lists 11 elements, the metals Pb, Cd, Cr, Co, Cu, Mo, Ni, Hg, Tl, Zn and the non metal F, explicitly as inorganic soil contaminants. A guide level is assigned to each of these elements which applies to the "amount of pollutant contained in a mixed sample taken from the upper 20 cm of mineral soil". These guide levels are based on either the "total contents" of the soil samples, extracted by boiling with 2 molar nitric acid or on the "soluble content", extracted with 0.1 M $NaNO_3$ for 2 hours. The ordinance prescribes that sites where there are indications that the concentrations of naturally-occurring pollutants are exceeded should be subject to closer observation. Such an investigation should show whether the soil fertility is affected.

One of the problems in the determination of inorganic soil contaminants is hidden in the definition of the guide levels. In contrast to the *organic* soil contaminants, which are by definition anthropogenic, most of the *inorganic* contaminants listed in the VSBo occur naturally in all soils in a wide range of concentrations. They can hardly be denoted simply as pollutants without regard to their origin, and it is also questionable whether critical guide levels of the amount of pollutants in soils should be defined without referring to the mineral and chemical composition of the parent material. This becomes immediately clear when comparing soils developed on different lithologies, rich or poor in the content of an element listed as a pollutant. Both the soil and its parent material affect concentrations of specific "pollutants" when the nitric acid extraction method is used. Consequently, the background concentrations originating from the dissolved parent material may contribute considerably and to an unknown extent to the element concentration in relation to the guide levels for the amount of pollutants.

Although Vogel *et al.* (1989) concluded that bedrock composition was of minor importance with respect to the concentration of metals in top-soil samples, this is not entirely true when analysing a large number of soil profiles developed on various parent materials (see below).

In this contribution some principles of the trace element dynamics in natural soils are discussed. The difficulties surrounding the interpretation of concentrations derived only from top-soil samples are discussed, using lead as an example. Methods and techniques by which some of the problems can be overcome are given.

Trace metal dynamics in natural forest soils

In an investigation of trace element concentrations in the soils of Switzerland Vogel *et al.* (1989) compared the concentrations of Pb, Cd, Cu, Ni and Zn in top-soil and sub-soil samples. In their statistical approach, the authors showed, for example, that the concentration of lead in the top-soil was considerably more than in the sub-soil, a finding which they attributed to anthropogenic pollution. In contrast, nickel was almost equally distributed in the top-soil and sub-soil although concentrations tended to be slightly larger in the sub-soil. From these findings the conclusion was drawn that nickel could probably not be considered as a pollutant in the soil investigated, although they mentioned that this element is mobile in acid forest soils and might therefore be translocated.

It is important to take into account the heterogeneous trace element distribution in natural, unpolluted soil profiles when assessing the extent of inorganic soil pollution. There is a quantitative relation between the metal content in the parent material and the corresponding concentrations in the fine earth fraction of a soil. The *distribution* in the soil profiles, however, is controlled primarily by pedogenetic processes acting through time.

Table1. Relative mobilities of metals in the weathering environment (from Mattigod & Page, 1983, with permission of the Academic Press.)

	Weathering conditions			
	Electron activity		Proton activity	
Relative mobilities	oxidizing	reducing	acid	neutral to alkaline
High	Mo, V, U, Ca, Na, Mg, Sr, Zn	Ca, Na, Mg, Sr	V, U, Ca, Na, Mg, Sr, Zn, Co, Cu, Ni, Hg, Cd	Mo, Ca, Na, Mg, Sr
Medium	Ca, Co, Ni, Hg, Cd,	----	Cd	Cd
Low	K, Pb	K, Fe, Mn	K, Pb, Fe , Mn	K, Pb, Fe, Mn
Very low to "immobile"	Fe, Mn, Al, Cr	Al, Cr, Mo, V, Zn, Co, Cu, Ni, Hg, Cd, Pb	Al, Cr	Al, Cr, Zn, Cu, Co, Ni, Hg

Enrichment or depletion of an element in a soil horizon, compared with the content in the parent material, can be estimated from its *relative mobility* under the conditions of the chemical environment in the pedosphere. The relative mobility of an element in a soil (Table 1) is affected primarily by the electron and proton activity conditions (Mattigod & Page, 1983). Information on the pH and the redox conditions in a soil are therefore helpful for the understanding of trace metal dynamics.

Beside the chemical conditions in the pedosphere, the distribution of elements is also governed by the soil matrix. Elements can be bound in an exchangeable, adsorbed or complexed form on the clay minerals, the organic matter and at the surface of oxides and hydroxides. The mobility of an element is therefore considerably diminished by its interaction with the solid phase of a soil. The nature and the extent of these inhibition reactions depend on the soil characteristics, almost unique for each soil. In addition to these capacity factors, all factors influencing water movement in the soil (e.g. precipitation regime, porosity, frost) as well as the effects of vegetation on nutrient cycling have to be considered when interpreting the distribution of a given element in soil profiles. This is schematically summarized in Figure 1.

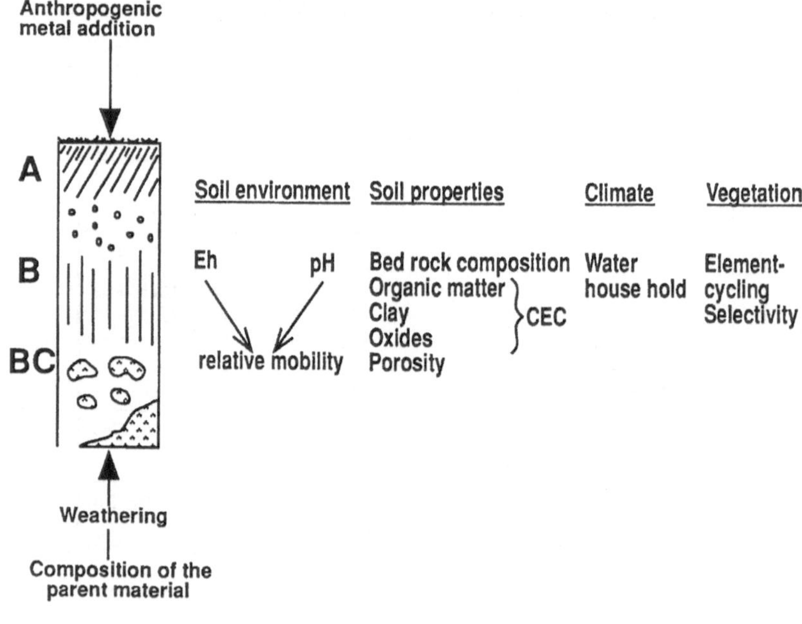

Figure 1. Factors affecting the element distribution in natural soils.

From the above it is evident that it is almost impossible to assess pollution by inorganic elements that occur naturally when analysing top-soil samples for total extractable elements only. To evaluate the natural background concentrations samples should be collected from the entire soil profile either by horizons or at fixed depths. Furthermore, in addition to the nitric acid method, a softer extraction procedure, not affecting the unweathered parent material, could provide a useful basis for the interpretation of trace element distribution in soils. This is demonstrated below, using results from a pilot study at the Swiss Federal Institute for Forest, Snow and Landscape Research to evaluate the feasibility of a national forest soil inventory.
Comparing the total element concentration in a soil with the concentration in the parent material, distribution patterns of mobile or non-mobile elements may be presented schematically as in Figure 2a. The measured concentration of an immobile element in the fine earth fraction is very similar to the bedrock concentration only at the bottom of a soil profile. Larger concentrations in the upper part of the profile are caused by residual enrichment resulting from weathering (curve 1).

When taking into account only the concentrations of the fine earth fraction, weathered and transformed by the soil-forming processes, the theoretical distribution pattern is likely to resemble more that in Figure 2b. In the latter the concentration of an immobile element should vary only slightly throughout the whole soil profile, depending only on the weathering intensity of the minerals in the fine earth fraction. Enrichment in the top soil might partly be compensated and masked through the amount of organic matter incorporated in this layer. This might result in smaller (curve 2) or only slightly larger concentrations (curve 3) in the upper part of the profile, depending primarily on the amount of organic matter.
Deviations from this general behaviour occur as a result of bioturbation in biologically active soil or in elements essential for plant nutrition. These elements tend to accumulate in the top soil by nutrient cycling, irrespective of their chemical mobility.

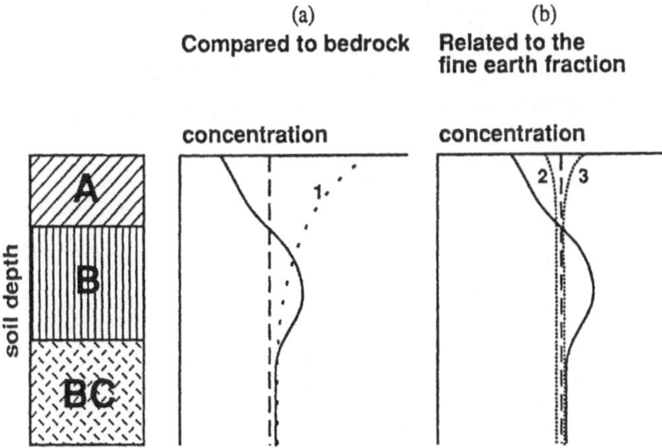

Figure 2. Relation between the element content in the bedrock material and in the soil in the case of in situ weathering. (a) compared to the bedrock; (b) compared to the fine earth fraction (<2 mm). Dashed line: concentration of bedrock. The distribution pattern of mobile and immobile elements are represented by the solid and dotted lines, respectively.

An attempt to assess the soil pollution with lead (Pb) using the "total concentrations" as an index.

Material and methods

In this study forest sites were chosen according to the Swiss national forest inventory in the five main wood-producting regions of Switzerland: Jura, Swiss Plateau, Prealps, Alps, and in Ticino in the Southern Alps (Figure 3). In each of these regions 20 soil profiles were exposed and samples were collected by horizons. Site and soil type characteristics are summarized in Table 2.

The fine earth (< 2 mm) of the dried samples was extracted and analysed according to the procedures listed in Table 3. Extraction was done in triplicate at an extractant/sample ratio of 10:1. The results were averaged as long as the deviation of independent measurements did not exceed 10 %. The results from samples collected in a desired range of soil depths were pooled for the soils of each region to form a single mean value for each soil depth. In this study the element distributions in the soil profiles are represented for fixed depths and not by genetically defined soil horizons.

Figure 3. Location of the sites investigated in the pilot study.

Table 2. Site and soil characteristics of the pilot study.

Region	Geology	Soils	Vegetation
Jura	Limestone; Malm and Dogger Sediments often combined with marl interlayers	Rendzines sometimes transitions to Cambisols	Eu-Fagion Associations
Swiss Plateau	Molasse, sometimes covered with moraines	Cambisols, Luvisols seldom gleyic Luvisols	Eu-Fagion Associations
Prealps	Flysch; partially sediments of Dogger and Cretaceous	Gleysols Planosols	Piceo-Abieton Associations
Alps	Paragneiss, Calc schists (Bündnerschiefer)	Podzols Rankers	Vaccinio-Piceion
Ticino	Ortho gneiss	Crypto Podzols	Carpinion

With the three extraction procedures elements of different binding intensities can be extracted. The 2 molar nitric acid method does not dissolve the soil matrix completely, and the measured concentrations therefore represent only about 70 % of the total element content that is extractable with more aggressive methods (Aitang & Häni, 1983). It is, however, adequate for practical purposes.

Table 3. Extraction procedures applied in the pilot study

Assignment	Abbreviation	Extractant	Procedure
Total element concentration	Me tot	2 M HNO_3	2 hours in water bath at 100 °C
Extractable elements	Me ext	0.5 M NH4-acetate 0.02 M EDTA buffered at pH 4.65	1 h under shaking in a water bath at 20°C
Exchangeable cations	Me exc	1 M NH_4Cl	1 h on an end-over-end shaker at room temp.

According to the concept of the relative mobility (Table 1), Pb can be considered as nearly immobile in all the soils that were investigated. In natural, non-polluted areas one would therefore expect a fairly uniform distribution of lead in the soil profiles when analysing the fine earth fraction. Support for this assumption can be obtained from fossil soils (Scherelis, 1989) and from soil protected, for example, by buildings erected before the start of the industrial epoch.

Results

In Figure 4 the mean values of the total Pb concentrations are presented as functions of the soil depth, together with the mean pH values for each region. For the interpretation of soil data note that the mean pH values of the profiles from the Swiss Plateau, the Alps and Ticino are almost identical. The dotted lines in the figures indicate the guide value of 50 ppm, listed in the VSBo, and the solid lines correspond to the mean Pb concentration of the samples collected at depths > 70 cm.

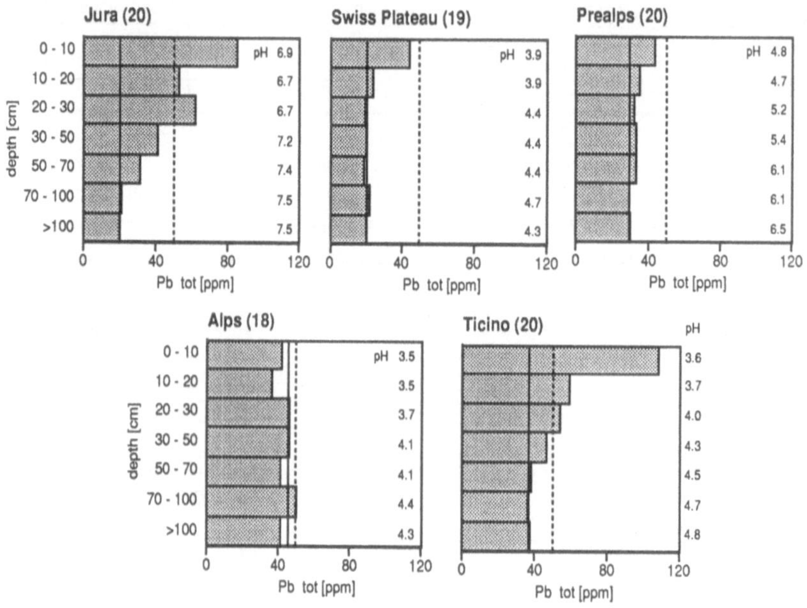

Figure 4. Distribution of "total"extractable lead in the profiles of the five regions. In brackets the number of investigated soil profiles.

From Figure 4 the following information can be obtained.

* The largest total extractable Pb concentrations are generally found in the top soil. This is not so in the alpine soils. For the interpretation of the alpine soils, the results from one profile out of 19 profiles with a soil depth > 70 cm were discarded because the sub-soil samples showed unusually large Pb concentrations, between 160 and 210 ppm.

* Very large concentrations, far exceeding the guide level, were found in the soils of the Jura and in Ticino, and, despite the different pH values at all soil depths, the distribution pattern in the profiles from the two areas is quite similar. The samples from the Swiss Plateau contain much Pb only in the top soil.
* In all profiles the lead concentration reaches a nearly constant value in the samples collected at a depth > 70 cm. Because of the very low mobility of lead in soils, this might reflect the background concentration of the unpolluted fine earth.
* According to the different chemical compositions of the parent materials of the five regions, different background concentrations in the fine earth fractions are observed in the soils (Table 4). Consequently the guide level of 50 ppm is reached with different amounts of pollutants in three of five regions. In some profiles in the alpine region Pb concentrations exceeding 200 ppm in the sub-soil were found. Some of this lead comes from the lead-rich granitic bedrock in this region and so these soils are the most susceptible to Pb pollution.

Table 4. Mean concentration of total extractable Pb in the soil samples collected at depth > 70 cm.

Region	ppm
Jura	20 ± 7.0
Swiss Plateau	21 ± 5.2
Prealps	29 ± 6.7
Alps	46 ± 18.9
Ticino	37 ± 15.8

To assess soil pollution we need to know whether the large Pb concentrations in the top-soils are entirely attributable to man and to what extent the composition of the parent material or the influence of pedogenic processes contribute to such an accumulation.

Since even in acid soils Pb is very immobile, it is generally assumed that deposited lead is fixed in the uppermost few centimetres of the soils, although in some circumstances it can be leached to deeper layers (Fisenne *et al.*, 1978.). In addition, it has been shown experimentally that Pb can be transported over several decimetres distance, probably as soluble complexes with dissolved organic matter (Stevenson & Welch, 1979). For the Ticino soils we know that water-soluble organic matter from the litter layer together with organo-metallic compounds are leached into the deeper soil horizons. It therefore seems possible that in the acid Ticino soils, lead is translocated downwards through the soil profiles, probably as organometallic complexes. The large precipitation (≈ 1800 mm /year), together with the porosity of these soils, may favour such a process (there is little bioturbation because earthworms are rare in these acid soils). In the calcareous soils of the Jura the lead distribution is caused by another process. The high pH together with the large concentrations of exchangeable Ca^{2+} inhibit both the translocation of trace metal cations and the translocation of organic matter or organo-metallic compounds. In these soils useful information for the interpretation of the lead distribution is available from the distribution pattern of aluminium or iron. Figures 5a and 5b show the similarity of the distribution pattern of aluminium and lead. At the pH measured in these soils aluminium is practically insoluble, and leaching of this element is unlikely. Because aluminium is not considered a pollutant, deposition of aluminium can safely be disregarded. The Al-distribution in these profiles may therefore be explained by *residual enrichment,* caused by limestone weathering and subsequent loss of $CaCO_3$. The aluminium is consequently of purely lithogenic origin, and its distribution pattern can be explained by a natural soil process. For the same reason, lead accumulates in the calcareous soil by limestone weathering. Using the Al-distribution as a reference, the lithogenic contribution to the lead concentration can be estimated. With the mean Al-concentration from the deepest soil horizons as a reference for the Al-content of the poorly weathered fine earth, enrichment factors can be calculated by comparison of the Al-concentration of a given soil horizon with that in the reference horizon (Table 5).

Table 5. Enrichment factors for aluminium calculated by comparison of the Al concentration of a given soil horizon with that of the deepest horizon as a reference.

Depth [cm]	0 - 10	10 - 20	20 - 30	30 - 50	50 - 70	70 - 100	>100
Relative Al-tot concentration	2.94	2.53	2.70	2.48	1.76	1.30	1.00

These enrichment factors can then be applied to calculate the lithogenic fraction of lead (Fig. 5c). This figure shows that the amount of Pb, exceeding the guide level of 50 ppm in the three top horizons, is entirely of lithogenic origin! Only the amount that cannot be explained by residual enrichment may be caused by pollution.

The application of the ammonium-acetate EDTA extract to assess soil pollution

Unfortunately, there is no extractant which is strictly selective for anthropogenic pollutants. For this reason the total metal concentrations are still considered to be the most useful index to estimate the degree of soil pollution. However, we feel that a milder extractant that does not affect the unweathered bedrock material but, instead, extracts the weathering and deposition products in exchangeable, adsorbed, precipitated or complexed form, would be better for the purpose. There are several extractants with such properties, most containing a complexing agent such as EDTA or DTPA. The extractant used in this study was proposed by Lakanen & Erviö (1971) who showed that the elements extracted with this method represent a "labile pool"

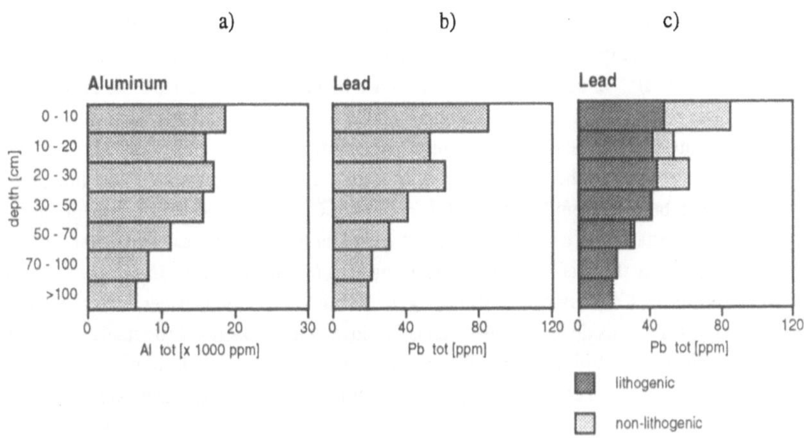

Figure 5. Estimation of the lithogenic fraction of lead by residual enrichment in the soil profiles of the Jura due to limestone weathering. a) Distribution of the total aluminium as a non-mobile reference element ; b) distribution of total lead; c) estimated lithogenic and non-lithogenic Pb-fractions, calculated with the enrichment factors from the Al-distribution.

of potentially plant available trace elements. Because of its composition it affects the exchangeable and organically complexed elements together with elements in pedogenetic oxides and hydroxides. In addition, elements of anthropogenic origin should be partially extracted by this procedure. For this reason we call the elements extracted with this procedure *extractable*. This amount is completely operationally defined and is certainly not selective in a strict chemical sense. It simply means that the extractable fraction contains most or all of the exchangeable elements, together with an unknown amount of the pedogenetic elements more strongly bound but not being part of the unweathered bedrock material. It is important to realize the conditional nature of the procedure using such an extractant. To obtain reproducable and comparable results the extraction must be performed under strictly defined and controlled conditions, and one should understand how the extractant works when interpreting the results.

- It acts as a cation exchanging agent the exchanging cation being NH_4^+.
- The strong complex-forming EDTA extracts barely soluble trace metals in various chemical forms.
- It acts as an acid because it contains acetic acid as a buffer.

The effect of the last property is important when working with calcareous soils. Because the extractant is acid, some of the limestone is dissolved, causing increased concentrations primarily of Ca^{2+} and Mg^{2+} in the extract. This is shown by the soil samples from the Prealps. In samples with pH > ≈ 6.5, free $CaCO_3$ occurs, and the amount of extractable Ca increases drastically (Figure 6a). This calcium was not previously fixed in an exchangeable form at the cation exchange sites of the soil, but it results from a dissolution reaction of the limestone grains. As can be seen from Figure 6b, the amount of Ca^{2+} extracted with the ammonium-acetate-EDTA extractant is identical with the amount of the exchangeable calcium extracted with NH_4Cl, as long as there is no free carbonate present in the fine earth. In the presence of calcium carbonate, Ca is dissolved, disturbing the strong correlation between extractable and exchangeable Ca^{2+}. This dissolution reaction is sensitive to the extraction conditions, especially time and temperature. With respect to the determination of trace metals, however, the effect is negligible because the concentration of these elements in the limestone is usually very small.

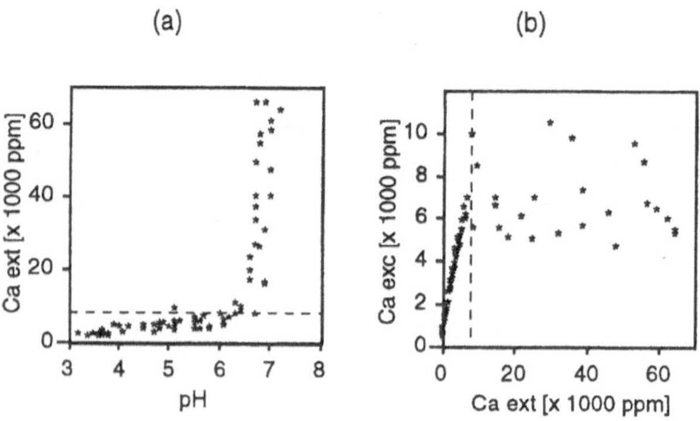

Figure 6. Dissolution reaction of $CaCO_3$ by the ammonium acetate-EDTA extractant demonstrated with the samples of the Prealps (see text). (a) Extractability as a function of the soil pH; (b) relation between the extractable (NH_4-acetate-EDTA) and the exchangeable (NH_4Cl) amount of calcium. Above the indicated limit the soils contain free $CaCO_3$, and the increased Ca concentration in the extracting solution is due to acid attack of the bedrock.

As the ammonium-acetate-EDTA extractant does not remove lead from the unweathered rock the distribution pattern of the extractable Pb-fraction in an unpolluted soil should be closely related to the weathering intensity in the various soil horizons, provided that only the pedogenic fraction is affected by this procedure. This means that in comparable horizons of soils developed under the same conditions the *relative* amount of extractable Pb is expected to be independent of the total Pb-concentration and is influenced only by the weathering intensity. Significant deviations from this relation can be considered to indicate the non-pedogenic origin of an element that may be caused by pollution. Relating the ammonium-acetate-EDTA-extractable fraction to the total extractable amount therefore offers a convenient way of assessing the extent of possible soil pollution. If in contrast only the absolute values are considered then the extent of pollution cannot be estimated because the extracted absolute amount depends on the Pb concentration of the weathered bedrock. Absolute values of the extractable amount, however, are more important for an assessment of possible phytotoxicity.

Results

Fig. 7 shows the distribution of the relative amounts of the extractable Pb in the soil profiles of the 5 areas. The bars represent the mean values of samples collected in the indicated soil depths.

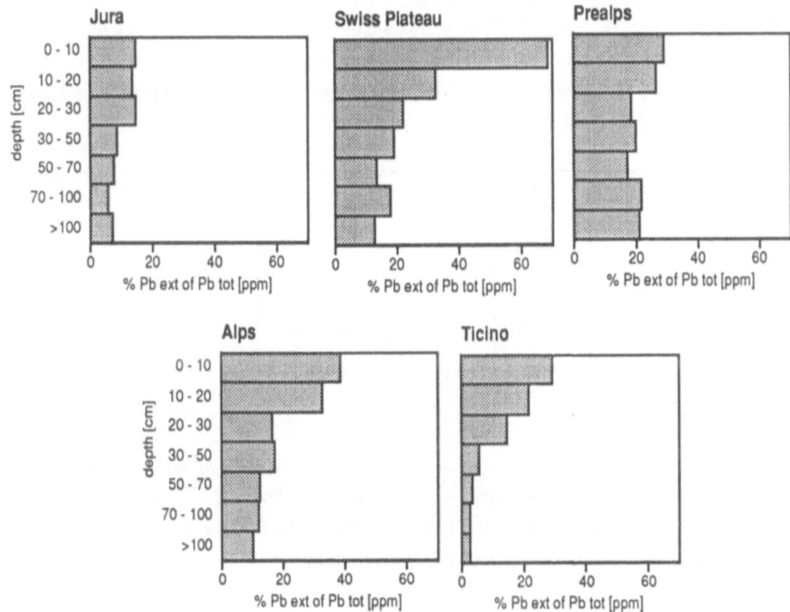

Figure 7. Distribution of the relative amounts of the extractable Pb-fraction in the soils of the 5 regions.

From this figure we can deduce the following.

* The relative amount of extractable lead in the soils of the Swiss Plateau, the Prealps and the Alps does not converge to zero at the bottom of the soil profiles. This indicates that these soils developed on pre-weathered material. In the Swiss Plateau the parent material is a Riss moraine, in the Prealps it is Flysch.
* A remarkable deviation from a smooth distribution curve is observed in the profiles of the Swiss Plateau, in the Alps and in Ticino. The large proportion of lead in the upper soil horizons cannot be explained by pedogenesis. Residual accumulation due to a weathering is unlikely to result in such large Pb concentrations, especially when the dilution effect of the organic matter in these horizons is considered. In addition the soils in these three regions are very acidic, and for this reason there are few if any earthworms. As lead is not an essential element for the plants, bio-accumulation due to nutrient cycling can be discounted. Consquently the large lead concentrations in the top soil samples in these regions may be caused by pollution.

An additional indication for possible Pb pollution is obtained when plotting the relative extractable Pb fraction against total amount of Pb of several different localities. As outlined above, in unpolluted samples from comparable soil horizons the relative extractable Pb fraction should not depend on the total amount of Pb, and consequently the variation of the corresponding values should be small.

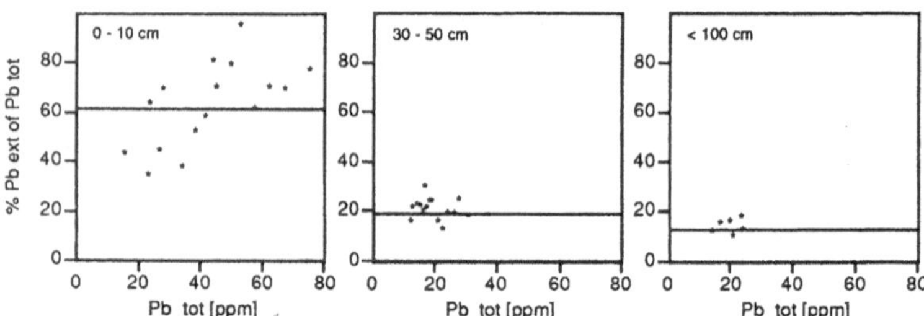

Figure 8. Relative amount of extractable Pb as a function of the total Pb concentration in unaffected soils, developed under the same conditions. This amount should be close to a constant value. Scatter in the individual measurements indicates an anthropogenic effect. Sample locality: Swiss Plateau.

Lead pollution does not follow the principles of soil formation and would therefore be evident in a large scattering of the values of the relative lead extractability in such a plot. This is shown for three soil horizons of the soils from the Swiss Plateau (Fig. 8). Scatter occurs only in the top soil samples, whereas the values from deeper horizons vary only a little. The decreasing mean values with increasing soil depth result from the decreasing weathering intensity.

With respect to toxicity the absolute values of the extractable fraction are more relevant than the relative values. Comparing Fig. 9 with the total Pb concentration in Fig. 4 leads to a completely different interpretation of the pollution situation.

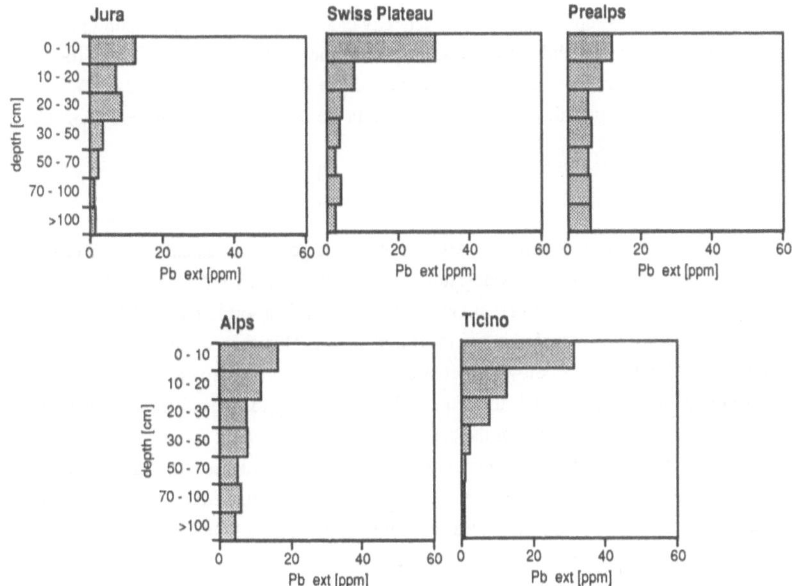

Figure 9. Distribution of the absolute amounts of extractable lead in the 5 regions.

* The *total* Pb concentration exceeded the guide levels in the soils of the Jura down to 30 cm (Fig. 4). Following from this result, these soils should be considered as severely polluted. In contrast, these soils have the smallest *extractable* Pb concentrations of all the regions investigated, indicating more reasonably a comparatively moderate degree of pollution (Fig. 9).
* For the soils of the Swiss Plateau the situation is different. Only the topsoil samples show large *total* concentrations, but the guide level is not exceeded. Compared to the soils of the other regions, however, these soils have the largest concentration of *extractable* lead, giving rise to the presumption of severe pollution.
* The soils of the Prealps and the Alps show only slightly larger *extractable* concentrations in the top horizons, and the distribution pattern in the profiles indicates only moderate anthropogenic influence.

* The distribution patterns of both the *total* and the *extractable* lead are similar for the Ticino soils. It seems that these soils are strongly influenced by "non lithogenic" lead and may be considered as heavily polluted. In addition there is some evidence that because of the special site and soil conditions Pb is leached to a greater soil depth than in the soils elsewhere.

Because of the relationship between the extractabilty of an element with the ammonium-acetate ETDA extract and its availability to plants, the presentation in Fig. 9 may be helpful for a risk assessment of the soils. Using this approach, the soils of the Swiss Plateau and of Ticino are the most severely endangered, whereas the situation in the other regions seems less dramatic.

Conclusions

A reliable assessment of soil pollution with the inorganic pollutants, as listed in the VSBo, is difficult or impossible when analysing top-soil samples with the recommended nitric acid method. With this method the "total" amounts of an element in a soil sample are dissolved, and the lithogenic fraction is included in the measured values. This consequently implies that the lithogenic fraction should be known when assessing pollution. The diversity in the chemical composition of the parent materials therefore makes it necessary to relate all element determinations to the chemical composition of the parent material. For this only soil measurements with identical, or at least similar, bedrock material are comparable. In addition, the background concentrations obtained with this procedure are usually large. This makes it difficult to evaluate pollutants, which in most cases occur in small amounts. Changes of the extractable concentration over time are consequently difficult to detect.

Results from top-soil samples alone are insufficient for interpretating analytical data obtained by the nitric acid method. Only the distribution of an element through the entire soil profile allows an estimate of the lithogenic fraction as well as of a possible soil pollution. A meaningful interpretation therefore has to be based on additional soil data and on a knowledge of the most important soil forming processes. This is especially true for soil developed on limestone. The specific weathering processes of calcareous material together with the residual enrichment of many elements in the soil results in a overestimation of the amount of metals of anthropogenic origin.

The assessment of soil pollution can be much improved using additional extraction procedures such as the proposed ammonium-acetate-EDTA method. Because this method does not affect the unweathered parent material, the results are more closely related to pedogenic and anthropogenic influences, and they may be interpreted additionally in respect to plant availability.

Acknowledgements

We are deeply indepted to Dr J. Innes for carefully revising our script and for his many suggestions.

References

Aitang,H., & Häni, H., 1983. Dissolving heavy metals from soils with acids in order to approximate total element content. *Zeitschrift für Pflanzenernährung und Bodenkunde* **146**, 481-493.

Fisenne, I.M., Welford, G.A., Perry, P., Baird, R. & Keller, H.W., 1978. Distributional uranium-234, uranium-238, radium-226, lead-210, and polonium-210 in soil. *Environmental International* **1**, 245-246.

Lakanen, E. & Erviö, R., 1971. A comparison of eight extractants for the determination of plant available micronutrients in soils. *Acta Agral. Fennica,* **123**, 223-232.

Mattigod, S.V. & Page, A.L., 1983. Assessment of metal pollution in soils. In: I.Thornton (ed.), *Applied environmental geochemistry,* Academic Press, London.

Scherelis, G. 1989. Untersuchungen zur profildifferenzierten Variabilität der Schwermetalle Cr, Mn, Fe, Ni, Cu, Zn und Pb in rezenten und fossilen Parabraunerden Baden-Württembergs. *Stuttgarter geographische Studien,* Band 112.

Stevenson, F.J. & Welch, L.F., 1979. Migration of applied lead in a field soil. *Environmental Science and Technology* **13**, 1255-1259.

Vogel, H., Desaules, A. & Häni, H., 1989. Schwermetallgehalte in den Böden der Schweiz. *Bericht 40 des Nationalen Fortschungsprogrammes "Boden";* Liebefeld-Bern.

P. Blaser and S. Zimmermann
Swiss Federal Institute for Forest, Snow and Landscape Research
Zürcherstr. 111
CH-8903 Birmensdorf
Switzerland

Discussion

Question by R. Mayer

In forest soils a large part of atmospheric input of contaminants can be retained in the litter and humus surface layer (forest floor). Wouldn't it be very helpful for distinction of anthropocenic and lithogenic origin of contaminant accumulation in soil horizons to take into account the amounts of contaminants stored in the organic surface layers?

Answer by P. Blaser

I agree that this would be very helpful. But in the Swiss Programme for Soil Testing (from which the data were reported here) it was decided, for reasons of simplicity and regional comparability, to analyse only mineral top soils.

Question by R. Kohl

Are there also forest soils included in your investigation, and do you look at the litter layer?

Answer by P. Blaser

The soils investigated in this study are exclusively forest soils, but because in the Swiss Ordinance we have no guide levels for the litter layer we did not sample it.

Question by A. Papritz

What did the frequency distribution of the lead values in each region and depth look like? Were the distributions strongly skewed?

Answer by P. Blaser

Apart from the values observed in the Alps, the distributions seem to be close to normal. In the Alps the values varied largely because of heterogeneities in the bedrock.

Question by K. Skark

Did you look at geological maps to correlate large Pb concentrations in the Alps?

Answer by P. Blaser

The selection of the plots was related to bedrock composition. However, there is further work to do, especially in Ticino where we know of two places where the bedrock contains a lot of lead.

Soil Monitoring, Monte Verità, © Birkhäuser Verlag Basel

ANALYTICAL PROBLEMS IN THE DETERMINATION OF ORGANIC SOIL CONTAMINANTS

M. D. Müller

The analytical process is described from sampling and preparation of samples to determination of compounds of interest and data evaluation. Emphasis is given to critical points, which are prone to errors. The key factors in soil contaminant analysis, the sensitivity, specificity and standard deviations of a particular analytical method together with the implementation of quality assurance programs to minimize the risk of false results are presented and discussed.

Introduction

The last hundred years have seen a continuously growing number of organic compounds being synthesized and, to some extent, being produced and marketed on a large scale. As well as the petroleum- and coal-related chemicals, there are now several thousands of chemicals manufactured and widely used. As a result, soil contamination by persistent organic compounds from various sources has become increasingly severe. Whereas at the beginning of industrialization it was primarily the sites of production which were known to be polluted, the last 30 years have taught as that global transport mechanisms carry contamination even into remote areas. In this paper the principal problems associated with the determination of organic soil contaminants are discussed and illustrated by a few selected examples.

Relevant Steps in the Analytical Process

In most cases the determination of organic compounds in soil is carried out by following a series of steps summarized below:
- site identification, sampling, transport and storage of the soil samples,
- preparation of the sample extracts,
- qualitative and quantitative determination of the analytes of interest using a suitable method,
- confirmation and quality assurance (GLP) of the data,

- interpretation of the analytical data.

All these steps to some extent depend on one another. Inadequate sampling or storage for example must lead to erroneous results, even if the laboratory part with extraction and analysis is carried out correctly. These steps are discussed and illustrated by selected examples below.

Site identification, sampling, transport and storage of the samples

As soil is a structured, solid material with a considerable degree of inhomogeneity, the collection of samples is crucial in soil monitoring. For larger areas, sampling arrays can be used. For example, in field studies to assess the accumulation of pesticides in orchard soil, soil cores are collected in the rows that are treated with herbicide. Sampling locations have to be carefully marked to prevent sampling the same spot more than once. Samples are bulked by blending several subsamples, taken at defined depths and frozen until analysis. These steps must be carried out in such a way that no sample alteration (losses and contaminations) can occur (e.g. no plastic gloves or bags used when plasticizers are the analytes of interest).

Extraction and clean up of the samples

In most cases the soil sample cannot be used directly for determination of the contaminants, which have to be isolated in a suitable way, depending on the physical-chemical properties of the contaminants. The governing factors are:

- volatility (vapour pressure),
- solubility,
- the acid-base constants,
- adsorption on organic soil matter (K_{oc}),
- distribution between octanol and water (P_{ow}),
- adsorption on inorganic soil matter.

The large vapour pressure of volatile compounds (e.g. chlorinated hydrocarbons) makes them amenable to collection by head-space analysis. For many other less volatile or non-volatile compounds which are not neutral molecules the acid-base constants of proton-acceptor or donor groups determine, together with the pH of the soil, the charge distribution on the molecule. Beside the lipophilic interaction of organic molecules expressed by the distribution between octanol and water (K_{ow}) or adsorption constant K_{oc} strong attracting and repellent electrostatic interactions between the charges on the molecule ion and soil particles can pose significant problems when attempts are made to isolate the organic molecules from the soil. For these compounds solvents are used which can compete for the adsorptive sites on soil particles and solvate the molecules of interest. For neutral (non-charged) molecules, mixtures of solvents with intermediate and low polarity are commonly used (e.g. acetone with hexane), whereas with charged molecules buffer solutions are used to influence the charge distribution at the molecule of interest or ionic interactions (see below).

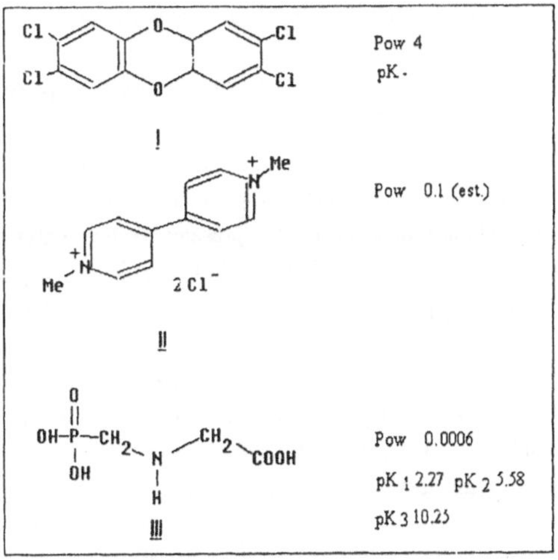

Figure 1. Structures and selected physico-chemical data of 2,3,7,8-tetrachlorodioxin (I), paraquat (II) and glyphosate (III).

Figure 1 shows the structures and relevant physico-chemical data of three potential soil contaminants. I, 2,3,7,8-tetrachloro-p-dibenzodioxin is a neutral molecule with lipophilic (and π-electron) interactions with soil humic acids dominating. II is the herbicide paraquat. The molecule carries two positive charges in a broad range of pH. III is also a herbicide, glyphosate. It is a zwitterion under normal soil pH-conditions, with a positive charge on the nitrogen and at least two negative charges on the carboxylate and phosphonic acid moieties (Worthing, 1991).

Compounds I and II are nonvolatile with long half lifes and they have a high potential for accumulation in soil. Paraquat has a half life of several years under field conditions when incorporated in soil (Domsch, 1992). The degradation of III is very dependent on soil micro-organisms and pH and its half life ranges from a few weeks to several months (Torstensson, 1985; Worthing, 1991).

Whereas the isolation of I from soil can be done using a fairly simple solvent extraction (see above) with adequate recovery, II is tightly bound to the soil matrix by its dicationic character. The extraction of II can be accomplished using boiling concentrated sulphuric acid, which breaks the strong electrostatic bonds between cationic adsorption sites of the soil colloids and the paraquat-divalent cation. In a similar way, the cation form of III is strongly adsorbed on soil particles. Shifting the pH value above 11 by adding a strong base leads to a

deprotonation of the nitrogen atom, and glyphosate, which is no longer zwitterionic, can easily be brought into aqueous solution.
The resulting extract often contains coextracted material which is likely to interfere in the analysis. So these extracts are purified to eliminate these substances as far as possible. In laboratory routine purification is standard, e.g. by automated gel permeation chromatography (Specht, 1987). Especially in trace residue analysis this cleaning step is essential to remove interfering compounds, which might mimick the presence of the analytes and lead to false positive results. On the other hand, the analytes might be lost in this step, resulting in a false negative result. An example is given with the determination of 2,3,7,8-substituted dioxins in soil (see below).
In many cases, the compounds of interest are not suitable for chromatographic analysis, either because of their polarity or because they cannot be detected. For example, compound III can be determined by liquid chromatography, but is difficult to detect because the molecule does not absorb light at wavelengths less than 200 nm. Therefore, III is transformed, in a simple reaction in buffered aqueous solution, into a stable, anionic and highly fluorescent derivative which can easily be detected (cf. Figure 2). There are many reactions which can be carried out on a microscale and which are ideal to convert polar or otherwise difficult-to-analyse compounds into derivatives which can then either be better separated or be determined with increased sensitivity or both (Knapp, 1979).

Determination of the analytes

Assuming the sample has been taken and extracted in a proper way yielding a purified extract, it is now ready for the determination of the analytes. The methods can be grouped according to the underlying process into:

- non-chromatographic methods
- chromatographic methods.

In the first group the immunological ones have received considerable interest in the last few years. They use the specific interaction of an organic compound with an antigen. Various enzyme-linked imunoassay (ELISA) kits are now commercially available and these offer high sample throughput at moderate price, but unfortunately have severe drawbacks: ELISA tests are available for only a limited number of compounds.

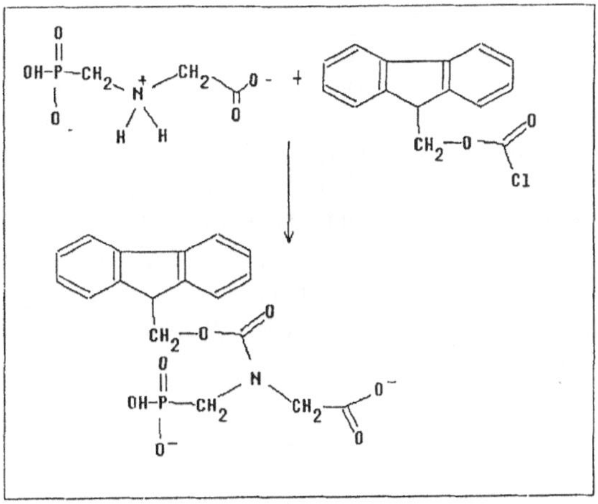

Figure 2: Derivatization of glyphosate with fluorenylmethylchloroformiate (FMOC-Cl) (Bardalaye, 1985).

They work well at moderate sensitivity with aqueous samples, but the analysis of soil extracts or biological samples can lead to erroneous results and must be carefully evaluated (Stanker, 1991); furthermore, there are many compounds such as the polychlorinated dioxins and furanes where many different isomers have to be determined, which cannot be analyzed with an ELISA test (see below).

The spectroscopic methods such as colorimetry, fluorimetry or nuclear magnetic resonance also belong in this group. For example, organic fluorescent dyes used in tracer experiments for solute transport in soil can be determined directly in an extract by measuring the fluorescence of an aqueous extract. The glyphosate derivative shown in Figure 2 could not be determined in this way, because the excess of reagent, which is also fluorescent, requires the HPLC-separation of the derivative from interfering compounds. The non-chromatographic methods are fast and economic, as no time-consuming chromatographic step is involved, but they are limited to such cases where the detection selectivity is sufficient.

For complex samples and analytes, a *chromatographic process* is often essential. As well as high-resolution gas and high-performance liquid chromatography (HRGC and HPLC, respectively), which are still the workhorses in the analytical laboratory, new techniques have been introduced in the last few years, such as supercritical fluid chromatography and capillary electrophoresis, extending the scope of the former techniques into the field of low volatility or charged molecules. In many cases, high-resolution gas chromatography is still the best method of choice (and also required for analyses carried out according to EPA regulations) coupled to one of the most specific and sensitive detection instruments, the mass spectrometer (MS). As both HRGC and mass spectrometry are gas phase techniques, coupling the two instruments is simple, in contrast to the combination of HPLC and MS. In the latter case, the large amount of solvent and often buffer substances require a complicated interface to the mass spectrometer. In general, coupled HPLC-MS provides a lower sensitivity (typically in the ng range) compared to the pg range for modern HRGC-MS instruments. The relative importance of HRGC and HPLC may be judged from a simple estimate of potential analytes (pesticides) which are amenable to residue analysis (Worthing, 1991) using one or other technique given in Table 1.

Table 1: Estimate of percentages of pesticides amenable to residue analysis using HPLC or HRGC

Technique	Non-derivatized	After Derivatization
HPLC	80	95
HRGC	15	30

Table 1 shows the broad applicability of HPLC techniques, whereas, even after derivatization, only a minor part of pesticides can be analysed by HRGC. This difference lies in the need for sufficient vapour pressure and thermostability for a successful HRGC analysis, which is not the case with HPLC. The sensitivity and specificity of HPLC routine detects such as UV-absorption or fluorescence systems may be insufficient when small concentrations of analytes are to be quantified.and many interfering compounds are present.

Chromatography is based on the interaction of the analytes, transported by a mobile phase. It has a stationary phase which produces a signal for observation with a suitable detector system. The assignment of the compound of interest and measurement is done via retention times and signal area of the analytes in the chromatographic system in comparison to that of standards. The processes are prone to various errors. In general, standard deviations of a few to 100 % were observed for various analytes in intercalibrations of environmental samples, depending on concentrations. It is therefore necessary to include an assessment of these possible errors in an analytical scheme.

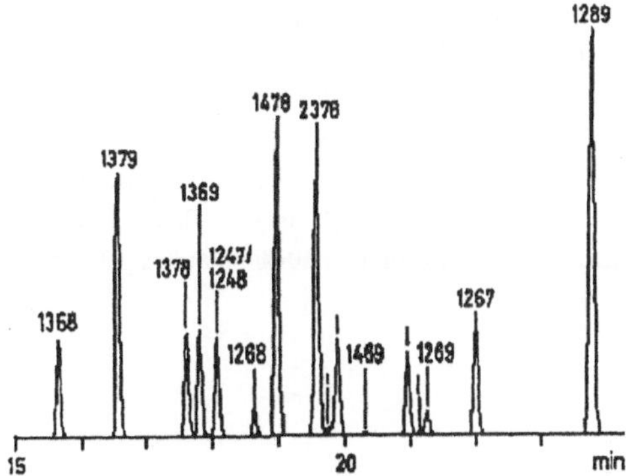

Figure 3: Separation of 2,3,7,8-tetrachlorodibenzodioxin from the other isomers with substitution indicated using HRGC-MS (Schmid, 1987)..

These facts are illustrated by the determination of 2,3,7,8-tetrachlorodibenzodioxin (I) using HRGC and detection by mass spectrometry. Figure 3 shows a chromatogram of the separation of most isomers of tetrachlorodibenzodioxin with substitution indicated. The HRGC column used allows the separation of I from the other tetrachloro-isomers. A proper clean-up of the sample, sufficient selectivity of the stationary phase to separate different isomers and the sensitivity and selectivity of the mass spectrometric detector are necessary for the correct determination of I in the presence of other isomers. The limit of detection is defined for the amount of analyte giving a signal/noise ratio greater than 3:1. Together with a certain sample aliquot taken from the extract, the minimum detectable concentrations can be estimated.

Confirmation and quality assurance (GLP) of the data

All the steps mentioned above are potentially subject to errors: from sampling the wrong sites, exchange of samples to extraction and clean up with partial or complete loss of the analyte to the analytical process, where false positive or negative results are produced. Therefore, a critical evalutation of data together with quality assurance are required. The following questions are important in this respect.

- What are the limits of detection, and are they adequate to solve the problem?
- What are the errors at a certain confidence level, and are the errors acceptable?
- In the case of positive results, (residues above detection limit being found), is there confirmation of these results?

These questions should be addressed in a parallel programme for quality assurance, as it is now a requirement for legal use of the data (good laboratory practice, GLP (Morris, 1987)). In general, these aspects are covered by running separate samples of known content or samples with some additional analyte added. The confirmation step can be carried out for a limited number of samples using an independent method. In residue analytical laboratories working with HRGC-mass spectrometry, samples are spiked prior to analysis with surrogates of the analytes labelled with stable isotopes. In the analysis of dioxins, all relevant 2,3,7,8-substituted congeners are available as either all ^{13}C- or ^{37}Cl- labelled compounds. In this way, comparably small standard deviations of 10 - 20 % at 95 % confidence interval in the low ng/kg range can be achieved.

References

Bardalaye, P. C., Wheeler, W. B. & Moye, H. A. 1985. Analytical techniques of glyphosate residue analysis. In *The herbicide glyphosate.* Grosbard, E. & Atkinson, D. Eds, Butterworths, London. pp. 263-285.

Domsch,K. H.1992. *Pestizide im Boden.* Verlag Chemie Weinheim, New York.

Hamaker, J. W. & Thompson, M. J. 1972. Adsorption. In *Organic Chemicals in the Soil Environment*, Goring, C. A. I. & Thompson, M.J. (Eds), Dekker, New York 51.

Knapp, D. R. 1979. *Handbook of analytical derivatization reactions.* Wiley, New York.

Morris, C. R. 1988. Good Laboratory Practices: Birth of a New Profession. In *Good Laboratory Practices*, Garner, W.Y. & Barge, M.S. (Eds). ACS Symposium Series 369, American Chemical Society, Washington, D.C.

Specht, W. 1987. Methode S 19 in Deutsche Forschungsgemeinschaft: Rückstandsanalytik von Pflanzenschutzmitteln. Mitteilung VI der Senatskommissionfür Pflanzenschutz-Pflanzenbehandlungs- und Vorratsschutzmittel, Methodensammlung der Gruppe "Analytik", 9. Lieferung .

Schmid, P. & Müller, M. D.,1987. Preparation and evaluation of OV-240-OH coated glass capillary columns for isomer specific determination of PCDDs and PCDFs. *Journal of hHgh rRsolution Chromatography and Chromatography Communications* **10,** 548-552.

Stanker, L. H., Watkins, B. E. & Vanderlaan, M. 1990. Environmental Monitoring by Immunoassay. In *Pesticide Chemistry*, Proc. of the 7th International Congress of Pesticide Chemistry (IUPAC), Frehse H. (Ed.). Verlag Chemie Weinheim, New York, pp. 397-407.

Torstensson, L. 1985. Behaviour of Glyphosate in soils and its degradation. In *The herbicide glyphosate*, Grosbard, E. & Atkinson, D. Eds, Butterworths, London, pp. 137-150.

Worthing, C. R. & Hance, R. J. (Eds) 1991. *The pesticide manual.* 9th edition. The British Crop Protection Council, Farnham.

Markus D. Müller, Eidgenössische Forschungsanstalt für Obst- Wein und Gartanbau,
CH-8820 Wädenswil, Switzerland.

Discussion

Question by S. Crivelli
Which analytical techniques are applied to determine volatile organic substances in soil?

Answer by M.D. Müller
In most cases, head space sampling with gas chromatography is used.

Question by P. Baccini
To clarify your last conclusion (the expertise of the analytical chemist being a key factor in a successful analytical programme), what is a good analytical chemist?

Answer by M.D. Müller
A chemist working in the field of analytical chemistry needs a certain integration in a project where analytical data are needed. In order to get correct results, analytical work carried out by contract laboratories may require careful and intense project management by the sponsor.

Question by M.A. Oliver
How many replicates of analyses do you need to determine accuracy and precision?

Answer by M.D. Müller
We analyse five or six replicates. Of course, a comparison between different laboratories is feasible only if several laboratories are available to analyse a certian compound.

Question by M. Flury
Did all the laboratories in the atrazine ring test use the same analytical technique?

Answer by M.D. Müller
No, they used different analytical methods. The spiked samples were sent to contract laboratories which used their routine technique for the determination of atrazine and other s-triazine derivatives

THE POTENTIAL OF MICROBIOLOGICAL PROPERTIES AS INDICATORS IN SOIL POLLUTION MONITORING

P. C. Brookes

Microbial properties can be used for monitoring soil pollutants. Pollution by organic pollutants is generally more difficult to evaluate than inorganic pollutants because more of the former can enter the soil ecosystem. Pollutants or stresses that cause small or transient changes to the soil microflora and its activity should be distinguished from those that have more persistent and presumably more serious effects.

No single microbial property can now be used universally for monitoring soil pollution. Microbial activity, such as microbial respiration, C and N mineralization, biological N_2-fixation and some soil enzymes, can be measured as can the total soil microbial biomass. Combining microbial activity and population measurements (e.g. biomass specific respiration) might be more sensitive to soil pollution than either activity or population measurements alone.

Measurements that have some form of "internal control", e.g. biomass as a percentage of soil organic matter, might be advantageous as it might then be possible to determine if the natural ecosystem is being altered by pollutants without recourse to expensive and long-running field experiments. Finally, new applications of molecular biology to soil pollution studies (e.g. genetic fingerprinting) might also have considerable value in the future.

The functioning of the soil ecosystem

Most ecosystems obtain the energy required for their functioning directly by photosynthesis. Soil, in contrast, normally contains few functioning photosynthetic organisms. Instead, it gains most of its energy from dead organic materials, i.e. plant and animal residues. During the mineralization of these residues, carbon dioxide and inorganic nutrients such as nitrate, phosphate and sulphate are released so that plants can utilize them again. The less labile fractions of the residues are ultimately incorporated into the soil organic matter pool, where mineralization also proceeds, albeit more slowly.

The soil micro-organisms (collectively the soil microbial biomass) and soil fauna mineralize both fresh inputs and the much larger pool of soil organic matter. The fertility of natural soil ecosystems therefore significantly depends on the rate of turnover of the soil organic matter, mediated by the soil microbial biomass. Any agent that suppresses or poisons the soil organisms or changes the quality or quantity of organic matter (either fresh inputs or the soil organic matter itself) can damage the functioning of the soil-plant ecosystem either in the short-term or over much longer periods.

In agricultural ecosystems soil fertility can be increased by applications of inorganic or organic fertilizer. The fertility of natural ecosystems however depends almost entirely on natural microbial processes, including N_2-fixation, the mineralization of organic forms of N, C, P and S, and organic matter transformations, all mediated by the soil microbial biomass. Any decline in natural fertility resulting from harmful chemicals will therefore have proportionately greater effects on natural ecosystems.

The nature of soil pollutants

Soil pollutants are of two main kinds, inorganic and organic. Of the inorganic pollutants, by far the most important are the heavy metals, e.g. Cu, Ni, Cd, Zn, Cr, Pb. Once they enter soil they remain for extremely long periods, having a half-life, depending upon the metal, of several thousands of years. In practice, the metals can be removed only by removal of the soil itself, seldom a practical proposition.

Heavy metals enter soil from several sources, including wastes from mines and smelters, atmospheric deposition (following release of metals into the atmosphere from metal smelting or other industries), animal manures, and sewage sludge. The last, while containing useful quantities of organic matter, N and P, is often contaminated with significant quantities of heavy metals, which are chelated by the organic matter in the sludge. When the sludge decomposes the metals are released and fixed on the soil, so that the metals accumulate as sludge is added.

Mandatory EC limits restrict the amounts of metals permitted to accumulate in agricultural soil. The limits are based upon known effects of metals on plant uptake and animal health. They take no account of effects on soil micro-organisms or important microbial processes, e.g. organic matter or soil N dynamics. This apparent oversight is presumably because, at the time the limits were originally set, the methods necessary to investigate the effects of metals on these properties either awaited development or had not yet been tested. Recently developed methods have indicated significant effects of heavy metals at concentrations around, or less than, current EC limits on both the standing crop of microbial biomass and its activity. The suitability of biomass-related measurements for use in pollution monitoring will be discussed, mainly by reference to heavy metals.

The potential of microbiological properties as indicators for monitoring soil pollution by organic pollutants is, in some ways, more difficult to evaluate than inorganic pollutants. This is first because there are literally thousands of possible organic chemicals which either continuously or spasmodically enter the soil ecosystem. Second, as indeed with heavy metals, there seems to be a need to distinguish between 'pollutants' or stresses that cause small or transient changes to the soil microflora or its activity and those that have more persistent, and presumably more serious, effects. One of the features of organic pollutants is that, unlike heavy metals, they will be eventually metabolized to CO_2 and other inorganic materials, which plants can use again. It is therefore important to arrive at some criteria by which the magnitude of their effects on the soil microflora and its activity can be assessed.

Among the many organic pollutants that could enter the soil ecosystem are: polychlorinated biphenyls (PCBs), polyaromatic hydrocarbons (PAHs), biocidal chemicals (e.g. insecticides, nematicides, herbicides) and organic solvents. As with heavy metals, it is mainly man's activities that introduce the organic pollutants into the soil and, indeed, most organic pollutants are man-made, or if they occur naturally, enter soil only in small amounts. For example, PAHs are always formed during forest fires (e.g. Matzner, 1984).

Assessing the significance of microbiological changes due to pollution

The use of microbiological properties as indicators of soil pollution has, in principle, much to recommend it. Microbes, having both mass and activity, and being in intimate contact with the soil micro-environment are, in many ways, ideal monitors of soil pollution. Probably the best criteria so far proposed were those of Domsch (1980) and Domsch *et al.* (1983). They first considered the following effects of naturally occurring stresses on soil microbial populations and their activities.

* Fluctuations of temperature.
* Extremes of water potential.
* Extremes of pH.
* Physical disturbances of soil, e.g. ploughing.
* Decreased gas exchange.
* Decreased supply of nutrients.
* Increases in inhibitors, predators and antagonists.

Any of these phenomena, singly or jointly, can markedly affect both the size and activity of the microbiological community. For example, Domsch *et al.* (1983) evaluated 55 documented fluctuations of bacterial populations under field conditions and showed that depressions to as little as 10% occurred frequently. On this basis they considered that any stress (or pollutant) that permitted a full recovery of the microbiological property studied within 30 days was normal, those resulting in delays of 60 days tolerable, while those taking longer than 60 days were considered to indicate that further investigation was required (Fig. 1).

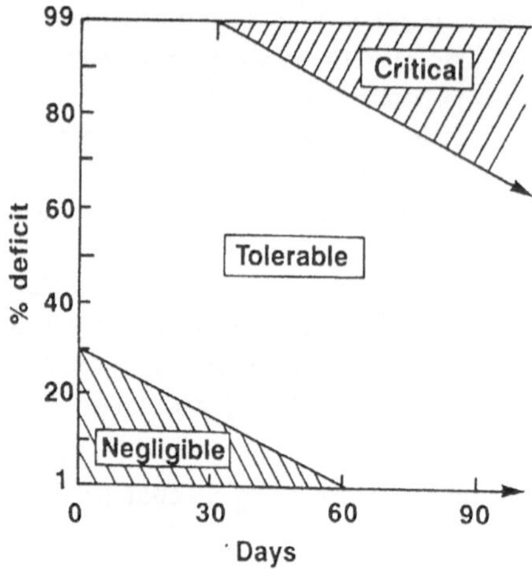

Fig. 1. Relation between monitoring period and change in biological activity (deficit) in three categories describing the ecological significance of injuries. From Domsch et al. (1983). Published with permission of Springer-Verlag.

Cook and Greaves (1987) have given an excellent example of the natural variation that might be expected within a single microbiological property over a year. They monitored soil CO_2 evolution directly from a field plot over 52 weeks in Oxfordshire, England, between April 1977 and May 1978. Carbon dioxide evolution was maximal at around 950 $\mu g\ cm^3$ soil in the warm wet conditions in the summer of 1987 but declined to less than half this by December of that year (Fig. 2). Similar patterns of N mineralization were also observed. With both properties there was a long recovery period of 15-20 weeks into the next spring, before increasing activity was observed again. These are natural phenomena, the result of changing microbial activity presumably as a result of changing moisture or temperature regimes or both, with no known extra stress, such as a pollutant, to confound the results. The potential of microbiological properties as indicators in soil pollution monitoring might thus appear small or non-existent when faced with such a wide range of natural microbiological activity. I hope to show that this is not actually the case.

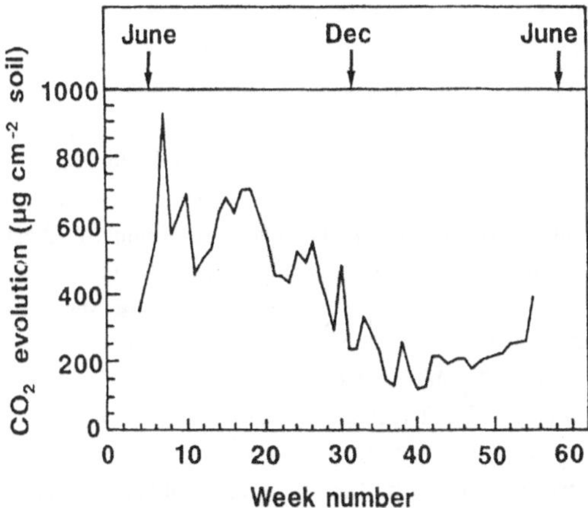

Fig. 2. Carbon dioxide evolved from an Oxfordshire field during one year. From Cook et al. (1983). Printed with permission of Taylor & Francis.

Evaluation of the potential of microbiological properties as indicators of soil pollution monitoring

The previous example illustrates the problems of field monitoring and interpretating the changes in the respiration of soil micro-organisms - apparently a simple characteristic to measure. If a pollutant was also introduced into the system, unless it had truly drastic effects, e.g. the near total kill produced by an efficient fumigant such as chloroform or methyl bromide, it would seem almost impossible to measure its effects under such conditions. It is indeed interesting that in this field experiment "Variation between samples was greatest when the rainfall was heavy ...". It is just at this time that microbiological activity would be maximal, which would also make effects due to pollutants very difficult to detect. Many other soil properties were also considered by Cook & Greaves (1987). These included soil physical and chemical properties and conditions, e.g. field moisture capacity, soil temperature, meteorological data, % organic C, soluble organic C, pH and biological properties, e.g. enumeration of micro-organisms by fluorescence staining, plate counts of bacteria, actinomycetes, fungi, yeast and algae, ATP, urease and O_2 uptake. Cook & Greaves (1987) concluded that the properties that appeared to be most responsive to environmental change were N transformations and soil respiration. However, they considered that the most striking feature of their field data was the considerable variability that they had encountered.

This work coincided with that of Domsch *et al.* (1983) in showing that measurements of changes in microbiological properties as indicators in soil pollution monitoring need to be considered against the background of natural variation and should not be interpreted solely on results from isolated laboratory tests.

This is not to say that only *field* measurements of microbiological properties are valid. Indeed, this is not what is meant at all (M. Greaves, personal communication). Instead, detection of small, transient changes in microbial population size or microbial activity *in the laboratory or field,* as a response to a real or apparent pollutant should not be accorded a great deal of weight simply because of the large natural variations observed in the field. Based on these criteria, only those approaching or exceeding a recovery period of around 30 - 60 days should be considered 'critical', while over the first 30 days at least a depression in activity of 90% *followed* by recovery, still falls within the range considered 'normal' (Somerville *et al.*, 1987). It should be noted that this theoretical framework was originally developed for testing the side effects of pesticides. However, it would seem sensible to extend the same approach to other soil pollutants.

Criteria to be used in selection of possible microbiological properties as indicators in soil pollution monitoring

Let us consider some basic criteria, by no means exhaustive, that a microbiological property might be expected to fulfil as an indicator in monitoring soil pollution.

1. The property needs to be accurately and precisely measurable across a wide range of soil types and soil conditions.
2. Because a large number of samples usually have to be analysed it is preferable that the property can be easily and economically measured.
3. The property needs to be of a nature that control or background measurements can also be made so that the effects of the pollutant can be precisely determined.
4. The property needs to be *sensitive enough* to indicate pollution but also *sufficiently robust* not to give false alarms.
5. The property needs general scientific validity based upon reliable and contemporary scientific knowledge.
6. Reliance upon a single property might be unsafe. Two or more, preferably independent, might be chosen. If so, their interrelations in non-polluted environments should be known.

Possible microbiological properties as indicators in monitoring soil pollution

There is probably no microbiological property that could be used universally for monitoring soil pollution. However, in attempting to select it is worth bearing in mind some of the main functions of the soil ecosystem itself. Two major functions are mineralization of soil organic carbon and nitrogen. Inhibition of either would have serious consequences for the ecosystem as a whole. It therefore seems reasonable to place particular emphasis upon these two biological processes in assessing the potential of microbiological properties as indicators in monitoring soil pollution.

In the following sections these and other possible microbiological properties will be discussed and illustrated. Although other pollutants may be referred to, the main references will be to heavy metals (e.g. Cu, Ni, Cd, Zn) and pesticides. The reasons are several. First, heavy metals persist in soil indefinitely once they enter it, with little, if any, decrease in their biological potency. Thus, any effects observed are unlikely to be transient. Pesticides, in contrast, decrease in potency with time as they are degraded. In each case, legislation, both on a national and European level, exists to limit both the amounts of heavy metals in soil and pesticides in the soil and aquifers. The effects on soil micro-organisms and microbial activity must also be evaluated for each new pesticide as it is developed and before it can be registered for general use. Thus there are considerable data both on the types of tests used and the results obtained.

Microbiological properties as indicators of soil pollution fall into two main groups. The first group contains those that measure the activity of the whole microbial population, e.g. microbial respiration, soil N mineralization, N_2-fixation, etc. The second contains those that measure the microbial population, at the single organism level, at the functional group level, e.g. N_2-fixers, or at the whole population level (i.e. the soil microbial biomass). A third possibility also exists by *combining* both activity and biomass data, giving specific activities of the microbial population; see below.

The use of pure cultures of micro-organisms isolated from soil as indicators of soil pollution is rejected. Pure cultures of isolated organisms may be atypical of their form in soil or may undergo undetected change during storage. Their metabolic rates may be very different from those in soil as they are removed from their normal ecological interactions. In view of this, interpretation of results and extrapolation to field conditions is impossible (Greaves *et al.* 1980).

Similarly the use of soil enzymes is also controversial because the total enzymic capacity of soil is made up of various fractions, i.e. growing micro-organisms, dead cells, extra cellular enzymes associated with the clay-humus complex (Greaves *et al.*, 1980). However, in view of the large amount of work done on the use of soil enzymes as pollution indicators, further discussion is warranted.

Microbial activity measurements

Any measurement *in isolation* is of little use for deciding whether an ecosystem is suffering from pollution. Microbial activity measurements are valuable in evaluating soil pollution, but the problem is always to decide what to compare the measurement with, i.e. what 'control' to use. The problem is removed if properly constructed field experiments with fully replicated treatments are available. A good example is the Woburn Market Garden Experiment at Rothamsted, which has provided unique data on the effects of heavy metals on microbial biomass and microbial activity (Brookes & McGrath, 1984). Interpretation of data from the natural environment is much more difficult as the determination of basal levels of microbial activity is a real problem. In view of this most, or all, of the following data are from field or laboratory experiments. However, ways must be found to study and to monitor the natural environment.

Microbial respiration

As shown previously, microbial respiration in the field is subject to enormous fluctuation (Cook & Greaves, 1980). However, under controlled laboratory conditions at suitable moisture, e.g. 40-50% of full water-holding capacity (WHC), and temperature, say 15°-25°C, the respiration of sieved (2-6.25 mm) soil can be determined accurately and precisely (e.g. Jenkinson & Powlson, 1976). This measurement is widely used by microbial ecologists as an index both of microbial activity and microbial biomass itself. Indeed, measurement of microbial respiration forms the basis both of the fumigation-incubation method for measuring soil microbial biomass (Jenkinson & Powlson, 1976) and the substrate-induced respiration method (SIR) for measuring the same property (Anderson & Domsch, 1978).
However, microbial respiration appears unaffected at heavy metal concentrations at around current EC mandatory limits (e.g. Brookes & McGrath, 1984; Brookes *et al.*, 1984). Only at very large concentrations is CO_2 evolution decreased. Thus Tyler (1981) reported that microbial respiration was decreased while concentrations of 1000 mg kg^{-1} each of Cu or Zn were exceeded.
Despite this, soil respiration (i.e. CO_2 evolved or O_2 consumed) coupled with total amounts of microbial biomass, and hence biomass specific respiration, is being shown to be a sensitive indicator of soil pollution.

Nitrogen and carbon mineralization

The conversion in soil of plant and animal residues, or soil organic matter itself, to its simple inorganic components is a microbially mediated process upon which the cycling of the major elements, i.e. C, N, P, S, ultimately depends. Many micro-organisms contribute to the release of NH from soil organic N. In contrast, only two genera of autotrophic bacteria, *Nitrosomonas* and *Nitrobacter*, are involved in the oxidation of NH to NO.
Again, as with microbial respiration, it seems that mineralization rates of soil organic N are unaffected at metal concentrations near current EC limits (e.g. Brookes *et al.*, 1984). This is supported by Doelman (1986) who reported that inhibition of both N mineralization and nitrification are inhibited at around 100 mg kg^{-1} Zn, Cu and Ni, around 100-500 mg kg^{-1} Pb and Cr and around 10-100 mg kg^{-1} Cd.

There is a general consensus that interpretations of the effects of heavy metals on N mineralization and nitrification is difficult for several reasons (e.g. Bååth, 1989; Brookes & Verstraete, 1989; Chander, 1991). These include lack of standardization of experimental procedures and because of variation in soil properties which may alter the relative or absolute toxicity of the metal (e.g. Tyler, 1981). Similarly, Duxbury (1985) concluded that information on the effects of pollutants such as heavy metals on these processes is conflicting and recommended caution in generalizing about the effects of heavy metals on microbially mediated processes in natural environments.

Goring & Laskowski (1982) extensively reviewed the effects of herbicides, insecticides, fungicides and nematicides on soil N transformations. They determined that most pesticides would either have no effect or cause less than 25% inhibition of N mineralization or nitrification, even when exceeding recommended doses.

More severe biocidal chemicals, e.g. chloroform, methyl bromide or chloropicrin, have much more substantial effects on N mineralization and nitrification in particular (e.g. Martin, 1972). Initially, such chemicals cause both the total microbial population and N mineralization to decrease. However, the population slowly increases again, although not to its original size, and N mineralization increases also. Often N mineralization in a previously fumigated soil exceeds that of a similarly incubated but non-fumigated control soil after a few days. This is attributed to the mineralization of that part of the microbial biomass killed by the fumigant, in addition to the mineralization of dead soil organic matter. However, nitrification may be depressed for several weeks after the fumigant is removed.

There is evidence that heavy metals introduced in sewage sludge, and giving soil metal concentrations several times the current EC mandatory limits, cause more accumulation of soil organic matter than does uncontaminated sewage sludge. Thus Chander & Brookes (1991c) reported that soil at Luddington (15% clay) which contained Cu at 3.7 times the limit contained 32% more organic matter than a soil receiving uncontaminated sludge. At Lee Valley (21% clay) soil contaminated with Zn at 3.4 times the permitted concentration contained 10% more organic matter than that treated with uncontaminated sludge. Similarly, plots containing Cu at 3.8 times the limit had about 14% more organic matter. These results strongly suggest that the heavy metals were decreasing the turnover rate of the organic matter, presumably because of inhibitory effects on the microbial biomass itself.

These results were obtained from well-maintained field experiments where all the treatments were known and where soil was reasonably homogeneous. Detection of such effects in natural environments would be more difficult, if not impossible. This is because proper control soils, essential for comparison, would not be available.

Nitrogen fixation

Atmospheric nitrogen (N_2) can be fixed biologically only by micro-organisms. All biological N_2-fixation depends upon the enzyme nitrogenase, which occurs in three main types of micro-organisms. These vary in the potential amounts of N_2 that they can fix in temperate regions as follows.

	kg N fixed ha^{-1} y^{-1}
Free-living heterotrophic bacteria	1-2
Cyanobacteria	5-30
Symbiotic associations (e.g. *Rhizobium* and clover)	100-200

(From McGrath 1993)

N_2-fixation, or more accurately nitrogenase activity, can be measured using the acetylene reduction assay. Nitrogenase converts acetylene to ethylene, which can be measured easily by gas chromatography. This provides a sensitive test for nitrogenase activity, but the results cannot be converted directly to the quantity of N_2 fixed. For this, measurement of conversion of $^{15}N_2$ to other forms of 15-labelled N is required (Giller & Day, 1985).
Nitrogen-fixation has frequently been suggested as a suitable test for pollution, particularly by heavy metals, in soil.

Free-living heterotrophs

N_2 is fixed slowly by free-living heterotrophs in most soils, and its rate of fixation is a poor indicator of soil pollution. Thus, Rother *et al.* (1982) found no relation between acetylene reduction and heavy metal pollution, even at metal concentrations many times in excess of current EC limits.
However, both Brookes *et al.* (1984) and Lorenz *et al.* (1992) reported that concentrations of heavy metals close to, or less than, current maximum EC limits decreased heterotrophic N_2-fixation (again measured by acetylene reduction) by as much as 90%. The reason for this apparent discrepancy is because the soil was incubated with glucose at up to 2000 μg glucose g^{-1} soil for up to 50 h before analysis. These conditions would have favoured the development of heterotrophic N_2-fixation and permitted more sensitive evaluation of the effects of metals on the nitrogenase enzyme system. More research is needed before this assay can be used as a standard test. Thus, Lorenz *et al.* (1992) showed that not all soils responded in a predictable way. Some English soils, apparently otherwise similar to those which developed measurable heterotrophic N_2-fixation, did not respond in the way expected. This difference could not be attributed to, for example, lower soil pH or more soil N. They concluded that the failure to determine standard conditions of incubation limits the usefulness of heterotrophic N_2-fixation as a biological indicator of soil pollution.
We know less about the effects of organic biocidal chemicals on N_2-fixation by heterotrophs. Depending upon the biocide used, effects may be stimulatory and depressive, with little clear

pattern that would appear helpful; see Simon-Sylvestre & Fournier (1979) for a review on this.

Autotrophic N_2-fixation

The cyanobacteria, or blue-green algae, contain nitrogenase and can grow on the soil surface. The amount of N_2 that they fix in the soils of temperate ecosystems is uncertain - estimates vary from between 10-50 kg N ha^{-1} (Witty *et al.*, 1979; Henriksson, 1971) to nil. Field measurements of N_2-fixation by blue-green algae as potential microbiological indicators of soil pollution are unlikely to be profitable. The variability between measurements, both temporal and spatial, if indeed N_2-fixation activity was detectable at all, would probably make such measurements meaningless. That, at least, was my experience when I tried to measure nitrogenase activity on the soil surface of plots of the Broadbalk Wheat Experiment weekly for a year; I found nothing. This is in contrast to Witty *et al.* (1979) who estimated (based on C_2H_2 reduction) that up to 28 kg N ha^{-1} y^{-1} was fixed from the same plots on the same field over 2 years of measurement. They also reported considerable variation (up to 5 fold) in amounts of N_2 estimated to be fixed on the same plots in different years.

Despite the uncertainty of detecting N_2 fixation in the field, it would seem that measurement of autotrophic N_2 fixation, under standard laboratory conditions, could be a useful indicator of soil pollution. However, the laboratory measurements would then reflect the potential of, for example, uncontaminated and polluted soils for N_2-fixation, rather than actual fixation in the field, which would be sensitive to environmental conditions.

Brookes *et al.* (1986) incubated moist fresh soil from the Woburn Market Garden Experiment at Rothamsted under laboratory conditions of 20°C day, 16°C night, 16 h day and 50% WHC. Colonies of cyanobacteria grew rapidly on soil that had received annual applications of farmyard manure (FYM) from 1942-67 (uncontaminated soil). In contrast, similarly incubated soils from the same experiment which had received metal-contaminated sewage sludge from 1942-61 (contaminated soil), showed little surface colonization by cyanobacteria, even by day 118 of incubation when the experiment was terminated.

In the uncontaminated soil there was an initial lag of about 14 days, then the rate of acetylene reduction increased rapidly, reaching a maximum at about day 28, declining slowly and regularly until day 118. In contrast, acetylene reduction had barely commenced by 50 days in the contaminated soil. It then increased regularly but much more slowly than in the uncontaminated soil, and was still increasing when the experiment ended. There was about 3 times more acetylene reduction in the uncontaminated than contaminated soil by day 118. Similarly, the uncontaminated soil fixed about 10 times more ^{15}N-labelled N in 24 h than did the contaminated one, although no differences in total N were discernible between the different treatments at the end of the experiment.

In a further experiment, soil was sampled at the same time and at 40 cm intervals along a gradient between the middle of an uncontaminated plot and a contaminated plot. Concentrations of EDTA-extractable Zn, Cu, Ni and Cd increased in a curvilinear manner between the uncontaminated and contaminated plot. In contrast, total C_2H_2 reduction decreased linearly (Fig. 3) with increasing soil metal concentration, during 60 days. It was halved at 50 μg Zn, 20 μg Cu, 2.5 μg Ni and 3 μg total Cd. Because the soils contained all these metals, it is not possible to say which metal, or metal interactions, produced the effects. However, apart from Cd, the maximum concentrations for individual metals were well within individual EC maximum limits (about 30% of the soils contained Cd at more than 3 μg Cd g^{-1} soil). These results therefore suggest that soil needs to contain only little metal for non-symbiotic N_2-fixation to be affected (Fig. 3).

Lorenz *et al.* (1992) also investigated the development of cyanobacteria on a gradient between an uncontaminated and contaminated soil from the Woburn experiment. However, instead of sampling along the middle of an uncontaminated and contaminated plot, as described by Brookes *et al.*, (1986), they produced their gradient by mixing the uncontaminated and contaminated soil to give different concentrations. They also reported inhibition of the growth of blue-green algae and decreased acetylene reduction in the contaminated soils, but only at the largest concentrations (i.e. a mixture of 83% sludge soil and 17% FYM or in 100% sludge soil).

The lag of 14 days that they observed for the contaminated soil was much shorter than the 50 days observed by Brookes *et al.* (1986), and significant differences in acetylene reduction between the least (i.e. 100% FYM) and most metal-contaminated (i.e. 100% sludged soil) remained only until day 28, compared to day 118 as reported by Brookes *et al.* (1986).

The major difference between the two experiments was that the metal gradient of Brookes *et al.* (1986) was obtained by sampling along a natural gradient between an uncontaminated and contaminated plot. Both were ploughed annually. In this case the metals would have been thoroughly dispersed throughout the soil, as more than 30 years had elapsed between the last sludge application and the measurements. In contrast, Lorenz *et al.* (1992) produced their gradient by mixing uncontaminated and contaminated soils in different proportions. It is likely, therefore, that the metals would not have been so intimately distributed within the mixed soil, however carefully the mixing was done. Thus 'islands' of uncontaminated and metal-contaminated soil particles might have existed side by side. Perhaps this is the explanation for the differences in the results obtained in the two experiments as all other incubation conditions were, as far as could be achieved, identical.

It may also be that cyanobacteria are very sensitive to incubation conditions in other ways. It is interesting that Lorenz *et al.* (1992) failed to get blue-green algae to grow even on uncontaminated Luddington soil (similar to Woburn in most respects). Similarly, they report on other experiments in Sweden where no cyanobacteria developed when freshly sampled, uncontaminated soils were incubated in the light. Much more research needs to be done to evaluate whether the cyanobacteria have the potential to act as indicators of heavy metal toxicity in soil and, if so, to standardize growth conditions.

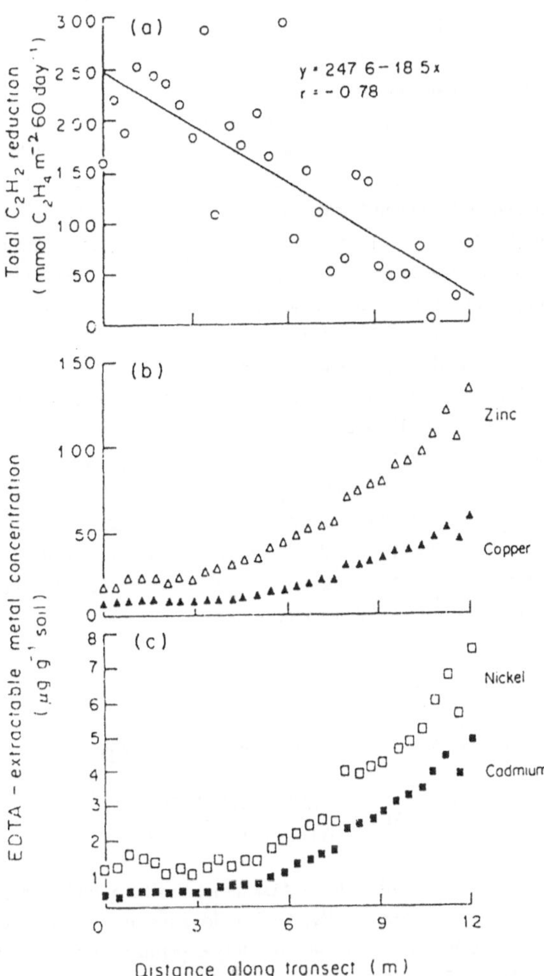

Fig. 3. *(a) Total soil $N_2[C_2H_2]$-fixation (mmol C_2H_2 m^2) during 60 days incubation in soils sampled at 40 cm intervals along a transect from the middle of an uncontaminated FYM plot to the middle of a contaminated sludge plot. (b) EDTA-extractable Zn (Δ) and Cu (▲) in soils sampled at 40 cm intervals along the transect (μg g^{-1} soil). (c) EDTA-extractable Ni (□) and Cd (■) in soils sampled at 40 cm intervals along the transect (μg g^{-1} soil). From Brookes et al. (1986). Published with permission of Pergamon Press.*

Symbiotic N_2-fixation

As with cyanobacteria, pot experiments in soil from the Woburn Market Garden Experiment revealed a drastic decline (i.e. 50% or more) in symbiotic N_2-fixation by *Rhizobium leguminosarum* biovar *trifolii* in symbiotic association with *Trifolium repens* (white clover) in soil containing more than 334 µg Zn, 99 µg Cu, 27 µg Ni and 10 µg Cd g^{-1} soil (McGrath *et al.*, 1988). It was not possible to identify which metal, or combination of metals, produced the effects. Addition of nitrogen fertilizer restored the yields of clover on the metal-contaminated soil to those of clover growing in uncontaminated soil, so that the effects were not due to direct phytotoxicity of the plants. Rather, McGrath *et al.* (1988) showed that the decreased N_2-fixation and decreased clover yields were because the nodules were not fixing N_2, although clover nodulation occurred in the metal-contaminated soil. Even when the bacteria from the ineffective nodules were isolated in the absence of heavy metals, they still could not fix of N_2 on white clover. Giller *et al.* (1989) concluded that ineffectiveness of *Rhizobium* in fixing N_2 in metal-contaminated soil was not due to direct metal toxicity. Rather, only a single genotype of *Rhizobium* sp, ineffective in fixing N_2, survived in the metal-contaminated soil.
Important as such experiments are, the *Rhizobium*-legume symbiosis is not simple to work with. It is unlikely that it could be easily developed as a routine indicator of soil pollution. The requirement to sample, prepare and to measure clover dry weights, % total N, ^{15}N-labelled N in the plants probably precludes it for routine bioassay.

Adenylate energy charge ratio

Atkinson (1977) introduced the concept of adenylate energy charge (AEC):

$$AEC = \frac{[ATP] + 0.5\,[ADP]}{[ATP] + [ADP] + [AMP]}$$

where ATP is adenosine 5-triphosphate, ADP is adenosine 5'-diphosphate and AMP is adenose 5'-monophosphate. Theoretically, AEC could range from 1.0, i.e. all ATP, to 0, i.e. all AMP. However, results from numerous laboratory experiments indicate that metabolically active organisms have AECs of between about 0.8 - 0.95, whereas resting or dormant organisms have AECs between about 0.7-0.5. Values of AEC much less than 0.4 are taken to indicate a stressed or senescent population, incapable of biosynthesis. Spores, for example, can have an AEC less than 0.1 (e.g. Atkinson, 1977).
Brookes & McGrath (1987) tested whether the AEC of the soil microbial biomass could be used as an indicator of environmental stress caused by heavy metals in soils from the Woburn Market Garden Experiment. Although other microbial indices, e.g. N_2 fixation and microbial biomass (see later), suggest stress-induced effects of metals on the soil microflora, AEC was similar in both the uncontaminated and contaminated soil; 0.85 and 0.89 respectively. This

suggests that AEC cannot be a valid indicator of stress produced by heavy metals. The effects of other pollutants upon soil AEC values await evaluation.

Soil enzyme activity

If we are to use enzymes as indicators, than we should differentiate between exocellular and endocellular enzymes. Brookes *et al.* (1984) reported less dehydrogenase activity in metal-contaminated soil than in similar uncontaminated soil while soil phosphatase was unaffected. Soil phosphatase can occur exocellularly as well as within the living cell. In contrast, dehydrogenase is active only inside intact, living cells. These results suggest that dehydrogenase is a better indicator than phosphatase of effects of metals on soil microbial activity. However, more recent work (Chander & Brookes, 1991a) suggests that the dehydrogenase assay itself is subject to interference from Cu in soil, because Cu prevents the red colour development of the end product (triphenyl-formazan) from the artificial substrate triphenyltetrazolium chloride. Thus, when Cu is present, this abiological reaction can be incorrectly interpreted as decreased dehydrogenase activity caused by Cu. In fact, it is simply an artefact of the method. Other common heavy metals, e.g. Ni, Cd, Zn, do not cause this effect. As the interference appears to be specific to Cu, dehydrogenase activity may be of value as an indicator of other heavy metals.

It would also be interesting to test the dehydrogenase assay with organic pollutants, e.g. pesticides, fumigants, etc., to see if a similar phenomenon occurs.

Microbial population measurements

Measuring effects of pollutants on the micro-organisms themselves, rather than their activity, is another possible means of monitoring soil pollution. There are two main ways in which pollutants can act upon the microbial population. One is by producing direct toxic effects, i.e. killing or biochemically disabling the organisms - antibiotics are good examples. The other way is by operating indirectly - for example by decreasing the availability of a substrate, e.g. plant root exudates. Thus the decreased energy available to the microbes could also result in a smaller population.

Repeated pesticide applications can cause shifts between sensitive and resistant populations of certain micro-organisms. However, processes such as crop residue decomposition and nutrient mineralization are less likely to be affected than are processes that depend on a few species of micro-organisms (Moorman, 1989).

There are serious problems in trying to relate *in vitro* studies of soil pollutants on micro-organisms to the soil environment, and this will not be considered here, unless it provides useful background information. For example, numerous papers refer to laboratory experiments where heavy metals were added to micro-organisms *in vitro* at concentrations far exceeding those present in soil solution.

Studies of single species or groups of micro-organisms in soil are also difficult because most organisms cannot be easily isolated from soil. It is generally considered that most pesticides, added at field rates, do not cause significant lasting effects on microbial populations, or microbial activities, related to soil fertility. In general, changes in the population due to pesticide activity are no more severe than changes caused by natural stresses (e.g. Moorman, 1989).

The effects of heavy metals on the microbial population can be virtually permanent, both because of their toxicity and persistence. This is illustrated by reference to effects of heavy metals upon the soil microbial biomass.

Effects of heavy metals from past applications of sewage sludge upon the soil microbial biomass

The soil microbial biomass comprises the total mass of micro-organisms that live in soil, defined as those that have volumes of less than 5000 μm^3. Methods now exist for measuring the biomass as a single compartment, and these have been reviewed elsewhere (e.g. Jenkinson & Ladd, 1981; Jenkinson, 1988). The fumigation-extraction procedure for measuring the biomass, coupled with automated analysis (Vance *et al.* 1987; Wu *et al.* 1991) permits rapid and routine measurements of large numbers of samples.

Biomass measurements certainly have their limitations in soil pollution studies. Being essentially 'black box' measurements they do not permit evaluation of changes in the community structure of the microbial population, e.g. shifts in the fungal: bacterial ratio in soil. Nevertheless, they have shown damaging effects to the microbial population near current maximum EC permitted limits of heavy metal concentrations in soil which other, apparently more sensitive, methods did not.

Relationship between biomass carbon and total soil organic carbon

Generally, biomass C comprises about 1-4% of total soil organic C (e.g. Jenkinson & Ladd, 1981; Insam *et al.*, 1989; Anderson & Domsch, 1989). Thus there is an approximate linear relationship between these two variables, although it may vary somewhat between soils of different physical characteristics or between soils under different managements. For example, clay soils contain several times more biomass than sandy soil under similar management and climate (e.g. Lynch & Panting, 1980; Van Veen *et al.* 1985). Also the microbial biomass in forest and grassland soil generally forms a larger proportion of total soil organic matter than in arable soil (e.g. Ayanaba *et al.* 1976; Adams & Laughlin, 1981; Gupta & Germida, 1988; Srivastrava & Singh, 1988). This accords with the statement of Jenkinson & Ladd (1981) that situations favouring the accumulation of organic matter in soil increase both the amount of biomass and the proportion of the total soil organic matter.

Changes in soil management cause the microbial biomass to increase or decrease much faster than the total amount of soil organic matter. Ayanaba *et al.* (1976) and Adams & Laughlin (1981) reported that changing from forest or grassland to arable management caused much greater decreases in biomass C than total soil organic C. Similarly, Powlson *et al.* (1987) reported that 18 years of straw incorporation in Danish soils caused about a 40-50% increase in biomass C whereas total soil organic C increased by only 5%. Similar results were also reported by Saffigna *et al.* (1989) for Australian soils. This, and much other similar research, supports the original idea of Powlson & Jenkinson (1976) that the biomass is a much more sensitive indicator of changing soil conditions than is total soil organic matter content, so that the biomass can serve as an 'early warning' of such changes long before they are detectable in other ways.

There is accumulating evidence that heavy metals at around, or a little in excess of, current permitted EC limits also decrease the proportion of biomass C in total soil organic matter. Thus, Brookes & McGrath (1984) reported that soil from the Woburn Market Garden Experiment which last received metal-contaminated sewage sludge more than 30 years ago (contaminated soils) contained about half as much microbial biomass as similar soil which received inorganic fertilizer or farmyard manure (uncontaminated soils) over this period. The contaminated soil now contains Cu, Ni and Zn at around current EC limits and Cd at up to 3 times. The biomass in the contaminated soil showed no correlation with total soil organic C, unlike the uncontaminated soil from the same experiment. Similarly, Chander & Brookes (1991c) recently reported that biomass C as a percentage of total soil organic C was twice as large (1.5-2.0%) in non-sludged soil or soil which received non-contaminated sewage sludge than in soil which received sludges contaminated with Cu or Zn (0.7-1.0%) in field experiments at Luddington and Lee Valley experimental farms.

Chander & Brookes (1993) also showed that amounts of biomass C as a percentage of total soil organic C in control soil and in soil given uncontaminated sludge in an experiment at Gleadthorpe ranged from 1.5 to 1.6 (Fig. 4). These values were within the range of those reported previously for other soils (see above), and are remarkably similar to those reported by Chander & Brookes (1991c). The values of this percentage in soils containing larger concentrations of either Zn or Cu or both were also less than half (0.4 to 0.7) those in control soils or in soils given uncontaminated sludge (1.5 to 1.6).

So far, such data have been obtained only from carefully designed and controlled field experiments, with full replication and containing plots that have never received sewage sludge or plots that have received uncontaminated sewage sludge, so that easily comparisons can be made. This research should be extended to the natural environment to determine the environmental impact of heavy metals from, for example, sewage sludge disposal, metal mining and smelting. The collection of such data is always beset by the difficulty of obtaining valid comparable analytical results from suitable uncontaminated control soils. I suggest that the link between biomass C and total soil organic C, discussed above, can itself constitute an 'internal control', so that when soils deviate much from (biomass C)/(total soil organic C) ratios perceived as normal, for the particular management, soil type and climate, it may be a prelimi-

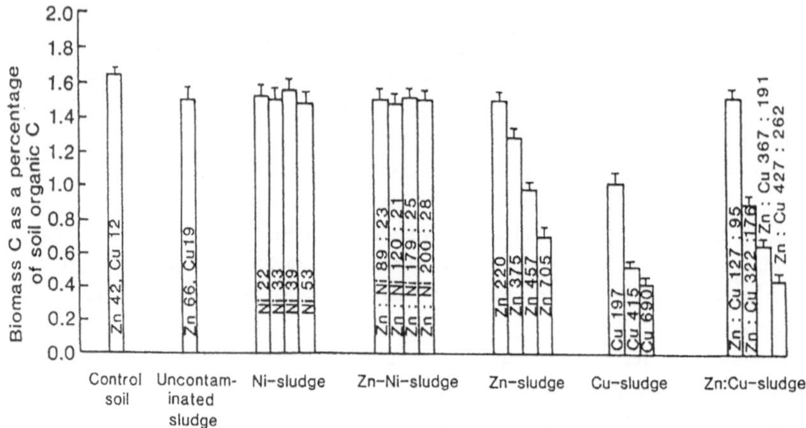

Fig. 4. Microbial biomass C expressed as a percentage of total soil organic C in Gleadthorpe soils (standard errors shown). Values given in boxes are the total soil metal concentrations ($\mu g\ g^{-1}$ soil). From Chander & Brookes (1991). Published with permission of Pergamon Press.

nary indication that some damage or change to the functioning of the soil ecosystem has occurred and indicate that further research should be done.

Effects of heavy metals on the soil microbial biomass

So far, because no experiment in Europe has soil contaminated with single metals at steadily increasing concentrations, it is impossible to establish the minimum soil metal concentrations at which effects on the soil microbial biomass are occurring. Summarizing results of Chander & Brookes (1991c, 1993) both Cu and Zn at 2 to 4 times EC limits decreased both the amounts of biomass and its proportion in soil organic matter, whereas similar increases in Ni had no observable effects. Similarly, Cd at twice the limit had no effect. There was also evidence that heavy metals in sandy loam soils affected the biomass at smaller metal concentrations than in soils containing more clay. It seems from these results that the most probable order of increasing metal toxicity (in terms of current EC limits) is Cu > Zn ≥ Ni or Cd.

Effects of heavy metals on the specific respiration of the biomass

I suggested above that the relation between biomass and soil organic matter might serve as an 'internal control' and help overcome difficulties of interpretating measurements made outside carefully controlled field experiments. However, this idea needs further testing before its validity can be assessed.

While measurements of soil respiration and microbial biomass made alone can, in certain circumstances, be useful indicators of environmental stress, combining the two measurements to give amounts of CO_2 evolved per unit of biomass (biomass specific respiration) has been shown to be a more subtle environmental indicator of stress. Because standard conditions of temperature, soil moisture and substrate availability are required, these measurements are probably valid only under laboratory conditions.

Environmental stress causes the soil microbial biomass to direct more energy from growth into maintenance (Killham & Firestone, 1984) so that an increased proportion of C taken up by the biomass will be respired as CO_2. Killham (1985) therefore suggested that determination of the relative proportions of ^{14}C-labelled glucose C going into microbial ^{14}C and $^{14}CO_2$ evolved might be useful for measuring the response of the microbial biomass to environmental stresses, such as pollution by heavy metals, salinity or changes in soil pH. Such stresses are known to induce this response in marine micro-organisms (Griffiths *et al.*, 1981). Accordingly, Killham (1985) developed a simple bioassay procedure based on proportioning ^{14}C-labelled glucose between biomass ^{14}C and $^{14}CO_2$ evolved. He showed that for a given increase in stress the ratio of respired $^{14}CO_2$: biomass ^{14}C formed was, on average, twice as great as the magnitude of the decrease in either respiration alone or dehydrogenase activity for a given increase in stress.

Killham (1985) estimated the proportion of added glucose ^{14}C entering the biomass indirectly because no other method was then available. As a result of improvements in methodology, biomass C can now be determined directly in soil by fumigation-extraction during the early decomposition of labile substrates (Ocio & Brookes, 1990). This would therefore appear to be a sensitive and practical approach.

Even without glucose amendment, biomass specific respiration was shown to be a good indicator of environmental stress due to heavy metals. Thus, Brookes & McGrath (1984) reported that rates of CO_2 evolution (μg CO_2-C g^{-1} soil) from both uncontaminated and metal-contaminated soil from the Woburn Market Garden Experiment were indistinguishable when the unamended soils were incubated at 50% WHC and 25°C. However, biomass specific respiration (measured as mg C respired g^{-1} biomass day^{-1}) was twice as fast in the metal-contaminated soils as in the uncontaminated ones.

In a later experiment more total and ^{14}C-labelled CO_2 was evolved from the contaminated than from uncontaminated soil (Chander & Brookes, 1991b) during the first 5 days after addition of ^{14}C-labelled glucose and maize (about 10% and 20% respectively). In contrast, about 30% less ^{14}C-labelled biomass was synthesised per unit of added substrate, which is in line with the findings of Killham (1985). Similarly, Chander & Brookes (1991d) showed that plant-derived inputs of organic ^{14}C were about 20% less in the contaminated soil than the uncontaminated soil. Also, the biomass in the contaminated soil contained about 35% less of this ^{14}C-labelled organic C than in the uncontaminated soil. These results indicate that two mechanisms cause smaller biomasses in metal-contaminated soils. These are decreased C inputs from growing plants and decreased efficiency of conversion of this C into new biomass C. The latter mechanism appears to be the more important.

Applications of molecular biology to soil pollution monitoring

Classical methods for identification of micro-organisms depend on culturing them following their isolation from soil. More rapid techniques, such as serology, have been used for decades for identifying micro-organisms on the basis of the surface properties of their cells. More recently, the use of monoclonal antibodies has improved the specificity of such methods. So far, such approaches have had only limited applicability because of the complex and heterogeneous nature of soil. However, attempts are now being made to combine monoclonal antibodies with other methods to enable isolation of specific microbes directly from soil (e.g. antibodies linked to magnetic beads for magnetic separation, or to fluorochromes to facilitate flow cytometry). These might offer great advances.

More recently, both individual species and strains of micro-organisms have been uniquely identified by methods exploiting differences in the nucleotide sequences of their DNA. These techniques, known colloquially as genetic fingerprinting, have become standard in studying population genetics. Indeed, because they are so specific, individual organisms within a species can be identified with certainty. They are therefore increasingly important in crime detection, as well as playing an ever-growing role in microbial ecology. Using such techniques, Giller *et al.* (1989) showed that isolates of *Rhizobium* sp. from white clover root nodules grown on metal-contaminated soil were virtually genetically indistinguishable whereas isolates from similar uncontaminated soil showed the expected genetic diversity.

There are also possibilities in monitoring the survival of introduced micro-organisms under field conditions. Thus Hirsch & Spokes (1988) introduced into the field a *Rhizobium* sp. which had been previously selected for chromosomally-located resistance to two antibiotics followed by insertion of a DNA sequence conferring additional antibiotic resistances. The micro-organism could still be detected five years later in the field on the basis of both its antibiotic resistance markers and its unique genetic fingerprint. Such techniques would appear to have great potential in investigating microbial survival under field conditions in polluted environments.

Conclusions

1. There is no single microbiological property that is ideal for the monitoring of soil pollution.
2. A major problem exists in interpreting results, particularly because of natural fluctuations in microbial populations and microbial activity. These fluctuations are often larger than effects due to pollution.
3. There are also problems in interpretation of environmental measurements because of lack of control or basal measurements.
4. There appear to be big advantages in using measurements that have some form of 'internal' control, e.g. biomass as a percentage of soil organic matter, as it is then easier to determine if the functioning of the ecosystem is being altered by factors such as pollution.
5. A 'watching brief' should be maintained on the newer methods of molecular biology as applied to microbial ecology. They might have considerable potential in monitoring soil pollution.

Acknowledgements

I thank Dr Penny Hirsch for her helpful comments on applications of molecular biology to soil pollution monitoring and Dr R. Webster for useful criticism of my script.

References

Adams, T.McM. & Laughlin, R.J. 1981. The effects of agronomy on the carbon and nitrogen contained in the soil biomass. *Journal of Agricultural Science, Cambridge* **97**, 319-327.

Anderson, J.P.E. & Domsch, K.H. 1978. A physiological method for the quantitative measurement of microbial biomass in soil. *Soil Biology and Biochemistry* **10**, 215-221.

Anderson, T.H. & Domsch, K.H. 1989. Ratios of microbial biomass carbon to total organic carbon in arable soils. *Soil Biology and Biochemistry* **21**, 471-479.

Atkinson, D.E. 1977. *Cellular Energy Metabolism and its Regulation.* Academic Press, New York.

Ayanaba, A., Tuckwell, S.B. & Jenkinson, D.S. 1976. The effects of clearing and cropping on the organic reserves and biomass of tropical forest soils. *Soil Biology and Biochemistry* **8**, 519-525.

Bååth, E. 1989. Effects of heavy metals in soil on microbial processes and population (a review). *Water, Air and Soil Pollution* **47**, 335-379.

Brookes, P.C. & McGrath, S.P. 1984. Effects of metal toxicity on the size of the soil microbial biomass. *Journal of Soil Science* **35**, 341-346.

Brookes, P.C. & McGrath, S.P. 1987. Adenylate energy charge in metal-contaminated soil. *Soil Biology and Biochemistry* **19**, 219-220.

Brookes, P.C., McGrath, S.P. & Heijnen, C. 1986. Metal residues in soils previously treated with sewage-sludge and their effects on growth and nitrogen fixation by blue-green algae. *Soil Biology and Biochemistry* **18**, 345-353.

Brookes, P.C., McGrath, S.P., Klein, D.A. & Elliott, E.T. 1984. Effects of heavy metals on microbial activity and biomass in field soils treated with sewage sludge. In: *Environmental Contamination* (International Conference, London, July 1984), pp. 574-583. CEP Ltd, Edinburgh.

Brookes, P.C. & Verstraete, W. 1989. The functioning of soil as an ecosystem. In: *Soil Quality Assessment.* State of the Art Report on Soil Quality. Report to Commission of the European Communities Directorate-General XII. Contract EV4A/0008/NL, 41 pp.

Chander, K. 1991. *The effects of heavy metals from past applications of sewage sludge on soil microbial biomass and microbial activity.* PhD Thesis. University of Reading.

Chander, K. & Brookes, P.C. 1991a. Is the dehydrogenase assay invalid as a method to estimate microbial activity in Cu-contaminated soils. *Soil Biology and Biochemistry* **23**, 901-915.

Chander, K. & Brookes, P.C. 1991b. Microbial biomass dynamics during the decomposition of glucose and maize in metal-contaminated and non-contaminated soils. *Soil Biology and Biochemistry* **23**, 917-925.

Chander, K. & Brookes, P.C. 1991c. Effects of heavy metals from past applications of sewage sludge on microbial biomass and organic matter accumulation in a sandy loam and a silty loam UK soil. *Soil Biology and Biochemistry* **23**, 927-932.

Chander, K. & Brookes, P.C. 1991d. Plant inputs of carbon to metal-contaminated soil and effects on the soil microbial biomass. *Soil Biology and Biochemistry* **23**, 1169-1177.

Chander, K. & Brookes, P.C. 1993. Effects of Zn, Cu and Ni in sewage sludge applied singly or in combination on microbial biomass in a sandy loam UK soil. *Soil Biology and Biochemistry* **25** (in press).

Cook, K.A. & Greaves, M.P. 1987. Natural variability in microbial activities. *Pesticide Effects on Soil Microflora.* Taylor and Francis, London, pp. 6-15.

Doelman, P. 1986. Resistance of soil microbial communities to heavy metals. In: *Microbial Communities in Soil.* FEMS Conference Publication, Copenhagen, Vol. **5**, Decker, New York, pp. 415-471.

Domsch, K.H. 1980. Interpretation and evaluation of data. In: *Recommended tests for assessing the side-effects of pesticides on the soil microflora.* Weed Research Organization Technical Report No.59, pp. 6-8.

Domsch, K.H., Jagnow, G. & Anderson, T.H. 1983. An ecological concept for the assessment of side-effects of agrochemicals on soil micro-organisms. *Residue Review* **86**, 65-105.

Duxbury, T. 1985. Ecological aspects of heavy metal responses in micro-organisms. *Advances in Microbial Ecology* **8**, Plenum Press, New York, pp. 185-235.

Giller, K.E. & Day, J.M. 1985. Nitrogen fixation in the rhizosphere; significance in natural and agricultural systems. In: *Ecological Interactions in Soil.* Special Publication No.4 of the British Ecological Society. Blackwell Scientific Publications, Oxford, pp. 127-147.

Giller, K.E., McGrath, S.P. & Hirsch, P.R. 1989. Absence of nitrogen fixation in clover grown in soil subject to long-term contamination with heavy metals is due to survival of only ineffective *Rhizobium. Soil Biology and Biochemistry* **21**, 841-848.

Greaves, M.P., Poole, N.J., Domsch, K.H., Jagnow, G. & Verstraete, W. 1980. Recommended tests for assessing the side-effects of pesticides on the soil microflora. In: *Weed Research Organization Technical Report No. 59.* Agricultural Research Council, Weed Research Organization, Oxford, p.9.

Griffiths, R.P., McNamara, T.M., Steven, S.S. & Morita, R.Y. 1981. Relative microbial activity and mineralization associated with water masses in the Lower Cook Inlet, Alaska. *Journal of the Oceaniographic Society of Japan* **37**, 227-233.

Goring, C.A.I. & Laskowski, D.A. 1982. The effects of pesticides on nitrogen transformations in soils. In: *Nitrogen in Agricultural Soils, Agronomy* **22**, 689-720.

Gupta V.V.S.R. and Germida J.J. (1988) Distribution of microbial biomass and its activity in different soil aggregate size classes as affected by cultivation. *Soil Biology & Biochemistry* **20**, 777-786.

Henriksson, E. 1971. Algal nitrogen fixation in temperate regions. In: *Biological Nitrogen Fixation in Natural and Agricultural Habitats. Plant and Soil Special Volume*, 415-419.

Hirsch, P.R. & Spokes, J.R. 1988. *Rhizobium leguminosarum* as a model for investigating gene transfer in soil. In: *Risk Assessment for Deliberate Releases.* Springer-Verlag, Berlin, pp. 10-17.

Insam, H., Parkinson, D. & Domsch, K.H. 1989. Influence of macroclimate on soil microbial biomass. *Soil Biology & Biochemistry* **21**, 211-221.

Jenkinson, D.S. 1988. Determination of microbial biomass carbon and nitrogen in soil. In *Advances in Nitrogen Cycling in Agricultural Ecosystems*, pp. 368-386. Commonwealth Agricultural Bureau International, Wallingford, pp. 368-386.

Jenkinson, D.S. & Ladd, J.N. 1981. Microbial biomass in soil: measurement and turnover. In: *Soil Biochemistry*, Vol. 5, Marcel Dekker, New York, pp. 415-471.

Jenkinson, D.S. & Powlson, D.S. 1976. The effects of biocidal treatments on metabolism in soil. V. A method for measuring soil biomass. *Soil Biology and Biochemistry* **8**, 209-213.

Killham, K. 1985. A physiological determination of the impact of environmental stress on the activity of microbial biomass. *Environmental Pollution* (Series A) **38**, 283-204.

Killham, K. & Firestone, M. 1984. Salt stress control of intracellular solutes in streptomycetes indigenous to saline soils. *Applied Environmental Microbiology* **47**, 301-306.

Lorenz, S.E., McGrath, S.P. & Giller, K.E. 1992. Assessment of free-living nitrogen fixation activity as a biological indicator of heavy metal toxicity in soil. *Soil Biology and Biochemistry* **24**, 601-606.

Lynch, J.M. & Panting, L.M. 1980. Variations in the size of the soil biomass. *Soil Biology and Biochemistry* **12**, 547-550.

Martin, J.P. 1972. Side effects of organic chemicals on soil properties and plant growth. In: *Organic Chemicals in the Soil Environment*, Vol.2, Marcel Dekker, New York, pp. 733-792.

Matzner, E. 1984. Annual rates of deposition of polycyclic aromatic hydrocarbons in different forest ecosystems. *Water, Air and Soil Pollution* **21**, 425-434.

McGrath, S.P. 1993. Effects of heavy metals from sewage sludge on soil microbes in agricultural ecosystems. In: *Toxic Metals in the Soil-Plant System*, John Wiley and Sons, Chichester, (in press).

McGrath, S.P., Brookes, P.C. & Giller, K.E. 1988. Effects of potentially toxic metals in soil derived from past applications of sewage sludge on nitrogen fixation by *Trifolium repens* L. *Soil Biology and Biochemistry* **20**, 415-424.

Moorman, T.M. 1989. A review of pesticide effects on microorganisms and microbial processes related to soil fertility. *Journal of Production Agriculture* **2**, 14-23.

Ocio, J.A. & Brookes, P.C. 1990. An evaluation of methods for measuring the microbial biomass in soils following recent additions of wheat straw and the characterization of the biomass that develops. *Soil Biology and Biochemistry* **22**, 685-694.

Powlson, D.S., Brookes, P.C. & Christensen, B.T. 1987. Measurement of soil microbial biomass provides an early indication of changes in total soil organic matter due to straw incorporation. *Soil Biology and Biochemistry* **19**, 159-164.

Powlson, D.S. & Jenkinson, D.S. 1976. The effects of biocidal treatments on metabolism in soil. II. Gamma irradiation, autoclaving, air-drying and fumigation. *Soil Biology and Biochemistry* **8**, 179-188.

Rother, J.A., Milbank, J.W. & Thornton, I. 1982. Seasonal fluctuations in nitrogen fixation (acetylene reduction) by free-living bacteria in soils contaminated with cadmium, lead and zinc. *Journal of Soil Science* **33**, 101-113.

Saffigna P.G., Powlson D.S., Brookes P.C. and Thomas G.A. (1989) Influence of sorghum residues and tillage on soil organic matter and soil microbial biomass in an Australian vertisol. *Soil Biology & Biochemistry* **21**, 759-765.

Simon-Sylvestre, G. & Fournier, J.C. 1979. Effects of pesticides on the soil microflora. *Advances in Agronomy* **31**, 1-92.

Somerville, L., Greaves, M.P., Domsch, K.H., Verstraete, W., Poole, N.J., Van Dijk, H. & Anderson, J.P.E. 1987. Recommended laboratory tests for assessing the side-effects of pesticides on the soil microflora. In: *Pesticide Effects on Soil Microflora*. Taylor and Francis, London, pp. 205-219.

Srivastava, S.C. & Singh, J.S. 1988. Carbon and phosphorus in the soil biomass of some tropical soils of India. *Soil Biology and Biochemistry* **20**, 743-747.

Tyler, G. 1981. Heavy metals in soil biology and biochemistry. In *Soil Biochemistry*, Vol.5, Marcel Dekker, New York, pp. 371-414.

Vance, E.D., Brookes, P.C. & Jenkinson, D.S. 1987. An extraction method for measuring microbial biomass C. *Soil Biology and Biochemistry* **19**, 703-707.

Van Veen, J.A., Ladd, J.N. & Amato, M. 1985. Turnover of carbon and nitrogen through the microbial biomass in a sandy loam and a clay soil incubated with [$^{14}C(u)$] glucose and [$^{15}N](NH_4)_2SO_4$ under different moisture regimes. *Soil Biology and Biochemistry* **17**, 747-756.

Witty, J.F., Keay, P.J., Fogatt, P.J. & Dart, P.J. 1979. Algal nitrogen fixation on temperate arable fields - the Broadbalk Experiment. *Plant and Soil* **52**, 151-164.

Wu, J., Joergensen, R.G., Pommerening, B., Chaussod, R. & Brookes, P.C. 1990. Measurement of soil microbial biomass C by fumigation-extraction - An automated procedure. *Soil Biology and Biochemistry* **22**, 1167-1169.

P. C. Brookes
Soil Science Department
AFRC Institute of Arable Crops Research
Rothamsted Experimental Station
Harpenden, Herts AL5 2JQ
Great Britain

Discussion

Question by R. Webster

Your results, which show that the biomass on the metal enriched soil is about half that on the uncontaminated soil and incidentally is about equally active microbially, represent a snapshot in time after 20 years of continuous application of sludge. In monitoring pollution we should like earlier indications of change so that we could act earlier to prevent undesirable accumulation.

Do you know of experiments in which the effects of adding metal contaminated sewage sludge sequentially to soil on the microbial biomass have been measured from the start? If so what were the results?

Answer by P.C. Brookes

As far as I know, there are no field experiments of the sort you describe, and I think you have made an important point. Krishan Chander at Rothamsted added metal-contaminated sewage-sludge to soil in laboratory studies. He showed that the biomass initially increased as it used the organic matter in the sludge as an energy source. Toxic effects on the biomass occurred later and the amount of biomass decreased to less than that in control soils. This suggests that it was the metals, rather than the sludges themselves, that were toxic. I also understand that the Ministry of Agriculture in Britain is going to commission field experiments where metal-contaminated sewage-sludge will be added to soil of different types. I hope very much that we can monitor some of these experiments.

Questions by P. Fry
I'm astonished that the microbial biomass shows a difference between heavy-metal polluted soil and soil that is not polluted, while the microbial activity doesn't show any difference. Is biomass generally insensitive to heavy metal pollution?

Answer by P.C. Brookes
Generally the biomass is not so sensitive, that is true. But in heavy metal polluted soil the biomass is working faster and producing more CO_2.
And after a long time (20 years in our study) the biomass does change.

Question by R. Schulin
You showed us very interesting data on the temporal fluctuation of microbial activity. Do you also have data on the spatial variability, and if so, how do they look?

Answer by P.C. Brookes
No, unfortunately we do not have many data on spatial variability. But I agree that this is a topic of paramount interest, in particular in soil monitoring. Richard Webster has some experience, I think, on which he might comment.

MAGNETIC SUSCEPTIBILITY OF SOILS IN THE AREAS INFLUENCED BY INDUSTRIAL EMISSIONS

Zygmunt Strzyszcz

The magnetic susceptibility in forest soils of Katowice province is largest in the litter layer, mainly in the Of and Oh horizons. In the arable soil near steel plants the magnetic susceptibility was greatest in the ploughed layer (20-25 cm). The magnetic susceptibility is due largely to industrial emissions of metallurgical dust, cement dust and fly ash from combustion of pulverized hard coal. The large magnetic susceptibility is accompanied by substantial contents of heavy metals. Measurements of the magnetic susceptibility of forest soil litter may be used in soil monitoring to ascertain the presence of ferromagnetics of anthropogenic origin.

Introduction

All substances are to some extent magnetic. Three forms of magnetic susceptibility are distinguished: diamagnetism, paramagnetism, and ferromagnetism, and ferromagnetic substances can be ferro-, ferri-, and antiferrimagnetic. According to Thompson and Oldfield (1986) the magnetic properties of soil depend on the magnetic properties of minerals that form it. Quartz, orthoclase, calcium carbonates, organic matter, and water have diamagnetic properties; olivine, pyroxene, garnet, biotite, iron and manganese carbonate are paramagnetic. The minerals magnetite and maghaemite are ferromagnetic, whereas goethite, lepidocrocite, and haematite are antiferromagnetic. Soil formed from basalt and other rocks rich in ferromagnesian minerals have large magnetic susceptibilities.

In other soils the magnetic susceptibility can be caused by soil-forming processes.

In particular it can result from the activity of specific bacteria that synthesize ferromagnetics from other iron compounds (Petersen *et al.*, 1989; Fassbinder *et al.*, 1990).

In addition to its having natural causes magnetic susceptibility is influenced by man's activities.

The following can contribute to an increase in magnetic susceptibility of top soil:

- forest and peat fires (Le Borgne, 1955; Schwertmann & Heinemann, 1959),
- addition of fly ash from the combustion of pulverized coal (Tölle, 1986),

- emissions from steel works, power and cement plants, and motor vehicles and dust from urban roads (Hopke *et al.*, 1980; Hunt et al., 1984; Strzyszcz *et al.*, 1988; Strzyszcz, 1989 a,b).

Thompson & Oldfield (1986) noted that the soil close to large towns and industrial centres can have a higher susceptibility than that elsewhere because magnetic spherules derived from burning coal are emitted into the atmosphere and fall onto the soil from it. In the industrial North of England the susceptibility of recent ombrotrophic peats are within the same range as that of strongly susceptible forest soils in rural areas. The same authors point out the gap in this field of investigation, writing "this atmospheric component has not been recognized in the literature on soil magnetism, but it must be taken into account in site selection for studies of true enhancement processes and products, and in any evaluation of soil magnetic properties close to major source of spherules".

There have been similar increases in magnetic susceptibility of peats in the USA, Finland, and Canada, all of which coincide with industrialization with its attendant increase in combustion of solid fuels, and increase in production of iron and steel (Richardson, 1986; Tolonen & Oldfield, 1986).

Metallurgical and industrial dusts and fly ash contain magnetite. Increasing their production has meant that more has fallen on to the soil, and so industry has been directly responsible for the increase in magnetic susceptibility of the top soil horizons (Hansen *et al.*, 1981; Mitchell & Sluskoter, 1976). This is true in Poland as elsewhere (Wilczynska-Michalik, 1981; Manecki *et al.*, 1981; Schejbal-Chwastek & Tarkowski, 1988). Magnetite is a main component of ferromagnetics emitted. It is characterized by very slow (if any) transformation in the soil. So, magnetite can be used as a tracer of past and current industrial pollution of the soil. The magnetite concentration in Polish ashes ranges from 0.9 to 12% (Laczny, 1983).

Modern filtration can diminish the emissions of dust. Nevertheless, emission can be substantial. For example, even with a 98% efficient electro-filter 2.6 t/h of fly ash (1.7 t/h within the aerosol fraction 0 to 10 µm) is emitted from a coal power plant of 1200 MW (Konieczynski, 1987). Tomza (1987) found that the aerosol fraction contains much iron — 1.76 to 30.9 µg Fe/m^3 of air in the Katowice area. In 1988 there were 1527 industrial plants in Poland of which 1339 were equipped with dust collectors, but only 241 trapped more than 90%, and the efficiency of 649 lay within the range of 70-90%.

Our data suggest that in many regions of Poland, and particularly in Katowice province, we should expect an increase in soil magnetic susceptibility caused by industrial emissions. For many years in Poland 100-160 Mt/a of hard coal and 30 to 75 Mt/a of brown coal have been burnt. High stacks have caused the fly ash to spread over the whole country, and other dusts are only of a local importance.

It seems therefore that by measuring the ferromagnetism of the soil we can monitor the accretions from industry. In particular we should be able to determine the contributions from iron and steel metallurgy, combustion of hard and brown coal, cement and ceramics production, and coking. Other authors have suggested the possibility of using the measurements of magnetic susceptibility in monitoring of various features and properties of soil too (Fine *et al.*, 1992; Hunt, 1984; Maher, 1986).

Recent studies in Poland

Investigation of the magnetic susceptibility of arable soils was started in Katowice province near the Pokoj steel plant in 1987. In the following years the investigations were extended to adjacent provinces (Czestochowa, Bielsko, Opole) and also to some national parks (Strzyszcz *et al.*, 1988; Strzyszcz, 1989 a,b; Strzyszcz, 1991).

Katowice province covers 6552 km^2, or 2.1% of Poland but 10.5% of the Polish population, about 4 M inhabitants, live in it. The population density is approximately 600 per km^2 in comparison with 120 per km^2 for the whole country. Katowice province produces 98% of the country's hard coal, and 100% of zinc and lead ores are mined. It produces 52% of the nation's steel, 32% of its coke, and about 25% of its electrical energy. In the central part of the province, the Upper Slesian Industrial Region, 1506 km^2 in area, are more than 2800 industrial plants, including 67 hard coal mines, 16 power plants and thermal-electric power stations, 17 iron and steel plants, 4 cement plants, 6 glassworks, 12 coke plants, and 3 zinc and lead plants. It has the greatest density of this kind of industry in Poland and probably in Europe. The result is ecological disaster due to a total disregard for the environment from the beginning of industrial development to our days.

To investigate the magnetic susceptibility of the soil in Katowice province more than 500 samples were taken from about 100 soil profiles, of which 72 were in forest. According to Vodyanitskyi *et al.* (1983) most ferromagnetics accumulate in the litter layer, and so sampling was from particular litter subhorizons. Strip pits were made 1.5 m from free trunks in 80-100 year-old forest stands. Podsols and acidic brown soils dominated. Additionally, magnetic susceptibility of metallurgical dusts (12 samples), cement dusts (18 samples), and fly ash from the combustion of brown (7 samples) and hard coal (35 samples) was measured.

The measurements were made by a home-made instrument (Strzyszcz *et al.*, 1988) following the method of Tölle (1986) for investigating soil erosion. The instrument consists of a feeder of direct current, a tension stabilizer, a generator, a coil, an amplifier and a frequency meter. A weighed amount of substance is slid into the holder and then into the magnetic field of the current-carrying solenoid. The high frequency current generated in the coil is a subject of considerable change of frequency, which is recorded in the system amplifier - frequency meter. The frequency of the generated current can be measured precisely within the limits of 10^{-1} Hz.

The results of soil magnetic susceptibility expressed in Hz/5g, were compared in the Institute of Geophysics of the ETH, Zürich, with its current method, and the results were converted to specific susceptibility in m^3/kg units. For the comparative purposes 127 samples of soil, metallurgical dust, cement dust, and fly ash were taken (Fig. 1).

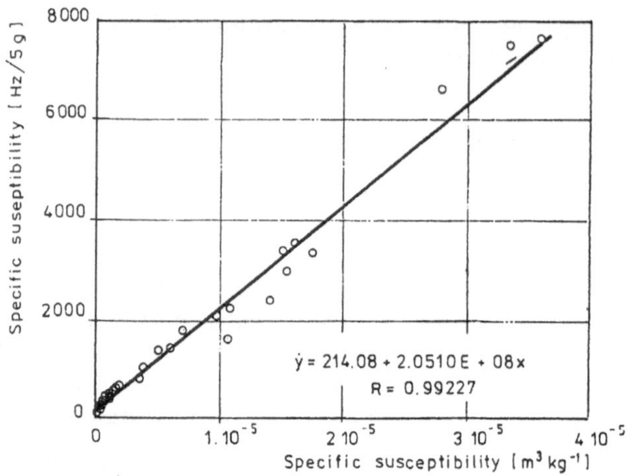

Fig. 1. Methodological comparison

In addition the contents of heavy metals in the soil, metallurgical dust, cement dust, and fly ash were determined by extraction with in 2M HNO_3 (soil) and 4M HNO_3 (dust and fly ash) followed by measurement by atomic absorption spectrometry.

Results and Discussion

The metallurgical dusts showed the largest magnetic susceptibility, probably from magnetite ores as well as the formation of ferromagnetics from other iron compounds, characterized by a significant magnetic susceptibility. It is interesting that metallurgical dusts show the greatest fluctuation in the susceptibility (Table 1).

Table 1. Magnetic susceptibility of dusts and fly ash

Type of dust	Number of samples	Specific susceptibility x $10^{-8}m^3/kg$		
		minimum	maximum	mean
Metallurgical	12	21	45816	13776
Cement	18	9	1620	363
Fly ash from hard coal	35	666	3605	2006
Fly ash from brown coal	7	508	1602	1047

The values of magnetic susceptibility for cement dusts are significantly less. The values depend mainly on additives used for cement production (fly ash, pyrite wastes). Upper values for cement dusts and fly ash from brown coal are similar, whereas the susceptibility of ashes from hard coal is significantly larger (Table 1). It probably results from iron content in Polish hard coals, which ranges from 6.8 to 13.2% Fe (Rozkowska, 1984). The content of iron in brown coal is less, only 0.70 to 1.24% Fe (Pacyna, 1980). Among iron compounds in hard coal about 1.5% of magnetite (Kuhl, 1961) and up to 15% of pyrite were found. Pyrite transforms during combustion into magnetite. Coal from the eastern part of the Upper Silesian Coal Basin the contains more pyrite, and therefore the magnetic susceptibility values of fly ash exceed 2000 x $10^{-8}m^3/kg$ from this area.

Strzyszcz (1988) found that the arable soils close to steel plants had a significant susceptibility over the whole ploughed layer (20-25 cm). Its value depended on the distance from plant and direction of prevailing winds. The susceptibility decreased with distance, and ploughing diluted it. Forest soils influenced by industrial emissions have an increase in magnetic susceptibility mainly in litter, especially in Of and Oh subhorizons. From the data presented in Fig. 2 it is evident that the susceptibility of the Of litter horizon in Katowice province ranges from 121 to 207 x 10^{-8} m^3/kg. Values between 121 and 500 are found in northern-western, western, and southern-western parts. In the central and eastern parts values exceeding 500 x 10^{-8} m^3/kg prevail. In the southern part the values range from 300 to 700 x 10^{-8} m^3/kg. These data in spatial terms practically overlap with iron fall values. (Fig. 3). Maximal value of iron fall in Katowice province amounted to 170 g m^{-2} a^{-1}, and the most often values are 10-35 g m^{-2} a^{-1}.

Table 2 lists the magnetic susceptibility in particular horizons of forest soils in Katowice province, its central part, i.e. USIR, Opole province, and Czestochowa province and in soils of some national parks in Poland. It also gives the annual dustfall.

Table 2. Magnetic susceptibility of particular horizons in forest soils

Locality	Number of strip pits	01	Of	Oh	Ah	E	B	Bbr	C	Dust-fall $t\ km^{-2}a^{-1}$
		Specific susceptibility x $10^{-8}m^3/kg$								
Slowinski NP	22	108	168	254	268	258	286		266	4
Czestochowa	15	197	400	294	223	206		225		27
Opole	29	222	363	317	265	347	376	266	210	73
Ojcowski NP	18	228	691	787	342	220	231	261	286	246
Katowice	22	425	922	725	305	198	232	207		328
USIR	27	1024	1729	1940	697	206	203	295	157	457

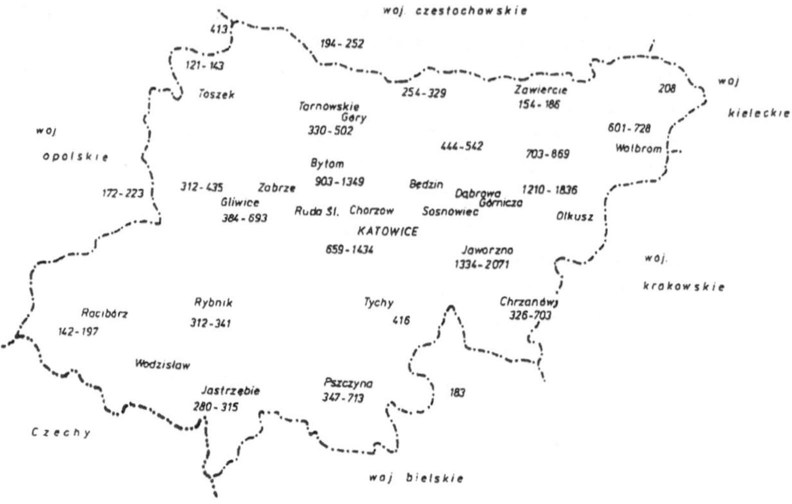

Fig. 2. Magnetic susceptibility (x 10^{-8} m^3 kg^{-1}) of the Of and Oh subhorizons in forest soil in the Katowice district

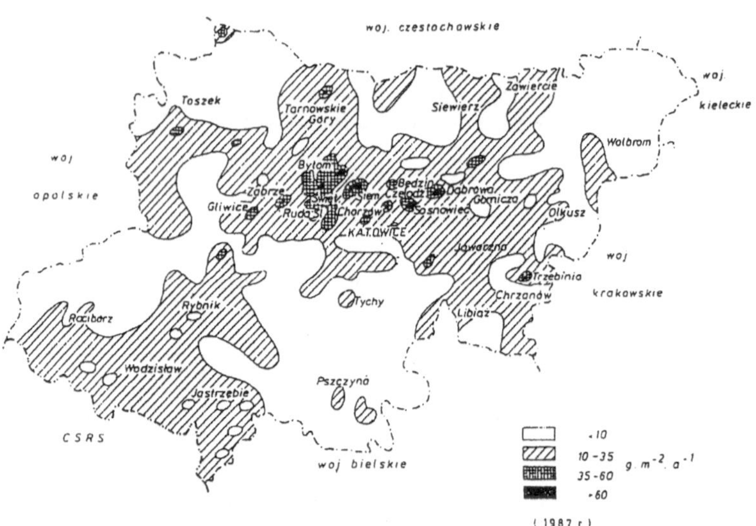

Fig. 3. Deposition of iron oxide (Fe_2O_3) in the Katowice district

The data suggest that accumulation of ferromagnetics takes place in the Of and Oh subhorizons. In the areas with pronounced emission (USIR, Ojcowski National Park), more ferromagnetics occur in the Oh subhorizon, whereas in areas with smaller emission more ferromagnetics are observed in the Of subhorizon. The greater accumulation of ferromagnetics in the Of subhorizon in the soil of Slowinski National Park might be because there is more precipitation there and the ferromagnetics are washed deeper into the soil.

These data confirm the observation of Vodyanitsky *et al.* (1983) that ferromagnetics in forest soil accumulate in the litter horizon, especially in the Of and Oh subhorizons. On the basis of the available evidence we attribute the large magnetic susceptibility found in upper soils layer to industrial emissions. This is also in accordance with results of Neumeister and Peschel (1968), who investigated soil close to Leipzig. On the basis of our own experience it can be stated that in the litter horizon near power plants the particles with diameters in the range 20 to 100 μm are rich in iron compounds, including ferromagnetics, and accumulate (Strzyszcz, 1991).

On the basis of the correlation coefficient value the conclusions can be drawn on dependence between soil magnetic susceptibility and some air pollutants in Katowice province. The results calculated from 13 sites are presented in Table 3.

Table 3. Product-moment correlation coefficient, *r*, between specific magnetic susceptibility of the soil and some air pollutants

SO_2	NO_x	F	dustfall	suspended dust	ironfall	iron conc. in air
0.110	0.340	0.220	0.677	0.656	0.440	0.656

The largest correlation coefficient was found for dustfall and content of Fe in the air. The larger *r* value for dustfall in comparison with suspended dust and iron fall should be noted. It can be explained by the fact that iron ferromagnetics falling from the air are accompanied by haematite. Its content in fly ash ranges from 2.0 to 9.5% and magnetite from 0.9 to 12.5% (Laczny, 1983).

Particles of industrial dust contain heavy metals in addition to ferromagnetics. As a consequence there is a correlation between magnetic susceptibility of soils and heavy metal contents in metallurgical and cement dusts, and fly ash. Values of correlation coefficients are given in Table 4.

Table 4. Product moment correlation coefficient between specific magnetic susceptibility and heavy metals content in dusts and soils

Type of the tested sample	n	Fe	Mn	Zn	Pb	Cd	Ni	Cu
Metallurgical dust	9	0.842	0.812	0.998	0.968		0.911	
Cement dust	13	0.377	0.940	0.990	0.999		0.809	
Fly ash	35	0.389	0.975	0.916	0.371		0.213	
Soils close to the steel work	18			0.618	0.843	0.175		
Soils close to the Jaworzno power plant	56			0.510	0.640	0.424		0.446
Horizon Of in 90 sites*	90			0.361	0.331			

The Table shows that the correlation coefficients are largest between magnetic susceptibility and heavy metal content in metallurgical dusts as well as in fly ash (except for Pb and Ni). Koniczynski (1982) found in the particular fractions of fly ash after combustion of hard coal the percent contribution of heavy metals listed in Table 5.

Table 5. Percent contribution of heavy metals in particular fractions of fly ash

	Fraction (μm)		
	0-10	10-40	>40
Pb	73-87	11-24	2-3
Ni	57-61	34-35	5-8
Zn	72-76	23-24	1-4

Lead in the aerosol fraction is the largest, and in soils far removed from industrial centres the correlation coefficient between magnetic susceptibility and content of lead in soil is larger than that with other heavy metals. In the soil of Ojcowski National Park (18 pits) and Opawski Landscape Park (13 pits) the correlation coefficient between magnetic susceptibility and content of heavy metals soluble in 2M HNO_3 was as listed in Table 6.

* forest soils from Katowice, Opole and Czestochowa provinces.

Table 6. Correlation coefficient between magnetic susceptibility of soil and content of some heavy metals in the soil of National Parks

	Fe	Mn	Zn	Pb
Ojcowski NP	0.697	-0.028	-0.028	0.794
Opawski NP	-0.081	-0.188	0.098	0.704

Bond (1992) found a similarly stronger correlation for Pb (0.575) than for Zn (0.490) in 56 soil samples collected close to Jaworzno power station.

Significant differences in correlation coefficients between the magnetic susceptibility and content of heavy metals in soil result both from granulation of emitted dusts, i.e. a different deposition rate and different filtration properties of various trees species and a distance from a trunk in which sampling of soil was made. So it seems that spatial dependence between the magnetic susceptibility and the content of heavy metals in soil needs further investigation.

Strong dependence between the magnetic susceptibility and heavy metal content within the pits was often found (Table 7). All the sites were influenced by significant industrial emissions. The relation between magnetic susceptibility and content of some heavy metals in soil of sites listed in Table 7 is presented in Fig. 4.

Table 7. Correlation coefficient, r, between magnetic susceptibility of soils and content of some heavy metals in strip pits in specific areas with much industrial emission.

Site of sampling	Number of strip pits	Zn	Pb	Cd
Kaziska	3	0.618	0.843	0.175
Chudow	2	0.704	0.724	0.581
Ruda Sl.	1	0.986	0.979	0.414
Szczakowa	5	0.075	0.599	0.280

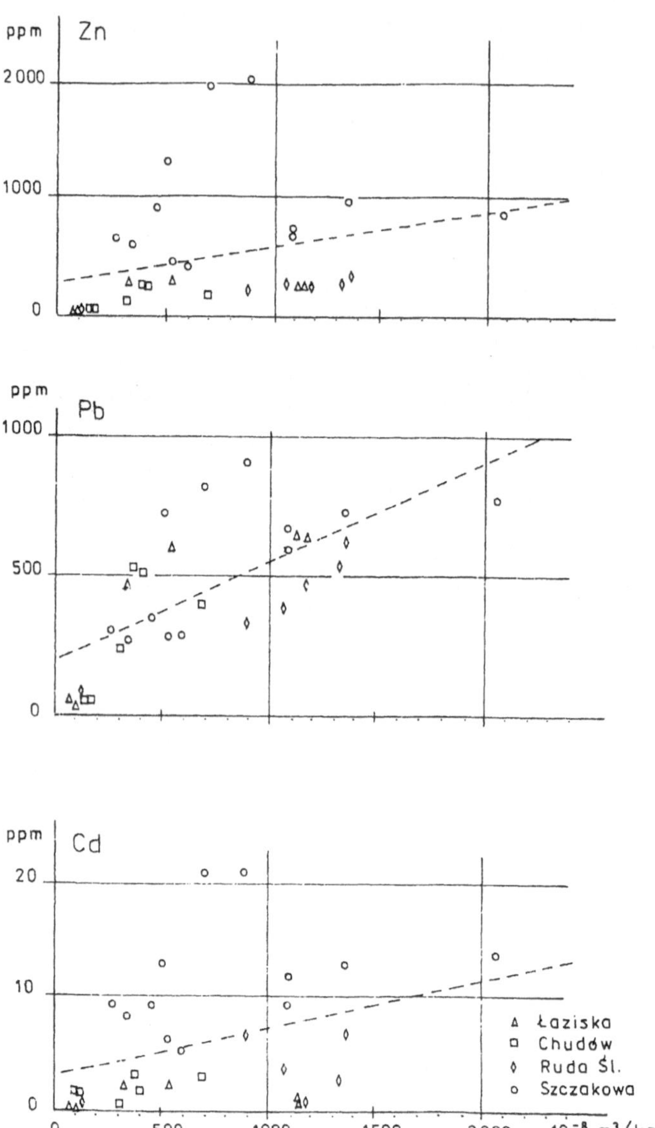

Figure 4. Relations between magnetic susceptibility (x 10^{-8} m^3/kg) and content of zinc, lead and cadmium (ppm) in soil

In the soil of the three national parks investigated the correlation between the magnetic susceptibility and content of heavy metals was as given in Table 8.

Table 8. Correlation coefficient between specific magnetic susceptibility of soil and content of some heavy metals in soil of pits in areas with little industrial emission.

Localization	Number of strip pits	Fe	Mn	Zn	Pb
Karkonoski NP	10	0.220	0.865	-0.153	0.181
Ojcowski NP	6	0.838	0.036	0.203	0.751
Opawski NP	5	-0.081	-0.188	0.098	0.704

Strong correlation between magnetic susceptibility of soil and the content of particular heavy metals (except Ni) was found in the pit close to a foundry in Dornach (P. Federer from ETH, Zürich), as follows:

Zn	Cu	Pb	Ni	Cd
0.96	0.92	0.95	0.10	0.91

The analysis shows that correlation is stronger near individual sources of emission and in forest soil than in arable soil.

In the litter horizon of the Silesian forest soil more than 5% of iron soluble in 10% HCl was found, out of its about half is characterized by magnetic properties (Strzyszcz, 1989). The question arises whether such increase in iron content can cause the ecological consequences and whether ferromagnetics should be treated as hazardous or toxic agents. Hansen *et al.* (1981) state: "Magnetite may also be a hazard to health because of its ability to occlude biologically active transition metal ions such as manganese and nickel by isomorphous substitution, and thus act as a slow release carrier agent for toxic elements". If the above proves true then the increase in amount of ferromagnetics in soil, the sources of which are industrial emissions, then we should treat the source of pollution as a time bomb with ecological consequences difficult to define nowadays.

Conclusions

The increase in magnetic susceptibility in Katowice province is caused by industrial emissions from metallurgical processes, cement production and the combustion of pulverized hard coal. Most ferromagnetics occur in metallurgical dust and fly ash. Ferromagnetics accumulate

mainly in the Of and Oh subhorizons of litter. Lower horizons of soil in the areas influenced show significantly less magnetic susceptibility. In the areas with small emissions (SNP) the ferromagnetics accumulate in the B subhorizon.
In almost all dusts tested there were strong correlations between magnetic susceptibility and content of heavy metals. The correlation between magnetic susceptibility and content of heavy metals in soil was weaker. This is probably caused by differences in granulation and mineralogical composition and conditions of dusts spreading and also by different filtration capacity of particular trees species. Within a soil profile stronger correlation was found only to areas influenced by industrial emissions.

Acknowledgements

I am very grateful to Professor F. Heller from the Geophysics Institute of the ETH in Zürich for the comparison of two methods for determining of magnetic susceptibility of soils, industrial dusts and fly ash.

References

Bond, C. 1992. *Immissionsbelastung in Oberschlesien. Bericht, Teil 5. Geostatische Zusammenhänge zwischen Schwermetallbelastung und Ferromagnetismus im Boden.* Diplomarbeit. Institut für terrestrische Ökologie, ETH Zürich, pp. 104-141.

Fassbinder, J.W.E., Stanjek H. & Vali, H. 1990. Occurence of magnetic bacteria in soil. *Nature, London* **343**, 161-163.

Fine, P., Singer, M.J., La Ven, R., Verosub, K. & Southard R.J. 1989. Role of pedogenesis in distribution of magnetic susceptibility in two California chronosequences. *Geoderma* **44**, 187-306.

Fine, P., Singer, M.J. & Verosub, K.L. 1992. Use of magnetic-susceptibility measurements in assessing soil uniformity in chronosequence studies. *Soil Science Society of America Journal* **56**, 1195-1199.

Hansen, L.D., Silberman, D. & Fisher, G.L. 1981. Crystalline components of stack-collected, size-fractionated coal fly ash. *Environmental Science and Technology* **15**, 1057-1062.

Hopke, P.K., Lamb R.E. & Natusch F.S., 1980. Multielemental characterization of urban roadway dust. *Environmental Science and Technology*, **14**, 164-172.

Hunt, A., Jones J. & Oldfield F., 1984. Magnetic measurements and heavy metals in atmospheric particulates of anthropogenic origin. *The Science of the Total Environment* **33**, 129-139.

Jarzebski, S. & Kapala J., 1976. Atlas zanieczyszczen wydzielanych przy procesach hutnictwa zelaza.Wyd. "Slask", p. 345.

Konieczynski, J. 1982. Skutecznosc pracy elektrofiltrow a emisja metali sladowych w spalinach elektrowni weglowych. *Ochrona Powietrza* **1-3**, 7-14.

Kuhl, J. 1961. Chemiczno-mineralna budowa nieorganicznej substacji mineralnej w weglu kamiennym. *Kwartalnik Geologiczny* **5**, 801-815.

Laczny, M.J., 1983. Model emisji zanieczyszcen ze skladowisk odpadow energetycznych do wod podziemnych. IKS Warszawa, p. 179.

Le Borgne, E. 1955. Susceptibilité magnetique anormale du sol superficiel. *Annales Géophysiques* **11**, 339-419.

Maher, B.A. 1986. Characterisation of soils by mineral magnetic measurements. *Physics of the Earth and Planetary Interiors* **42**, 76-92.

Manecki, A., Klapyta, Z., Schejbal-Chwastek, M., Skowronski, A., Tarkowski, J. & Tokarz, M. 1981. Wplyw przemyslowych zanieczyszczen atmosfery na zmiany geochemii srodowiska przyrodniczego Puszczy Niepolomickiej. *Prace Mineralogiczne* **71**, 1-58.

Mitchell, R.S. & Sluskoter, H.J. 1976. Mineralogy of ash of some American coals. *Fuel* **55**, 90-96.

Pacyna, J.M. 1980. *Coal-fired power plants as a source of environmental contamination by trace metals and radionucloides*. Scientific Papers of the Instiute of Environment Protection Engineering of Wroclaw Technical University No 47. (in Polish).

Petersen, N., Weiss, D. & Vali H. 1989. Magnetic bacteria in lake sediments. *Geomagnetism and Palaeomagnetism* 231-241.

Richardson N., 1986. The mineral magnetic record in recent ombrotrophic peat synchronised by fine resolution pollen analysis. *Physics of the Earth and Planetary Interiors* **42**, 48-65.

Rozkowska, A. 1984. Zawartosch pierwiastkow sladowych w weglach kamiennych z centralnej i poludniowej czesci GZW. In Proceedings of National Conference on: *Problemy badan wegla w pracach geologiczno-zlozo-owych w aspekcie nowych technologii jego utylizacji*. Uniwersytet Slaski, Katowice.

Schejbal-Chwastek, M. & Tarkowski J. 1988. Mineralogia przemyslowych pylow atmosferycznych i ich wplyw na zmiany geochemii srodowiska w parkach narodowych poludnieowej Polski. *Pracer Mineralogiczne* **8**, 1-92.

Strzyszcz, Z., Tölli, R. & Katzur J. 1988. Zur Anwendung eines hochfrequenten Messverfahrens für den Nachweis von ferromagnetischen Eisen in der Umwelt. *Archiwum Ochrony Srodowiska* **3-4**, 137-143.

Strzyszcz, Z., 1989 a. Anwesenheit des ferromagnetischen Eisen in oberschlesischen Waldböden und deren Ursachen. *Mitteilungen Deutsche Bodenkundliche Gesellschaft* **59**, 1197-1202.

Strzyszcz, Z., 1989 b. Ferromagnetic properties of forest soil being under influence of industrial pollution. Air pollution and forest decline. Proc. 14th Int Meeting for Specialist in Air Pollution Effects on Forest Ecosystems. IUFRO P2 05, 210-207, Interlaken.

Strzyszcz, Z., 1991. Ferromagnetism of soil in some Polish National Parks. Mitteilungen *Deutsche Bodenkundliche Gesellschaft* **66**, 1119-1122.

Schwertmann, U. & Heinemann B. 1959. Über das Vorkommen und die Entstehung von Maghemit in nordwestdeutschen Böden. *Neues Jahresbuch Mineralogie* **8**, 174-181.

Taylor, R.M. & Schwertmann U. 1974. Maghemite in soils and its origin. I Properties and observations on soil maghemites. *Clay Minerals* **10**, 189-297.

Tolonen, K. & Oldfield F. 1986. The record of magnetic-mineral and heavy metal deposition at Regent Street Bog, Fredricton, New Brunswick, Canada. *Physics of the Earth and Planetary Interiors* **42**, 57-66.

Tölle, R.,1986. Zum Einsatz weichmagnetischer Ferrite in der Bodenerosionsforschung. Dissertation Humboldt Universität, Berlin.

Thompson, R. & Oldfield F. 1986. *Environmental Magnetism*. Allen and Unwin London.

Tomza U., 1987. Trace elements pattern in atmospheric aerosols at Katowice. *Pracey Naukowe Universytat slaski* **924**, 120.

Vodyanitskiy, Y.N., Bagin, V.I. & Mymrin, V.A. 1983. Ferromagnetic minerals distribution in the podzolic soil profile. *Pocvoviedenie* **3**, 104-111.

Wilczynska-Michalik W., 1981. Z badan mineralogicznych pylow emitowanych przez Hute im. Lenina w Krakowie. *Prace Mineralogiczne* **68**, 1-52.

Z. Strzyszcz
Institute of Environmental Engineering
Polish Academy of Science
Zabrze
Poland.

Discussion

Question by R. Schulin

a) How strong was the correlation you found between your measurement signal of ferromagnetism and the content of heavy metals in your soils?

b) Is it site-specific?

Answer by Z. Strzyszcz

a) We found a rather close correlation of 0.87-0.90 (for different horizons)

b) Yes, the correlation gets worse if you make the regression over a larger area.

SAMPLING AND ANALYTICAL TECHNIQUES AS LIMITING FACTORS IN SOIL MONITORING

J.-P. Dubois and R. Schulin

Most properties of the soil vary substantially in space whereas their changes in time are rather slow. For soil monitoring this creates a problem because sampling is in most cases destructive and therefore repetitive samples have to be taken at different locations. As a consequence the 'signal' of interest, i.e. change in time, can be obscured by spatial variation which constitutes undesired background variation or 'noise'. Another source of noise which can seriously limit the conclusiveness (power) of a soil monitoring campaign is the variation due to procedural and analytical errors. This is a general problem in soil surveying, but it requires particular attention in long-term studies. The reason is that references and methods for quality control are themselves subject to temporal change, and that such changes are difficult to control and become increasingly likely and important as the time spanned by a series of observations increases.

The information provided by an investigation is given by its reliability and its comparability to existing knowledge. A way to control the reliability of the results of a study is to reproduce them by another study which is as independent as possible from the one to be controlled. To assure the reproducibility of the results of an entire investigation it should – at least in principle – be reproducible in each of its steps. Comparability is achieved by standardizing procedures and by calibrating the methods for measuring and observing to common reference systems or standards. In statistical terminology reproducibility is related to the precision and comparability to the unbiasedness of measurements and observations.

Measuring soil properties usually requires preparations before measurements can be made. The standard procedures for determining pollutant concentrations, for example, consist of the following main steps: locating and characterizing the sampling positions, excavating or coring to obtain soil samples, transport and storage in containers, conditioning (e. g. drying) and fractionation (e. g. sieving), homogenization and subsampling, extraction of subsamples, and finally chemical analysis of extracts. Each step contains a source of errors. The control of these errors is especially difficult in the first steps, because repetitive sampling at exactly the same location is impossible and because partitioning a sample into (more or less) equal subsamples is not feasible before the samples have been homogenized. Thus, short-range spatial variation of soil properties is inseparable from sampling errors.

Detailed protocols are necessary to control errors due to sampling in addition to a strict standardization of the sampling procedure. It is also important to record the positions of samples with respect to the morphological soil horizons because the soil varies with depth. Horizons with different ecological dynamics should not be mixed. In particular organic layers should be sampled separately from mineral horizons. Some problems still remain to be solved, for example, how a sample should be collected in the thin topsoil layer of a grassland where a separation of soil from plants is almost impossible.

Procedural errors, such as those from contamination by the sampling devices and losses during transport and storage, can be controlled by taking proper care and can be detected and estimated by analysing blank samples.

To assess analytical errors rigorously we need to be able sample repeatedly from homogeneous material. These replicates can then be analyzed in different batches, in different calibration series, in different laboratories, and in other different procedural units and facilities and the results compared. With the help of checks and standards individual components of the analytical error can be identified and quantified.

The chemical analysis of a soil extract is usually the least problematic part in measuring soil pollutants, although matrix effects can still cause significant difficulties. Much improvement is still possible in the extraction procedures, especially for organic substances. To minimize errors in laboratory analyses GLP ('good laboratory practice') rules have been formulated, and these should always be followed.

Many procedures of sampling and extraction can change the chemical and physical properties of the soil drastically. Whereas artefacts may not be a problem of the technical quality of the measurements, it is certainly of concern with respect to the meaning of the results.

Soil monitoring programmes must be designed to cover time spans of several decades. Given the rapid development of technology of today this means that analytical instruments and techniques will completely change during such a programme, with probably a trend to more *in situ* measurement. Moreover, there will be a considerable turnover of personnel involved in a programme. The criterion of overall reproducibility, in the classical sense used for laboratory experiments in the natural sciences, is obsolete and almost meaningless under these conditions. A long-term monitoring programme is a unique historical process. Only individual steps can be reproducible.

To be of value for assessing temporal changes observations made at different times during such a programme must be comparable to each other. This comparability can be achieved only by calibrating measurements to common reference standards, that is standard methods as well as standard samples. These methods must be as reproducible as possible. Because overall reproducibility is unattainable historical information is vital.

Experience tells us that the need for detailed documentation cannot be exaggerated. Matters that appear to be obvious and self-evident today might be hard to understand or be ambiguous tomorrow, especially when there is no longer anyone with first-hand knowledge to ask. Since no documentation will suffice to replace the real soil, materials, tools, and samples must be placed in archives wherever possible. This task is often underestimated in soil monitoring. It requires specialized and qualified personnel and facilities. In particular, however, cost-effective methods of archiving need to be developed and used. At present the effects of long-term storage on the composition and structure of soil samples and how they can be avoided or minimized through adequate preparation is insufficient, in particular with regard to organic matter and biological properties.

Conclusions

In conclusion the maintenance of constant quality standards over long times is a major challenge in soil monitoring. There are difficulties that go far beyond quality control of short-term surveys for which there already exist elaborate methods such as inter-laboratory testing. For long-term environmental monitoring archiving and documentation play a prime role and need to be much more developed as they are today, in particular with respect to soil structure, organic chemicals and biological components of soil. For this task the experiences of the historical sciences could prove valuable.

J.-P. Dubois
Institute for Land and Water Management
EPFL
1015 Lausanne
Switzerland

Prof. Dr. R. Schulin
Institut für terrestrische Ökologie (ITÖ)
ETHZ
Grabenstrasse 3/11a
CH-8952 Schlieren
Switzerland

SECTION 4

Survey and Data Analysis

Soil Monitoring, Monte Verità, © Birkhäuser Verlag Basel

Sources, propagation and control of errors in soil monitoring

Peter Lischer

Monitoring (by sampling) the amount of pollutants in soil, water, sewage sludge, etc., is important in today's society, especially when concentrations in samples significantly exceed officially tolerated limits. To determine whether some quantity surpasses a given threshold is a seemingly simple statistical problem. But how (apart from technical details like level and choice of test) is *significance* determined?
This paper shows how to estimate the different sources of errors, e.g. the variation between different soil samples, the interlaboratory variation, the variation between replicates and how to test if the concentration in a sample is significantly too large.

Introduction

Monitoring the amount of pollutants in soil, water, sewage sludge, etc., is important nowadays. To judge soil contamination by heavy metals and fluorine the Ordinance Relating to Pollutants in Soil (VSBo, 1986) sets threshold values of the total content (HNO_3-extract) and soluble content ($NaNO_3$-extract) that will not interfere with the fertility of soils in the long term. In the event that one of these values for a pollutant is exceeded, it is necessary to examine the source of contamination and to decide whether, according to Swiss federal regulations, measures should be taken.

As in many branches of science, soil monitoring involves many people and many groups. A good field description of soil and its environment requires the expertise of a pedologist who has not only a broad knowledge of geology, geomorphology and soil biology, but also an appreciation of present and past agricultural, horticultural and silvicultural practice. Equally important are soil chemists with experience. Laboratory determinations, however carefully carried out, will have no relevance if the samples do not represent the class or body of soil intended. Because there is a limit to the size of sample that can be carried from the field or dealt with in the laboratory, small samples are inevitable to represent huge volumes of soil. Much depends how the samples are selected. At least one member of the soil monitoring team should have experience in statistical design and analysis of experiments.

The monitoring problem

Let us consider an area where contamination by heavy metals is suspected. Samples are taken and analysed to determine the content of a heavy metal. Even if tests are performed on a single sample under identical conditions they do not, in general, yield identical results. This is attributed to unavoidable errors inherent in every test procedure since the factors that influence the outcome of a test cannot all be controlled fully. This variability has to be taken into account when interpreting test data. For instance, the difference between a test result and some specified value may be within the range of unavoidable random errors, in which case a real deviation from the specified value cannot be established. Similarly, comparing test results from two batches of material will not indicate a fundamental quality difference if the difference between them can be attributed to variation inherent in the test procedure. Errors arise in several ways and each group of scientists has its own system or *model* to describe and control them and so to reduce the total error.

Soil scientists know that soil is heterogenous so that its properties vary from one place to another. This variability is a source of error that is not easy to estimate. Data collected at known locations in space may be correlated, especially when the distance between sampling points is small. The theory of *regionalized variables* and its applications is an appropiate tool to evaluate such data (Webster & Oliver, 1990).

Chemists know from experience that the intralaboratory error tends to be smaller than the interlaboratory error. They try to assess the difference between by interlaboratory experiments in which the same material is analysed repeatedly by several laboratories. *Interlaboratory tests* enable the different kinds of analytical errors to be estimated and controlled (ISO-5725, 1986; SLB, 1989).

Statisticians are aware that many common assumptions are at best approximations to reality and that a minor error in the mathematical model can sometimes cause a much larger error in the final results. *Robust statistics* take into account deviations from the assumed ideal model (Hampel *et al.*, 1986; Huber, 1981).

The statistical model

Modelling spatial data

Just as data collected in time may be correlated, data collected at known localities in space may also be correlated. For example, the concentrations of lead contamination in the soil around a smelter are likely to be correlated. Soil samples are collected at various locations in space, and the data from them can be analysed by methods of spatial analysis. However, there is no single way of dealing with spatial variation in soil monitoring and in many instances no obvious best way. Each situation should be studied, and a sampling scheme designed for it (Webster, 1993).

Modelling laboratory errors

Interlaboratory trials are used in chemistry to control laboratory errors (ISO, 1986). We can distinguish four different types of interlaboratory tests, as follows.

1 *Collaborative trials* used for validating a standard method.

2 *Certification trials* used to establish the true value of an analyte concentration in a reference material.

3 *Co-operative trials* used to test the performance of laboratories, both as group and individually. Each laboratory is free to use any method that is consisered appropriate for the test material and the application.

4 *Proficiency testing* used for internal and external quality control.

Conducting interlaboratory tests is laborious, and so the organizers often try to combine different aims. This is possible, and I show an example of such an experiment below.

Many different factors may contribute to the variation in results of a test procedure, including

(a) the operator;

(b) the equipment;

(c) the calibration of the equipment;

(d) the environment (temperature, humidity, air pollution, etc.).

Such variation between tests performed by different operators or with different equipment, or both will usually be greater than that between tests carried out by a single operator using the same equipment.

The general term for the variation between repeated tests is *precision*. Two measures of precision, namely *repeatability* and *reproducibility* are necessary and are generally sufficient for describing the variability of a test measurement. Under repeatability conditions factors (a) to (d) listed above are considered constant and do not contribute to the variability, while under reproducibility conditions they vary and do contribute to the test results. Thus repeatability and reproducibility are the two extremes of precision, the first describing the minimum and the second the maximum variation in results. For some test methods there may be an accepted reference value for the property: for example if suitable reference materials are available, or if the accepted reference value can be established in relation to another test method or by preparing a known sample. The *trueness* of the test method can then be investigated by comparing the accepted reference value with the analytical results. Trueness is normally expressed in terms of *bias*. Bias can arise in chemical analysis, for example, if the test method fails to extract all of an element, or if the presence of one element interferes with determination of another. The general term *accuracy* is used in ISO (1986) to refer to trueness and precision combined.

For estimating the precision of a test method, we can assume that every single test result, y, is the sum of three components:

$$y = m + b + e, \tag{1}$$

where m is the general average, b is the between-laboratory variation and e is the random error occurring in every test.

The general average m of the material tested in an accuracy experiment involving many laboratories is called "the level of the test" (ISO,1986). In many technical situations the level of the test is exclusively defined by the test method, and the notion of an independent true value does not apply, e.g. the measurement of pH in soils which may lead to varying results depending on the extractant used (H_2O, KCl, $CaCl_2$). In other situations, however, the concept of a true value μ of the test property may hold good, such as the true concentration of a material in a solution that is being titrated. The level m is not necessarily equal to the true value μ ; the difference $(m - \mu)$, if it exists, is the bias of the method.

The term b is considered constant during any series of tests performed under repeatability conditions, but differs for tests carried out in different laboratories (that is, under reproducibility conditions). The value of b for a given laboratory is called the laboratory component of bias. When test results are always compared between the same two laboratories, they must determine their relative bias, either from their individual bias values, or by a trial between themselves. To make general statements about differences between two unspecified laboratories, or between two laboratories that have not determined their own bias, the distribution of laboratory components of bias must be considered. This was the reasoning behind the concept of reproducibility.
The variance of b is called the between-laboratory variance and is expressed as

$$\mathrm{var}[b] = \sigma_L^2 , \tag{2}$$

where σ_L^2 includes the between-operator and the between-equipment variabilities. In general, b can be considered as the sum of both random and systematic components.

The term e is the random error in every single test result. Within a single laboratory its variance is called the within-laboratory variance and is expressed as σ_w^2. This will have different values in different laboratories arising from the skills of the operators for example, but for a properly standardized method such differencies should be small. Therefore it is justifiable to establish a common value of within-laboratory variance, which is estimated from the average of all the laboratories taking part and is called the repeatability variance. It is given by

$$\sigma_r^2 = \text{var}[e] = \text{the average of } \sigma_w^2 . \tag{3}$$

The two quantities used as measures of precision are
the *repeatability standard deviation* σ_r and
the *reproducibility standard deviation* $\sigma_R = \sqrt{\sigma_L^2 + \sigma_r^2}$.
Usually b and e are further assumed to be independent and normally distributed with mean 0, $\text{var}[b] = \sigma_L^2$ and $\text{var}[e] = \sigma_r^2$.

However, suspect results occur much more frequently than the normal distribution would predict, for instance if one aspect of the analytical process is not completely under control. Since very few suspect values deviate by an order of magnitude, it is often difficult to decide whether the suspect value should be regarded as valid. It is the job of the statistician to identify suspect values and to point out that they are markedly different from most of the other, but he can neither give a reason nor say whether the majority of the data is correct. Distant outliers can have very serious effects on classical methods like mean, variance and least squares. Any reasonable procedure for rejecting outliers will prevent the worst from happening. Therefore, in evaluating collaborative trials, extreme results or all results obtained from a suspect laboratory are often eliminated before the components of variance are determined. But any such elimination inevitably entails the risk of overestimating the precision. An international convention suggesting suitable test statistics (e.g. Cochran and Grubbs tests) and decision limits (e.g. P = 0.01 or 'at least 40% reduction of the relative standard deviation') can be adopted (Horwitz, 1988). It does not, however, change the unsatisfactory *either-or situation*, which is typical for all outlier tests: as soon as the conditions for elimination are fulfilled the value of the desired quantity changes abruptly. Moreover, the proposed outlier tests are far from the best possible ones; e.g. the Grubbs test cannot even safely reject 2 distant outliers out of 20 (Hampel, 1985). On the other hand, Horwitz *et al.* (1986) found that the number of outliers in collaborative studies is of the order of 10% to 30%! These and other unsatisfactory features of outlier tests led the Swiss Federal Committee for Official Methods in Food Analysis (Lischer, 1987; SLB, 1989) and the Analytical Methods Committee of the Royal Society of Chemistry (AMC, 1989b) to suggest robust statistical methods for calculating repeatability and reproducibility.

Approximate statistical models, robust statistics (Huber, 1981; Hampel *et al.*, 1986)

Almost all statistical analysis has at its heart a model. Estimation, hypothesis testing, and inference are generally based on the data available and a conjectured model which may be defined implicitly or explicitly. Even in the simplest case, there are assumptions about randomness, independence and the type of the underlying distribution. These assumptions are not supposed to be exactly true. Possible deviations from the assumed parametric model (e.g. normal distribution) are the occurrence of gross errors, rounding and grouping effects and serial correlations. Rationalizations or simplifications are vital in applied mathematics, and one justifies their use by appealing to a vague continuity or stability principle: a minor error in the mathematical model should cause only a small error in the final conclusions. Unfortunately, this does not always hold true. Some of the most common statistical procedures (in particular those optimized for an underlying normal distribution) are very sensitive to seemingly small deviations from the assumptions. Nevertheless the normal distribution pervades statistical methodology, and its very name suggests widespread applicability. Users of statistics point to a theoretical result of mathematics to justify the assumptions, whereas pure mathematicians believe its applicability to have been proved empirically.

It may be more appropriate to describe real data as *contaminated normal distributions*. They have the form $F = (1 - \varepsilon)\,\Phi + \varepsilon H$, $0 < \varepsilon < 1$, where Φ is the normal distribution with mean μ and variance σ^2 and H is an arbitrary distribution.

$\varepsilon = 0.05$ means that 95% of the data are normally and 5% are arbitrarily distributed. Real laboratory data usually are better described by contaminated normal distributions with ε equal to a few percent. Moreover, transcription errors are surprisingly common. 5-10% of wrong values in a data set seem to be the rule rather than the exception (Hampel, 1980).

Robust statistical procedures possess the following features (Huber, 1981).

- They have a reasonably good (nearly 100%) *efficiency* at the assumed model. The percentage α efficiency of a procedure means that an optimal procedure needs only α% of the observations to get the same precision.
- They are robust in the sense that small deviations from the model assumptions impair the performance only slightly.
- Somewhat larger deviations from the ideal model are not catastrophic (high *breakdown point*).

The *breakdown point* (Hampel, 1980) of an estimator is the smallest fraction of free contamination that can carry the estimated value beyond all bounds. The breakdown point is a global measure of the safety or reliability of the estimator. The mean has a breakdown point of 0, because it does not tolerate a single outlier. The median has a breakdown point of 50%. However, it is not very efficient (in the ideal case of a normal distribution ≈ 60%).

Remark.

The notion of robust procedures is also used in ISO (1986). A standardized analytical test method is called robust if small variations in the procedure do not produce unexpectedly large changes in the results.

The soil-monitoring model

A statistical model for soil monitoring takes into account the views of the pedologist, the chemist and of the statistician. Every single result $y(\mathbf{x})$ of a specimen S taken at a point $\mathbf{x}$ is the sum of four components

$$y(\mathbf{x}) = z(\mathbf{x}) + b + e + a\,. \tag{4}$$

The true value of a specimen taken at $\mathbf{x}_k$, $z_k = z(\mathbf{x}_k)$, is regarded as random with mean μ and variance σ_s^2. The quantities b and e are the interlaboratory and the intralaboratory errors mentioned above, and a is an eventual bias. When examining the difference between test results obtained by the same test method the bias will have no influence and can be ignored. However, when comparing results produced using different test methods the bias has to be taken into account.

We assume further that b and e have contaminated normal distributions:

distribution of $b \in \{F\!: F = (1\text{-}\varepsilon)\,\Phi(\,0, \sigma_L^2\,) + \varepsilon\, H, 0 < \varepsilon < 1, H$ arbitrary;

distribution of $e \in \{G\!: G = (1\text{-}\varepsilon)\,\Phi(\,0, \sigma_r^2\,) + \varepsilon\, H, 0 < \varepsilon < 1, H$ arbitrary $\}$

Now let us consider a single additional measurement. This is made up of the true value + spatial effect + laboratory bias + measurement error and has so variance $\sigma_s^2 + \sigma_L^2 + \sigma_r^2$. Let

$$d = f\sqrt{\sigma_s^2 + \sigma_L^2 + \sigma_r^2}\,, \tag{5}$$

where f is a factor the value of which depends both on the number of test results available for estimating each of the variances and on the shape of the distributions. However, if the distributions are approximately normal and the number of test results is not too small, then at the level of 95% the factor f will be close to 2, and its use is recommended in international standards.

Then

$$\text{P}\,(|\text{measurement - true value}\,| < d\,) = 0.95, \tag{6}$$

and d measures the attainable precision.

A similar formula may be given if two additional measurements taken independently at the points $\mathbf{x}$ and $\mathbf{x} + \mathbf{h}$ are considered and if it is assumed that the *intrinsic hypothesis* (Matheron, 1965) hold true. Let

$$d(\mathbf{h}) = f\sqrt{2}\sqrt{\gamma(\mathbf{h}) + \sigma_L^2 + \sigma_r^2}\,, \tag{7}$$

where

$$\gamma(\mathbf{h}) = \frac{1}{2}\text{var}[z(\mathbf{x}) - z\,(\mathbf{x} + \mathbf{h})] \tag{8}$$

is the variogram. Then

$$\text{P}\,[|\,y(\,\mathbf{x} + \mathbf{h}) - y(\,\mathbf{x}\,)\,| < d(\mathbf{h})\,] = 0.95 \tag{9}$$

Computation of robust estimates

Consider n data points $x_1, x_2, ..., x_n$. We assume that the underlying distribution is contaminated normal, i.e. that μ is the mean and σ^2 the variance of the 'reliable' results, but not of the whole error distribution, and we want to find estimates of μ and σ^2.
Sample mean and sample variance are not the only possible estimators for population mean ('true value') and population variance ('precision'). Two other estimators are the median, M, and the median absolute deviation, MAD:

$$\mathrm{M} = \mathrm{med}_i\, x_i \; ; \tag{10}$$

$$\mathrm{MAD}_n = 1.4826\, \mathrm{med}_i\, \{| x_i - M |\} = 1.4826\, \mathrm{med}_i\, \{| x_i - \mathrm{med}_j\, x_j |\}. \tag{11}$$

Thr value 1.4826 is a normalizing factor. For normal distributions with mean μ and variance σ^2, $\mathrm{MAD_n}$ is a consistent estimator of σ. These estimators are highly resistant; they lead to results that change only slightly when a small part of the data is replaced by new numbers, possibly very different from the original ones, and have the best possible breakdown points (50%, if $n \rightarrow \infty$). The sample median and the MAD can be used to screen the data for outliers quickly, by computing

$$\frac{|x_i - \mathrm{med}_j x_j|}{\mathrm{MAD}_n} \tag{12}$$

for each x_i, and flagging as dubious those x_i for which this statistic exceeds a certain cutoff (say 2.5 or 3.0). However the MAD also has some disadvantages. Its efficiency for a gaussian distribution is only 37%. Furthermore, the MAD takes a symmetric view of dispersion because one first estimates a central value (the median) and then attaches equal importance to positive and negative deviations from it. The MAD corresponds to finding the symmetric interval (around the median) which contains 50% of the data, which does not seem to be a natural approach for asymmetric distributions.
Rousseeuw (1991) proposed two estimators of scale which are more efficient and not slanted towards symmetric distributions. They can be expressed by an explicit formula and have breakdown points of 50%. These estimators are

$$S_n = 1.1926\, \mathrm{med}_i\, \{\, \mathrm{med}_j\, | x_i - x_j |\, \} \tag{13}$$

and

$$Q_n = 2.2219\, \{\, | x_i - x_j |\, ;\, i < j\, \}_{(k)}\, , \tag{14}$$

where $k = \binom{h}{2} \approx \frac{1}{4}\binom{n}{4}$ and $h = [n/2] + 1$ is roughly half the number of observations, i.e. we take the k-th order statistic of the $\binom{n}{2}$ interpoint distances.

Unlike the MAD, S_n and Q_n do not need any location estimate. Instead of measuring how far away the observations are from the central value, S_n and Q_n look at a typical distance between observations, which is still valid for asymmetric distributions. The gaussian efficiency of S_n is 58%, whereas Q_n attains 82%.

Multivariate point clouds, robust distance.

Suppose we have data $\mathbf{X} = \{\mathbf{x}_1, \mathbf{x}_2, ..., \mathbf{x}_n\} = \{(x_{11}, x_{12}, ..., x_{1p}), ..., (x_{n1}, x_{n2}, ..., x_{np})\}$ of n points in p dimensions, (for example n laboratories, each of which analyses p samples) and we want to diagnose outlying points (laboratories). The word *outlier* is applied here to any vector (laboratory) that is markedly different from most of the other. It can happen that a point must be declared as outlying even if none of its elements is outlying. Figure 1a shows such a situation with an outlying point and the classical 95% tolerance ellipse obtained from the usual mean and variance-covariance matrix. This outlying point is characterized by a large (squared) *Mahalanobis distance*

$$D^2 = (\mathbf{x} - \mathbf{m})\, \mathbf{V}^{-1}(\mathbf{x} - \mathbf{m})^T, \tag{16}$$

where **m** is the mean of the group and **V** is the variance-covariance matrix.

The Mahalanobis distance (Flury & Riedwyl, 1983) is a generalized distance function that takes account of the correlation structure of the data set. Points with the same Mahalanobis distance from the mean vector lie on ellipsoids (ellipses for $p = 2$).

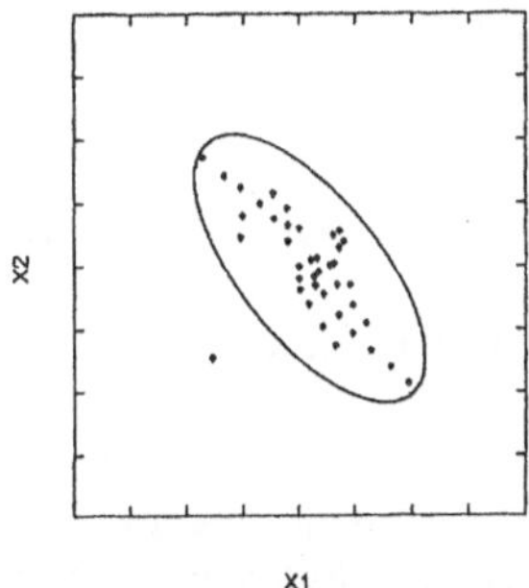

Figure 1a. Artificial data with 95% tolerance ellipse obtained with the classical estimator and an extreme point which is not outlying in either of the coordinates.

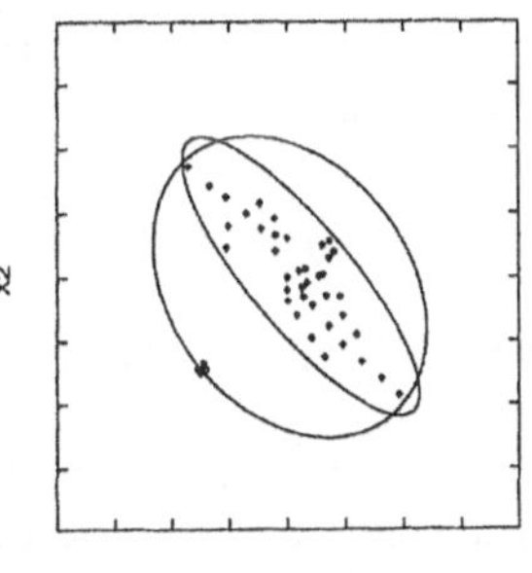

Figure 1b. Artificial data with 95% tolerance ellipses obtained with the classical estimator (big ellipse) and based on the MVE-estimator (small ellipse) and a group of 5 outlying points.

If **X** has a multivariate normal distribution then D^2 is distributed approximately as χ^2_p. Therefore it was originally suggested that one compare D^2 with a quantile of the χ^2_p distribution. This approach works well if only one single outlier is present, but suffers from the masking effect otherwise, because one distant outlier can make all other outliers have small D^2. Worse, the method fails completely in the presence of groups of outliers. Rousseeuw & van Zomeren (1990) suggest calculating robust distances based on estimators of location and scale with large breakdown points. Figure 1b shows a point cloud with a group of five markedly different points. The 95%-tolerance ellipse based on the Mahalanobis distance includes all points. Note that this ellipse is very big, because it is attracted by the five outliers. On the other hand, the ellipse based on the robust distance is much smaller and essentially fits the main group of points.
The software S-PLUS (1991) has an algorithm based on Rousseeuw's method for detecting multivariate outliers.
Robust distances summarise the overall performance of a laboratory. The robust distance is compared with the 95%-quantile of the χ^2_p distribution. Outlying laboratories have big robust distances. Robust distances seem to characterize outlying laboratories better than the sum of squared scores proposed in AMC (1992).

Experimental

Design of an experiment to assess the laboratory errors and a raw estimate of the spatial variation

As a single result y is a linear combination of independent quantities $z(\mathbf{x})$, b, e and a, the variance of y is the sum of the variances of the components:

$$\mathrm{var}[y] = \mathrm{var}[z(\mathbf{x}) + b + e + a] = \mathrm{var}[z(\mathbf{x})] + \mathrm{var}[b] + \mathrm{var}[e] \qquad (17)$$

An estimate of var[y] is needed to determine if a given limit value is exceeded.

It is not realistic to design an experiment to obtain reliable estimates of all possible variations because too many laboratories would have to collaborate and too many samples would have to be analysed. It is therefore necessary to proceed in stages, each of which will be subject to error. The over-all error in the final result can be determined from the errors in the separate experimental steps. The systematic and random laboratory errors need to be determined at the outset in monitoring. The information will be used to select laboratories with a high level of performance for further analysis, for example the estimation of the variogram and the variation in time. As a by-product of a well planned interlaboratory test we can get reference material with a consensus value and even a raw estimate of the spatial variation. The reference material can then be used for quality control and for estimating the long term variation of a laboratory.

At the second stage analysis we can get a more accurate estimate of spatial variation. This requires only one or a few well qualified laboratories.

Let the sampled points be distributed equally over the area (Figure 2).

Each laboratory L_i, $i = 1, 2, \ldots, p$ has to analyse 5 samples S_{i0}, S_{i1}, S_{i2}, S_{i3}, S_{i4}.

S_{i0} is taken from a bulked sample S_0 taken from the central part of the parcel.

S_{i1} and S_{i2} are taken at points $\mathbf{x}_{i1}$ and $\mathbf{x}_{i2}$ such that $\forall\, i : \mathbf{x}_{i2} - \mathbf{x}_{i1} = \mathbf{h}_i = \mathbf{h}$.

S_{i3} and S_{i4} are mixtures: $S_{i3} = 50\%\ S_{i0} + 50\%\ S_{i1}$, $S_{i4} = 50\%\ S_{i0} + 50\%\ S_{i2}$.

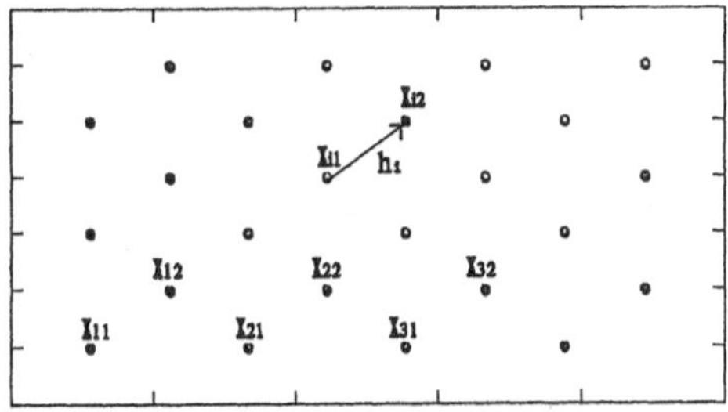

Figure 2. A systematic sample of 2n points (n = number of laboratories)

Let y_{ij}, $i = 1, 2, \ldots, p$; $j = 0, 1, \ldots, 4$, denote the result for the sample S_{ij}.

We have $y_{ij} = z_{ij} + b_i + e_{ij}$, where

$z_{ij} = z(\mathbf{x}_{ij})$ is the 'true value' of sample S_{ij};

b_i is the laboratory bias of L_i; $E[\,b_i\,] = 0$, $\mathrm{var}[b_i\,] = \sigma_L^2$

e_{ij} is the measurement error of L_i at S_{ij}; $E[\,e_{ij}\,] = 0$, $\mathrm{var}[e_{ij}\,] = \sigma_r^2$.

Let $w_{i1} = 2y_{i3} - y_{i0}$ and $w_{i2} = 2y_{i4} - y_{i0}$.

y_{i0}, w_{i1} and w_{i2} are three independent determinations by the same laboratory of the true content of S_{i0}.

We have $\forall\, i$:

$$\mathrm{var}[y_{i0}] = \sigma_L^2 + \sigma_r^2\,;$$

$$\mathrm{var}[w_{i1}] = \mathrm{var}[2\,y_{i3} - y_{i0}] = \mathrm{var}[b_i + 2\,e_{i3} - e_{i0}] = \sigma_L^2 + 5\sigma_r^2 = \mathrm{var}[w_{i2}] \qquad (18)$$

$$\mathrm{var}[y_{ij}] = \sigma_s^2 + \sigma_L^2 + \sigma_r^2\,;\ j = 1, 2, 3, 4\,.$$

Assessment of the performance of the laboratories

The laboratory results

	$\mathbf{y}$	$\mathbf{w}_1$	$\mathbf{w}_2$
L_1 :	y_{10}	w_{11}	w_{12}
L_2 :	y_{20}	w_{21}	w_{22}
....			
L_p :	y_{p0}	w_{p1}	w_{p2}

can be considered as a point cloud in 3 dimension. Outlying laboratories have robust distances $> \chi^2_{3,.975} = 9.348$.

Estimation of the interlaboratory and intralaboratory errors

The values $w_{i1} = 2y_{i3} - y_{i0}$ and $w_{i2} = 2y_{i4} - y_{i0}$ can be considered as results of a uniform-level interlaboratory trial.
We have $\forall\, i$:

$$\text{var}[\, w_{i1}\,] = \text{var}[\, w_{i2}\,] = \sigma_L^2 + 5\sigma_r^2. \qquad (19)$$

Algorithms to get robust estimates of σ_L^2 and σ_r^2 are given in SLB (1990).

Estimation of a consensus value of the reference material

Our experimental design allows two independent estimates of the 'true value' of S_0 :
(1) from $\{ y_{i0}\,;\, i=1, 2,..., p\}$ and
(2) from $\{ w_{i1}, w_{i2}\,;\, i = 1, 2,...., p\}$.
If the number of participating laboratories is not too small (> 7) then the estimates give a consensus value of S_0, and the sample can be used as reference material for further analysis.

Estimation of var[z(x)]

Let the intrinsic hypothesis (Matheron, 1965) hold true and consider the results

$$\mathbf{y}_j = \{y_{ij}\,;\, i = 1, 2,..., p\}\;, j = 1, 2, 3, 4. \qquad (20)$$

If we apply the scale estimator S_n (or Q_n) given above to the vectors $\mathbf{y}_j$ we get a robust estimate of the dispersion.

We have

$$S_n^2(y_2 - y_1) \approx 2\gamma(\mathbf{h}) + 2\sigma_r^2, \qquad (21)$$
$$S_n^2(y_4 - y_3) \approx \gamma(\mathbf{h}) + 2\sigma_r^2. \qquad (22)$$

Since the $z(\mathbf{x})$ are correlated we have further

$$S_n^2(y_1) \approx S_n^2(y_2) \approx \alpha\sigma_s^2 + \sigma_L^2 + \sigma_r^2, \qquad (23)$$
$$S_n^2(y_3) \approx S_n^2(y_4) \approx \tfrac{1}{4}\alpha\sigma_s^2 + \sigma_L^2 + \sigma_r^2 \qquad (24)$$

with $0 < \alpha \leq 1$. Thus

$$\gamma(\mathbf{h}) \approx S_n^2(y_2 - y_1) - S_n^2(y_4 - y_3) \qquad (25)$$

and

$$\alpha\sigma_s^2 \approx \tfrac{2}{3}[S_n^2(y_1) - S_n^2(y_3) + S_n^2(y_2) - S_n^2(y_4)] \qquad (26)$$

The actual design allows only a raw estimate of α.

A numerical example for estimating the laboratory errors

In an interlaboratory study carried out 1989 at the Swiss Federal Research Station for Agricultural Chemistry and Hygiene of the Environement (FAC) with 21 participating laboratories, several dried sewage sludges were analysed for different heavy metals and P and K. The primary goal was to compare laboratories using specified extraction techniques and optional analytical procedures (HCl-extraction and flame atomic absorption spectrometry). As far as this was possible a comparison of methods was also carried out. The results of the Cu-analyses are presented here.

Each laboratory L_i, $i = 1, 2, \ldots, 21$, had to analyse 10 specimens $S_1, S_2, \ldots, S_{10}$. The specimens $S_1, S_2, \ldots, S_5$ were taken from 5 different homogenized samples of dried sludges, S_6, S_7, S_8 and S_9 were mixtures: (S_6: 50% S_1 + 50% S_2; S_7: 50% S_3 + 50% S_2; S_8: 50% S_4 + 50% S_2; S_9: 50% S_5 + 50% S_2), and S_{10} was an extracted solution of S_1, produced in a laboratory of the FAC. This specimen was analysed to obtain an estimate of the effect of the preparation for different laboratories.

Four laboratories did not use F-AAS (Flame atomic emission spectrometry) but an other method. Laboratories L_{02} and L_{06} used ICP-AES (Inductively coupled plasma atomic emission spectrometry) and L_{15} and L_{21} used polarography.

Let y_{ij}, $i = 1, 2, \ldots, 21$; $j = 1, 2, \ldots, 10$; denote the result for the jth specimen S_j at laboratory L_i (Table 1) and w_{ij}, $i = 1, 2, \ldots, 21$; $j = 1, 2, 3, 4$; the calculated results of specimen S_2 (Table 2).

Table 1. Cu-concentrations in dried sewage sludges [mg/kg]

Labora-tory	Specimen S_1	S_2	S_3	S_4	S_5	S_6	S_7	S_8	S_9	S_{10}
L_{01}	233	491	428	376	200	364	474	444	351	234
L_{02}	253	530	453	389	191	402	495	453	371	244
L_{03}	240	530	460	390	200	390	480	440	350	274
L_{04}	243	518	456	389	186	385	483	459	369	240
L_{05}	244	521	446	392	193	385	481	451	357	245
L_{06}	257	547	459	407	198	382	502	462	466	243
L_{07}	252	477	414	391	190	366	456	387	339	243
L_{08}	263	530	458	420	221	389	493	475	378	242
L_{09}	262	530	454	408	182	388	496	468	382	242
L_{10}	260	554	472	426	211	404	488	483	383	260
L_{11}	243	527	444	369	163	393	492	456	351	246
L_{12}	226	470	416	373	178	347	445	323	387	233
L_{13}	239	529	425	375	188	357	460	430	335	229
L_{14}	239	501	422	402	171	337	444	416	363	266
L_{15}	296	732	548	562	167	552	642	639	431	333
L_{16}	252	521	460	405	208	416	497	467	373	242
L_{17}	218	489	430	369	187	366	456	414	338	226
L_{18}	216	480	416	357	179	369	450	423	319	242
L_{19}	250	500	470	390	210	390	490	450	380	230
L_{20}	223	495	403	361	173	353	442	416	325	216
L_{21}	260	268	378	223	157	557	110	493	819	287
mean	246.1	511.4	443.4	389.2	188.2	394.9	465.5	450.0	388.9	248.4
median	244.0	521.0	446.0	390.0	188.0	385.0	481.0	451.0	369.0	242.0
m	245.2	512.6	441.4	388.9	188.2	382.7	474.9	447.4	365.4	243.4
s	16.3	31.1	26.7	25.2	17.8	28.2	23.7	31.1	26.7	11.9

The quantities m and s are robust estimates of location μ and scale σ determined with the algorithm described in SLB (1989).

Outlying laboratories:

The critical value is $\chi^2_{5\ .975} = 12.833$. The robust distances rd^2, Table 2, show that laboratories 6, 7, 12, 15 and 21 are outlying, while the Mahalanobis distance D^2 fails to detect the seventh laboratory which has a marked outlying value (383).

Table 2. Measured and calculated Cu-concentrations of S_2 .

Laboratory	S_2 y_2	(S_6,S_1) w_1	(S_7,S_3) w_2	(S_8,S_4) w_3	(S_9,S_5) w_4	D^2	rd^2
L_{01}	491	495	520	512	502	3.5	1.0
L_{02}	530	551	537	517	551	0.4	1.1
L_{03}	530	540	500	490	500	2.8	1.2
L_{04}	518	527	510	529	552	0.5	1.0
L_{05}	521	526	516	510	521	0.3	0.1
L_{06}	547	507	545	517	734	11.4	19.0
L_{07}	477	480	498	383	488	3.2	18.1
L_{08}	530	515	528	530	535	0.9	0.4
L_{09}	530	514	538	528	582	1.9	1.0
L_{10}	554	548	504	540	555	4.0	3.9
L_{11}	527	543	540	543	539	1.3	0.8
L_{12}	470	468	474	273	596	14.4	85.5
L_{13}	529	475	495	485	482	4.9	1.0
L_{14}	501	435	466	430	555	5.0	8.9
L_{15}	732	808	736	716	695	16.9	76.2
L_{16}	521	580	534	529	538	2.4	1.0
L_{17}	489	514	482	459	489	1.5	1.0
L_{18}	480	522	484	489	459	3.2	2.1
L_{19}	500	530	510	510	550	1.4	0.7
L_{20}	495	483	481	471	477	1.3	0.6
L_{21}	268	854	-158	763	1481	19.0	2581.8
mean	511.4	543.8	487.6	510.7	589.6		
median	521.0	522.0	510.0	512.0	539.0		
m	512.6	520.8	508.5	506.7	540.6		
s	31.1	38.5	38.5	34.1	57.8		

Precision:

For this experiment we can also estimate the two measures of precision, the repeatability standard deviation σ_r and the reproducibility standard deviation σ_R .

Two methods are available: either a split-level experiment using the four pairs of samples (S_2,S_6), (S_2,S_7), (S_2,S_8), (S_2,S_9) or a uniform-level experiment with the four calculated values w_1, w_2, w_3 and w_4.

The four laboratories using methods other than F-AAS are not included in the calculations. The definitions for repeatability and reproducibility assume a closely specified method (e.g. HCl-extraction and F-AAS).

Table 3. Estimated inter- and intralaboratory standard deviations
(without the laboratories L_{02}, L_{06}, L_{15} and L_{21})

	s_L	s_r	s_R
(S_2,S_6):	20.38	13.44	24.41
(S_2,S_7):	21.18	12.40	24.54
(S_2,S_8):	24.56	13.55	28.05
(S_2,S_9):	16.52	14.78	22.17
w_1, w_2, w_3, w_4:	29.33	6.72	30.09

The estimates do not differ much.

True Cu-content of S_2 *:*
Our experimental design gives two independent estimates of Cu-content of S_2 :

- From sample S_2 : 512.6 (s.e.m. ≈ 6.8)
- From w_1, w_2, w_3, w_4: 517.9 (s.e.m. ≈ 4.5)
- Consensus value 515.3 (s.e.m. ≈ 5.8)

The sample S_2 can be used as reference material for further analysis.

Conclusions

The fascinating thing about analytical chemical measurements is that they can quantify chemical contents objectively. The sad aspect is that they suffer from a lack of comparability. It is, no doubt, common knowledge amongst those who practise analysis for trade and commerce that analysts can obtain different results on the same material. Obviously it may not be in their interest to expose this fact. It is in the field of public health and environmental monitoring, where determinand concentrations are often small and where slight differences may be significant, that interlaboratory variation has received most attention. The disturbing thing is the suggestion of unreliability and its possible diffusion to the general public as well as to the governments responsible in cases where important decisions must be made on the basis of chemical measurements. However, the situation is not as bad as it seems. If soil scientists, chemists and statisticians collaborate, try to understand each other's problems and use realistic models, representative samples, standardized methods, reference materials, interlaboratory tests and good robust statistics, errors can be controlled.

References

AMC 1989a: Analytical Methods Committee. Robust Statistics - How Not to Reject Outliers, Part 1: Basic Concepts. *Analyst* **114,** 1693-1697.

AMC 1989b: Analytical Methods Committee. Robust Statistics - How Not to Reject Outliers, Part 2: Interlaboratory Trials. *Analyst* **114**,1699-1702.

AMC, 1992: Analytical Methods Committee. Proficiency Testing of Analytical Laboratories: Organisation and Statistical Assessment. *Analyst* **117**, 97-117.

Flury, B., & Riedwyl, H. 1983. *Angewandte multivariate Statistik.* Gustav Fischer Verlag, Stuttgart.

Hampel, F.R. 1980. Robuste Schätzungen: Ein anwendungsorientierter Ueberblick. *Biometrical Journal* **22**, 3-21.

Hampel, F.R. 1985. The Breakdown Points of the Mean Combined With Some Rejection Rules. *Technometrics* **27**, 95-107.

Hampel, F.R., Ronchetti, E.M., Rousseeuw, P.J. & Stahel, W.A. 1986. *Robust Statistics, The Approach Based on Influence Functions.* John Wiley & Sons, New York .

Horwitz, W. & Albert, R. 1986. Significance of the Coefficient of Variation in the Analysis of Pesticide. Abstract SS 1-01. The Sixth International Congress of Pesticide Chemistry, IUPAC, Ottawa, Canada.

Horwitz, W. ed. 1988. Protocol for the Design, Conduct and Interpretation of Collaborative Studies. *Pure & Applied Chemistry* **60**, 855-864 .

Huber, P. J. 1981. *Robust Statistics*. John Wiley & Sons, New York.

ISO Guide 35. 1985. *Certification of Reference Materials, General and Statistical Principles*. International Organization for Standardization, Geneva, Switzerland.

ISO-5725. 1986. *Precision of Test Methods - Determination of Repeatability and Reproducibility for a Standard Test Method by Inter-Laboratory Tests.* International Organization for Standardization, Geneva.

Lischer, P. 1987. Robuste Ringversuchsauswertung. *Lebensmittel-Technologie* **20**, 167-172.

Matheron, G. 1965. *Les variables régionalisées et leur estimation.* Masson, Paris.

Rousseeuw, P.J. & Leroy, A.M. 1986. *Robust Regression and Outlier Detection.* John Wiley & Sons, New York.

Rousseeuw, P.J. & van Zomeren, B. C. 1990. Robust Distances: Simulations and Cutoff Values, *The IMA Volumes in Mathematics and its Applications.* **34**, 195-203. Springer Verlag, Berlin.

Rousseeuw,P.J. & Croux, C. 1991. Alternatives to the Median Absolute Deviation, Technical Report 91- 43. Department of Mathematics, University of Antwerpen.

SLB 1989: Schweizerisches Lebensmittelbuch, Kapitel 60. *Statistik und Ringversuche,* EDMZ, Bern.

S-PLUS. 1991. Statistical Sciences, Inc. Seattle.

VSBo 1986: *Ordinance Relating to Pollutants in Soil of June 9, 1986. SR 814.12*, Government of Switzerland, Bern.

Webster, R. & Oliver, M. A. 1990. *Statistical methods in soil and land resource survey.* Oxford University Press.

Webster, R. 1993. Dealing with spatial variation. In *Soil monitoring*, eds R. Schulin, A. Desaules, B. von Steiger and R. Webster. Birkhäuser Verlag, Basel, this volume.

Discussion

Question by B. von Steiger

Why are pedologists often afraid to formulate assumptions on spatial variation in the field?

Answer by P. Lischer

Statisticians are not in the position to make assumptions about things only visible in the field. Pedologists should not be afraid to formulate these assumptions provided they have read and understood the textbooks such as that by Webster & Oliver (1990).

P. Lischer,
Swiss Federal Research Station for Agricultural Chemistry and Hygiene of Environment, CH-3097 Liebefeld-Bern, Switzerland.

Soil Monitoring, Monte Verità, © Birkhäuser Verlag Basel

DEALING WITH SPATIAL VARIATION

R. Webster

Linear geostatistics can provide, by kriging from soil and environmental data, local estimates that are unbiased and have minimum and known variance, and are in this sense optimal. The procedure assumes a model, usually that the variables are realizations of random processes and intrinsic *sensu* Matheron. It can also be used for global estimation. Variograms, or covariance functions, are vital and must themselves be estimated from data. That demands fairly heavy sampling. Classical sampling also provides global estimates. It needs no assumptions about the nature of the variation, but to avoid bias and to estimate the errors sampling must be randomized.

The bases of both techniques are described. The advantages of kriging for global estimation are by no means clear. Environmental scientists should judge the relative merits of the two approaches in relation to their aims and the resources they have for sampling.

Introduction

Scientists have recognized variation in soil for a long time. Some have been fascinated by it. They have been motivated to investigate its causes and to devise schemes for classifying it. Others have found it a cause of frustration. Estimating values of the soil's properties has meant replicate sampling, and the magnitude of the variation has meant that surveys for the purpose have demanded more resources than were affordable. This is especially so for monitoring for which sites must be maintained and equipped with expensive collectors and instruments. Yet other investigators have been dismayed that the classifications devised by the systematists seemed to confer little practical benefit.

Means for describing variation statistically and for using the descriptions for estimating means, variances, and confidence intervals were developed in the 1920s and 1930s. They were based on sound sampling designs. Later they were used with some success for spatial prediction in civil and military engineering (e.g. Morse & Thorburn, 1961; Kantey & Williams, 1962; Beckett & Webster, 1971). In the last dozen years or so geostatistics has brought a fresh appreciation of soil variation and new analytical technique for prediction. By treating soil properties as autocorrelated random variables it has become possible to apply regionalized variable theory and estimate soil properties by kriging. Kriging, in its various forms, gives unbiased estimates (predictions) with minimum variance, and in this sense it is best. The kriging variances can themselves be estimated, and under certain reasonable assumptions confidence can be judged. Further, since the kriging variances depend only on the sampling configuration and the variogram (see below) and not on the sample values, future surveys can be planned to predetermined tolerances. Geostatistics can therefore be used either to gain precision or to reduce sampling or both. It is eminently applicable to soil, and scientists who wish to monitor the soil, to determine pollutants in it, and to detect changes in time naturally want to take advantage of it if possible.

In this paper I describe briefly the two approaches to spatial estimation. I summarize the assumptions underlying them and steps to be taken in applying them. And I try to show in what circumstances we should use them for soil monitoring and the resources that we should make available in survey.

Geostatistical estimation: kriging

Kriging will be familiar to some readers, but not to all. For the latter I give a brief résumé. At its simplest an ordinarily kriged estimate is no more than a weighted average of data. Suppose that we want the spatial mean of some property in a region B of area $|B|$. We compute

$$\hat{Z}(B) = \sum_{i=1}^{N} \lambda_i z(\mathbf{x}_i) . \tag{1}$$

In the conventional notation $\hat{Z}(B)$ is the estimate of the integral of a random variable Z (note the upper case) in B:

$$Z(B) = \frac{1}{|B|} \int_B Z(\mathbf{x}) \mathrm{d}\mathbf{x} . \tag{2}$$

The $z(\mathbf{x}_i)$ are the data (lower case z) at points $\mathbf{x}_1, \mathbf{x}_2, \ldots, \mathbf{x}_N$, and the $\lambda_i, i = 1, 2, \ldots, N$ are the weights associated with them.

From the practical point of view kriging looks no different from many other schemes of weighted averaging, such as moving averages and inverse functions of distance for interpolation and the unweighted mean for estimation. What makes kriging unique is the choice of weights. First, they sum to 1 to assure unbiasedness, as for the global mean. Then, subject to this constraint, they are chosen to minimize the estimation variance:

$$\begin{aligned} \sigma^2(B) &= \mathrm{E}\left[\{\hat{Z}(B) - Z(B)\}^2\right] \\ &= 2\sum_{i=1}^{N} \lambda_i \bar{\gamma}(\mathbf{x}_i, B) - \sum_{i=1}^{N}\sum_{j=1}^{N} \lambda_i \lambda_j \gamma(\mathbf{x}_i - \mathbf{x}_j) - \bar{\gamma}(B, B) , \end{aligned} \tag{3}$$

in which E denotes the expected value, here the expected squared difference between the true but unknown value $Z(B)$ and its estimate $\hat{Z}(B)$. The quantity $\gamma(\mathbf{x}_i - \mathbf{x}_j)$ is the semivariance of Z between the sampling points $\mathbf{x}_i$ and $\mathbf{x}_j$, $\bar{\gamma}(\mathbf{x}_i, B)$ is the average semivariance between point $\mathbf{x}_i$ and the block B for which we want the estimate, and $\bar{\gamma}(B,B)$ is the average semivariance of Z within B. The variance is minimized when

$$\begin{aligned} \sum_{i=1}^{N} \lambda_i \gamma(\mathbf{x}_i - \mathbf{x}_j) + \psi &= \bar{\gamma}(\mathbf{x}_i, B) \quad \text{for all } j \\ \text{and} \qquad \sum_{i=1}^{N} \lambda_i &= 1 , \end{aligned} \tag{4}$$

where ψ is a Lagrange multiplier to achieve the minimization. These are the kriging equations that must be solved. The kriging variance is also obtained by

$$\sigma^2(B) = \sum_{i=1}^{N} \lambda_i \bar{\gamma}(\mathbf{x}_i, B) + \psi - \bar{\gamma}(B,B) . \tag{5}$$

The problem can be formulated in terms of covariances instead of semivariances, or of regression (see de Gruijter & ter Braak, 1990), but the result is the same.

The geostatistical model

The above summarizes the technology of kriging. It requires data for equation (1), of course. Just as important are the semivariances for equations (3) to (5). They must be obtained from the variogram, the function that relates the variance of Z to the separation between points, and this in turn demands that we define a model.

Geostatistics assumes that we are dealing with random processes and that the actual values of a variable $Z(\mathbf{x})$ that we might observe in a region represent just one realization of such a process. If we take the strict definition then what we estimate from a set of observations, $z(\mathbf{x}_i)$, is another random variable, $Z(B)$, which is the spatial integral defined in equation (2). In any actual piece of ground, however, we have only one realization of the variable ever, and the spatial average of Z over B is single valued but unknown. To justify our approach we have to invoke certain other assumptions. The commonest of these and the one underlying ordinary kriging is Matheron's (1965) *intrinsic hypothesis*:

(a) the expected difference $\mathrm{E}[Z(\mathbf{x}) - Z(\mathbf{x}+\mathbf{h})] = u(\mathbf{x})$ is constant, and

(b) the squared difference, $\mathrm{E}[\{Z(\mathbf{x}) - Z(\mathbf{x}+\mathbf{h})\}^2] = 2\gamma(\mathbf{h})$, depends on the separating vector $\mathbf{h}$ and not on position $\mathbf{x}$.

In practice we can often assume that a property is locally stationary within some neighbourhood V, a condition known as *quasi stationarity*. In these circumstances our model of variation is given by

$$Z(\mathbf{x}) = \mu_V + \varepsilon(\mathbf{x}) , \tag{6}$$

in which μ_V is the local mean, and $\varepsilon(\mathbf{x})$ is a random variable with a mean of zero and variance defined by

$$\frac{1}{2}\mathrm{var}\left[\varepsilon(\mathbf{x}) - \varepsilon(\mathbf{x}+\mathbf{h})\right] = \frac{1}{2}\mathrm{E}\left[\{\varepsilon(\mathbf{x}) - \varepsilon(\mathbf{x}+\mathbf{h})\}^2\right] = \gamma(\mathbf{h}) . \tag{7}$$

This is equivalent to

$$\begin{aligned}\frac{1}{2}\text{var}\left[Z(\mathbf{x}) - Z(\mathbf{x}+\mathbf{h})\right] &= \frac{1}{2}\mathrm{E}\left[\{Z(\mathbf{x}) - Z(\mathbf{x}+\mathbf{h})\}^2\right] \\ &= \gamma(\mathbf{h})\,. \end{aligned} \tag{8}$$

The quantity $\gamma(\mathbf{h})$ is the semivariance, namely the quantity required in the kriging equations.

Matheron (1989) explains why it is appropriate to treat the real world in this way, and I shall not pursue his argument further here. Our concern is whether such a model is reasonable in the situations with which we have to deal. Much of our experience comes from soil survey and suggests that it is reasonable. In monitoring pollution, however, a more elaborate model may often be necessary.

The commonest departure from intrinsic stationarity that we are likely to encounter involves some kind of deterministic trend. For example, incinerators, factories, electric power stations and motor vehicles create patterns of pollution. The fallout depends regionally on the distance and direction from them and on the weather. The configuration is unique: it can hardly be considered the outcome of a random process. There is some randomness in the wind, the rain and atmospheric conditions generally, and to this must be added the original random nature of the soil. Trend is not necessarily man-made, however. Overland or subsurface flow on a slope and water rising from the ground water can redistribute pollutants in a deterministic way, just as in nature the soil can become increasingly wet, salty, or deeper down a slope.

We can modify our model to take this into account, replacing μ_V with a more elaborate term, $q(\mathbf{x})$, which depends on the position $\mathbf{x}$ and is known as the *drift*:

$$Z(\mathbf{x}) = q(\mathbf{x}) + \varepsilon(\mathbf{x})\,. \tag{9}$$

The term $q(\mathbf{x})$ is frequently taken as some polynomial:

$$q(\mathbf{x}) = \sum_{k=0}^{K} a_k f_k(\mathbf{x})\,, \tag{10}$$

in which the f_k are known functions of $\mathbf{x}$ and the a_k are unknown coefficients. The variogram of $\varepsilon(\mathbf{x})$ is no longer the same as that of $Z(\mathbf{x})$, but that of the squared residuals from the drift. Both must appear in the kriging equations. This takes us into the realm of universal kriging, which Olea (1975) has described in detail.

Another way of dealing with the problem is to filter out polynomials of degree k, so that what remains is intrinsic in the above sense. These are the intrinsic random functions of order k, with $k > 0$ (Matheron, 1973), and to work with them we must replace the variogram with the generalized covariance function.

The drift is not necessarily continuous. For example, the soil changes abruptly in many instances along cliffs, old river bluffs, and above the outcropping boundaries between contrasting rocks, and so on. We should not want to krige across sharp discontinuities of this kind. Voltz & Webster (1990) describe such a situation. They removed the effects of the discontinuities before computing their variograms; they kriged the residuals, and finally they added the discontinuous effects to obtain their final estimates. Further, the morphogenesis might not be the same on the two sides of a boundary, and almost certainly the variogram would change there.

From a statistical view there is nothing absolute in deciding what is trend and what is random. It all depends on the scale. The trend at one scale might appear as stochastic at another. For example, the soil around each individual tree in a forest might have a trend of pH radiating from the base of its trunk. At the scale of the whole forest, however, with trees distributed at random, this trend might be of no consequence. Soil boundaries might seem deterministic when viewed from a given place in the landscape, but the ensemble in a large region can quite easily be regarded as random (Burgess & Webster, 1984).

Estimating regional means: the classical approach

In monitoring we are often more concerned with change, change in time, than with spatial variation. For example, if we want to know how heavy metals in the atmosphere of a region are affecting the load in the soil there or the soil's response to global warming or general increase in CO_2 then typically we shall want to compare regional means from time to time. We shall want to know whether on average things are getting worse, because if they are then we ought to do something about it. We should also like to know if on average they are getting better, in which case we might relax controls. So, what we want in most instances is a single value to represent the soil's condition at any one time. We should really like to know the global average, the regional mean of, say, the soil's pH or its lead content. We might want some refinement, especially where there are contrasts in land use or soil type. The effect of acid rain on soil that is already acid will not be the same as that on calcareous soil, and it makes no sense to average the pH over both. Similarly, we should monitor forests separately from agricultural land. This does not alter the principle: for each type of soil or land use we want a single figure, an average for that class, on each occasion.

We might still use geostatistics to provide such estimates. We could treat the region of interest as a single block and krige. The estimated variogram is often unreliable at lag distances approaching that across a region, and so this treatment is usually inadvisable. An alternative is to average numerous local kriged estimates. The result is unbiased as before, and the estimation variance can still be estimated (Journel & Huijbregts, 1978). McBratney & Webster (1983) illustrate the technique's application to soil.

There is a more practical disadvantage of kriging in these circumstances. We need values of semivariance in the kriging equations. These can come only from a model fitted to an experimental variogram, which in turn derives from data. The model must be reasonably accurate, and for that so must be the experimental values. We now know from Monte Carlo studies that an investigator must expect to sample at some 150 places to estimate the variogram accurately, and certainly more than the 50 sometimes thought sufficient (Webster & Oliver, 1992). Perhaps that should be extended to 300 where variation is anisotropic.

We have to ask ourselves whether such effort is justified solely to estimate a regional mean. We could obtain a very precise value using the classical approach, and it is to this that I now turn.

Classical sampling

Classical estimation is based on different assumptions from that in geostatistics. First, it has no underlying model of variation. It treats the world as we find it, however it got that way. It could be the realization of a random process, but that is immaterial. That world, for example, the soil of a region, constitutes a *population*, consisting of *units*. Classical statistics is almost always concerned with mean values. In the present context these are again the spatial means. However, we should rewrite equation (2) for the integral using lower case for the variable because for every unit, i.e. each point $\mathbf{x}$ of which there are infinitely many, there is a fixed value, $z(\mathbf{x})$, and their mean is a constant:

$$z(B) = \frac{1}{|B|}\int_B z(\mathbf{x})\mathrm{d}\mathbf{x}\ . \tag{11}$$

The mean is unknown, and the aim is to estimate it.

Of course we cannot measure the soil everywhere, and so we cannot determine $z(B)$ exactly. The best we can do is to find an approximation, and this we must do by measuring z on some sample of the units. Further, we have to randomize the sampling to put the estimation on a probabilistic footing and thereby enable ourselves to estimate the error in our approximation.

So, whereas in geostatistics we assume the variables to be random, and we may choose the sampling points in any convenient way, in classical estimation we do not care how the variable is distributed, we regard it as fixed, but we must select our sampling points at random from the population, both to avoid bias and to estimate the errors. This is perhaps the most important theoretical difference between geostatistical and classical estimation.

We can estimate the regional mean with a formula analogous to equation (1):

$$\hat{z}(B) = \sum_{i=1}^{N} \lambda_i z(\mathbf{X}_i)\ . \tag{12}$$

Note that I use the capital $\mathbf{X}$ to signify that the units sampled are random. It is still a vector of coordinates, of course, not a matrix. The estimate is random, but since it is an estimate of a fixed quantity I retain the lower case z beneath the circumflex. It is usually denoted by $\bar{z}$, following the same reasoning and meaning the average of several fixed values. In the simplest cases (simple random sampling) all the units are chosen independently of one another. The weights are equal with value $\lambda_i = 1/N$, and this assures unbiasedness.

The variance of the population is estimated by

$$s^2 = \frac{1}{N-1}\sum_{i=1}^{N}\{z(\mathbf{X}_i) - \bar{z}\}^2 . \tag{13}$$

From this we can obtain an unbiased estimate of the estimation variance of the mean as $s^2(B) = s^2/N$. In fact this is the same result as that obtained from equation (5) where there is no autocorrelation, and the Lagrange multiplier equals $s^2(B)$.

Increasing precision and efficiency

The estimation variance, and its square root, the standard error, measure the precision of the estimate, and as in kriging we should like them to be as small as possible. It is well known that simple random sampling of a spatially dependent variable is imprecise, and also inefficient in the sense that more sampling is needed to achieve a particular standard error than almost any other scheme. The reason is that when points are placed entirely at random they tend to cluster in some parts of the space, there duplicating information, while they are sparse elsewhere, so that those parts of the space are unrepresented.

There are several ways in which precision and efficiency can be improved, all of which involve sampling the space more evenly. I shall mention two.

Stratified random sampling. Here the region of interest is divided into tessellating strata, or cells. Let there be K of them. They are usually squares, but they can be elongated rectangles, triangles, or hexagons. Sampling points are then chosen at random in each of the cells. If the cells are of equal size and we choose the same number of sampling points, n, in each then we estimate the mean without bias as before. However, the estimation variance is given by

$$s^2(B)_{st} = \frac{1}{K}\sum_{k=1}^{K}\frac{s_k^2}{n} , \tag{14}$$

where s_k^2 is the variance of n observations in the kth stratum, and the subscript st signifies that our estimate is from a stratified sample.

The quantity s_k^2/n represents the pooled variance within strata. It replaces s^2, the variance of the whole sample in the formula. Thus, the sampling error in stratified sampling derives solely from variation within the strata, and differences between strata make no contribution. So wherever there is spatial dependence, and this is almost always so at some scale, the estimation variance of stratified sampling will be less than that of simple random sampling. This represents a gain in precision. Alternatively we can economize by taking a smaller sample to obtain the same precision, and so be more efficient.

The strata need not be equally represented. If there are n_k observations in stratum k then equation (14) becomes

$$s^2(B)_{st} = \frac{1}{K}\sum_{k=1}^{K}\frac{s_k^2}{n_k} . \tag{15}$$

Further, the strata may vary in area, and in this circumstance we assign a weight to each stratum proportional to its area:

$$w_k = \frac{\text{area of stratum } k}{\text{total area}} .$$

The global mean is then estimated by

$$\hat{z}(B) = \sum_{k=1}^{K} w_k \bar{z}_k , \tag{16}$$

where $\bar{z}_k$ is the mean of the kth stratum. The global estimation variance is

$$s^2(B)_{w,st} = \sum_{k=1}^{K} \frac{w_k^2 s_k^2}{n_k} . \tag{17}$$

The strata need not be simply geometric: they can be the classes on a soil map or parcels of different kinds of land use. It is eminently sensible to sample these separately, not only because their within-class variances are likely to be smaller than that of the population as a whole, but also because their means are of interest.

Systematic sampling. We can obtain the most even coverage by sampling at regular intervals on a grid, and following the above reasoning this should give the most precise estimates. Yates (1948) and Quenouille (1949) showed this to be so with certain kinds of spatial correlation, and Berry (1962), Webster & Oliver (1990) and, most recently, Papritz (1993) have illustrated it with examples.

The main disadvantage from our point of view is that no entirely valid estimate of the estimation variance can be obtained from the data, since the sampling points are not located randomly within the grid. To overcome this Yates (1981) devised his scheme of balanced differences, and in our experience this gets close to the true estimation variance (Papritz, 1993).

A hazard of sytematic sampling often cited in the statistical texts is the risk of bias. The worry is that the grid might be so placed that it is offset in relation to some trend, or that its interval is in tune with some periodicity in the landscape. Periodic variation is clear in an orchard, a vinyard, or a planted forest, and it is easy to plan a sampling scheme unrelated to it. It is not quite so clear on land that has been underdrained, and when surveying a single field or farm it is a matter to beware of. However, periodicity in larger landscapes is very unlikely indeed, and the statisticians' taboo on systematic sampling seems quite unjustified. Here we can agree with Matheron (1989). Nevertheless, if you are still concerned you can avoid all risk by using a modification in which the sampling points are deliberately shifted away from the grid nodes in a random but controlled way. The result, known as *systematic unaligned sampling*, is almost as precise as fully systematic sampling and suffers the same disadvantage of not yielding an estimation variance. It is described in Webster & Oliver (1990).

The full theory of classical sampling is extensive, and I cannot dwell on it further here. It is well documented in standard texts such as those by Cochran (1977) and Yates (1981), where you can pursue it.

What of the future ?

Oliver & Webster (1991) have argued cogently for using geostatistics. If you want to predict locally then you are unlikely to do better, except by good fortune. If you want to reveal the spatial variation in a region and to map it then you will need many sampling points, and you should have ample data from which to compute a variogram precisely. If you can accept reasonable assumptions of stationarity needed to treat the variable as random and model the variogram or covariance satisfactorily in that framework then you should krige: the estimates you get will, in principle, be the best. This has been borne out in practice. Laslett *et al.* (1987) and Laslett & McBratney (1990) showed kriging to be the most reliable of the procedures that they tried for mapping soil variables.

If on the other hand you want only global estimates then consider the classical approach using randomized sampling. A simple random design is inefficient, but efficiency and precision can be gained by stratification, especially if the region of interest can be classified sensibly for the particular variables to be estimated. A further advantage of it is that by randomizing the sampling no assumption about the spatial distribution of the variables is needed. The technique is 'safe' in this respect, even if somewhat conservative.

Some studies of pollution require local estimates repeated in time. Geostatistics is being used to provide them, e.g. Soares *et al.* (1993). The studies run into two kinds of problem, neither of which is adequately understood. The first concerns the distribution which is to some extent deterministic in both a physical and a statistical sense. The second is that the monitoring requires fairly elaborate and expensive apparatus. Few agencies can afford to equip more than a few tens of sites, and once they have done so they want to maintain their stations at those sites: it is even more expensive to keep moving the equipment. As a result there are usually too few stations producing data from which to compute a reliable variogram for any one time. However, because the stations often return values every day or every week through one or more years there are many data altogether. Some geostatisticians are using the data from all occasions to compute pooled and much less erratic spatial variograms. It is not at all certain whether these are valid for kriging on a single particular occasion. The problem needs a thorough theoretical investigation.

In conclusion, there is no single way of dealing with spatial variation in soil monitoring, and in many instances no obvious best way. Each situation should be studied, and a sampling scheme designed for it. In some instances sampling and estimation should be based on classical design, in others a geostatistical approach should be adopted. New theory may be needed for sequential local estimation.

References

Beckett, P. H. T. & Webster, R. 1971. The development of a system of terrain evaluation over large areas. *Royal Engineers' Journal* **85**, 243-258.

Berry, B. J. L. 1962. *Sampling, coding and storing flood plain data.* Agricultural handbook no 237, U S Department of Agriculture, Washington, DC.

Burgess, T. M. & Webster, R. 1984. Optimal sampling strategies for mapping soil types. I. Distribution of boundary spacings. *Journal of Soil Science* **35**, 641-654.

Cochran, W. G. 1977. *Sampling techniques.* 3rd edition. John Wiley & Sons, New York.

Gruijter, J. J. de & ter Braak, C. J. F. 1990. Model-free estimation from spatial samples: a reappraisal of classical sampling theory. *Mathematical Geology* **22**, 407-415.

Journel, A. G. & Huijbregts, C. J. 1978. *Mining geostatistics.* Academic Press, London.

Kantey, B. A. & Williams, A. B. A. 1962. The use of soil engineering maps for road projects. *Transactions of the South African Institution of Civil Engineers* **4**, 149-159.

Laslett, G. M., McBratney, A. B., Pahl, P. J. & Hutchinson, M. F. 1987. Comparison of several spatial prediction methods of soil pH. *Journal of Soil Science* **38**, 325-341.

Laslett, G. M. & McBratney, A. B. 1990. Further comparison of spatial methods for predicting soil pH. *Soil Science Society of America Journal* **54**, 1553-1558.

Matheron, G. 1965. *Les variables régionalisées et leur estimation.* Masson, Paris.

Matheron, G. 1973. The intrinsic random functions and their applications. *Advances in Applied Probability* **5**, 439-468.

Matheron, G. 1989. *Estimating and choosing.* Springer Verlag, Berlin.

McBratney, A. B. & Webster, R. 1983. How many samples are needed for regional estimation of soil properties? *Soil Science* **135**, 177-183.

Morse, R. K. & Thornburn, T. H. 1961. Reliability of soil maps. In: *Proceedings of the 5th International Conference on Soil Mechanics and Foundation Engineering*, Paris, pp 259-262.

Olea, R. A. 1975. *Optimum mapping techniques using regionalized variable theory.* Series on Spatial Analysis, no 2. Kansas Geological Survey, Lawrence, Kansas.

Oliver, M. A. & Webster, R. 1991. How geostatistics can help you. *Soil Use and Management* **7**, 206-217.

Papritz, A. 1993. Interference between spatial variability and the detection of temporal changes in soil. In *Soil Monitoring*, eds R. Schulin, A. Desaules, R. Webster and B. von Steiger, Birkhäuser Verlag, Basel, this volume.

Quenouille, M. H. 1949. Problems in plane sampling. *Annals of Mathematical Statistics* **20**, 355-375.

Soares, A., Távora, J., Pinheiro, L., Freitas, C. & Almeida, J. A. 1993. Predicting probability maps of air pollution concentration: a case study on the Barreiro/Seixal industrial area. In: *Geostatistics Tróia '92*, Vol. 2, ed. A. Soares, Kluwer Academic Publishers, Dordrecht, pp 625-635.

Voltz, M. & Webster, R. 1990. A comparison of kriging, cubic splines and classification for predicting soil properties from sample information. *Journal of Soil Science* **41**, 473-490.

Webster, R. & Oliver, M. A. 1990. *Statistical methods in soil and land resource survey.* Oxford University Press.

Webster, R. & Oliver, M. A. 1992. Sample adequately to estimate variograms of soil properties. *Journal of Soil Science* **43**, 177-192.

Yates, F. 1948. Systematic sampling. *Philosophical Transactions of the Royal Society of London* **A241**, 345-377.

Yates, F. 1981. *Sampling methods for censuses and surveys.* 4th edition, Griffin, London.

R. Webster
ETH Zürich
Institut für terrestrische Ökologie
Grabenstrasse 3
8952 Schlieren
Switzerland

Discussion

Question by C. Lajaunie

The sampling fluctuations of the experimental variograms are in fact very large. The consequences are likely to be less dramatic, however, for the variograms have similar forms, and so their differences will have little effect on the kriging estimates. The estimation variances will be strongly affected, of course.

Estimating the variogram in two steps, namely drift estimation plus estimation of the variogram of residuals, leads to biased experimental variograms. IRF*k* theory was developed to deal with this situation.

Answer by R. Webster

Yes, many of the experimental variograms that I showed had similar bounded forms and differed mainly in their sill variances. When modelled and used for kriging they would lead to similar estimates. However, there were some that appeared unbounded and others that seemed to be almost entirely nugget variance, and kriging with them would produce different results. A. Papritz (in this volume) shows the fluctuation that can happen with small samples.

I agree with your second remark.

Question by S. van der Zee

In the geostatistical approach two points draw my attention.

1. You assume randomness in geostatistics (i.e. no structure) and then look for structure.
2. You subtract a more or less arbitrary trend from a signal of which you want to know the structure, and then look for structure in the (more or less arbitrary) residuals.

   Don't you think that these two points contain contradictions?

Answer by R. Webster

1. The rationale for using geostatistics is that the reality we observe is the outcome of one or more autocorrelated random processes. Our analysis seeks to identify and measure this autocorrelation, which is the structure and is unknown *a priori*, and to take it into account in estimation.
2. Dealing with trend is to some extent arbitrary, partly because what we regard as trend and what random fluctuation depends on the scale. Nevertheless, by modelling the two simultaneously (for example, as drift+random as in universal kriging, and by IRFk and generalized covariances) we can avoid some of the arbitrariness. Where we know the trend from other sources of information such as a physically based model of deposition or the step-wise change of soil type that we can learn from a soil surveyor, we should use that information, or at least consider it before deciding what to do with it.

Question by P. Lagas

Have you investigated the difference between random and systematic sampling in agricultural fields where crops are grown systematically in rows?

Answer by R. Webster

No; but if you align the systematic grid in some direction different from that of the rows then there should be no serious risk of bias, which I think is the concern behind your question. Also in that situation systematic sampling usually gives the best results, best being in the sense of most precise.

Question by R. Schulin

This is more a comment with respect to geostatistical terminology than a question. When we talk about 'true' values of, for example, the mean or the variance we refer to a fictitious concept. In a given case we have a particular realization of the hypothetical underlying random process, and we are interested only in real values of this specific case. The 'true' values are only imagined. Their 'truth', which by the way we cannot prove without additional assumptions, is a purely theoretical concept, and it can mislead since it suggests that we should interest ourselves in these 'true' parameters of the spatial process.

Answer by R. Webster
Clearly, any serious user of geostatistics must understand the underlying theory and the significance of its assumptions. Following your train of thought, there will be instances where the expectations and the variances of the random processes are of little consequence, and assuming them to exist is purely a means to an end. However, there are other instances where such parameters have crucial roles, and we should not dismiss the concepts lightly. The matter is discussed at length by G. Matheron (1989) in his *Estimating and Choosing*, Springer-Verlag, Berlin.

Soil Monitoring, Monte Verità, © Birkhäuser Verlag Basel

INTERFERENCE BETWEEN SPATIAL VARIABILITY AND THE DETECTION OF TEMPORAL CHANGES IN SOILS

A. Papritz

Temporal changes of storage of chemical elements in the soil may be assessed either by input-output analysis or by repeated inventories of the soil pool. The latter method gives information only *a posteriori*, but usually it is cheaper and more accurate than the first. Both geostatistical and classical sampling techniques can be used to estimate the mean temporal change within an area. The performance of the two have been compared by sampling from a large autocorrelated random field. Although theoretically more precise than the classical methods, the geostatistical global estimation scheme did not perform as well. To estimate the mean temporal change of a soil property over an area a paired stratified random or a paired systematic sampling scheme is recommended.

Introduction

Temporal changes of storage of chemical elements in the soil can be estimated by two methods. First, an analysis of the fluxes into and out from a soil compartment within a given time allows an indirect assessment of the change of the elements in the pool (Bader & Baccini, 1993; Schulin, 1993; von Steiger & Obrist, 1993). If a compound is chemically unstable then in addition to fluxes the degradation rate must be estimated. Second, direct estimates of the change can be obtained *a posteriori* from repeated inventories of an element in the soil. Both approaches have their merits and disadvantages.

The spatial variation of the element fluxes and the accuracy of their measurement determines the quality of the estimate in the first approach. The estimate is not affected by spatial fluctuations of the element content itself. Furthermore, the changes can be predicted *a priori* without having to wait till the change is measurable. Very often estimating fluxes is difficult and costly. The uncertainty of estimates of areal fluxes—although probably large—usually remains unknown, and the estimation error of the change cannot be quantified.

In the second approach the spatial variation of the soil property directly affects the estimation if sampling is destructive. A sample of the soil must then be taken away to measure its element content, and subsequent measurements have to be made on material taken elsewhere. Thus, the temporal change and the spatial variation are confounded. If soil sampling is destructive then the temporal trend of a property cannot be assessed for a single point in space but must be based on an analysis of mean values in its locality. The change to be expected within a certain time is typically small compared to the quantity stored in the soil. The more variable the soil is and the less precise the statistical sampling and estimation techniques are, the longer will it take to detect a temporal change that occurs at a constant rate. Within a given interval of time the change might even be smaller than the measurement error of a device. Thus, not only spatial variation but analytical precision, too, limits the ability to detect changes by repeatedly sampling the soil pool. The problem of measurement errors is addressed in the articles by Laslett & McBratney (1990) and Lischer (1993). I shall review some of the statistical methods to analyze data on temporal change of element pools in the soil. Different methods are evaluated by sampling from simulated data fields representing the spatial distribution of a soil property at two different times.

Theory

Estimation methods may be divided into two main categories: the *model-based* and the *design-based* techniques (de Gruijter & ter Braak, 1990; Webster, 1993). Model-based methods consider the spatio-temporal distribution of a soil property to be a realization of a random process that depends both on space and time. An observer then postulates a statistical model of the so-called *superpopulation* which includes certain assumptions about the stationarity and ergodicity of the random process. The latter must always be assumed because in real problems an observer knows only a part of a single realization of a spatially continuous random process, and inference about its probabilistic properties must be based on spatial averages.

According to the structure of a given set of data, the general model of a spatio-temporal random process may be simplified. If many repeated observations have been measured at a few locations then the data may be modelled as a multivariate time series. This approach, however, can be used only if the soil property has been measured at the same locations or if local mean values are to be analyzed. More common to studies monitoring soil pollution are data consisting of observations that were simultaneously measured at many locations but only at two or a few sampling times. A multivariate spatial random process is then a suitable simplified model, and *geostatistics* provides the tools for dealing with it.

In contrast to geostatistical techniques, design-based or—synonymously—*classical* or *probability sampling* methods (e.g. Cochran, 1977; Yates, 1981) are usually free from any assumptions about a statistical model. Central to classical sampling theory is the concept of the *population*. In spatial sampling the population is identical to the set of all possible sampling locations in an area. A given sampling location is called a *unit* of the population. A set of characteristics, the soil properties averaged over the volume of a single core sample, is attached to each unit. The spatial distribution of a soil property is then considered to be an unknown but fixed, deterministic function of the spatial coordinates. Thus, a set of observations is not considered to be a realization of a spatio-temporal random process. This concept implies that one would know—apart from measurement errors—the true mean change if a given area, the population, could be sampled exhaustively at two times. This contrasts the concepts of model-based estimation where a local mean value over a finite area is still considered to be a random quantity.

Of course, repeated exhaustive sampling is impossible, and only a small part of an area can be sampled. The mean change must then be computed from the observed values, using a suitable *estimator* such as the difference of the arithmetic sample means. If the sampling locations are selected by some random scheme, i.e. if every sampling location has a fixed and known chance to be included in a sample then the quality of the estimated change can be assessed in terms of its *bias* and *variance*. Both the bias and the variance of an estimator are defined as the mean behaviour over all samples that can be drawn under a given design from the population. This notion is again different from the concept of model-based estimation where bias and variance refer to the average behaviour of a statistic over all realizations of a random process. To make a distinction between the differing use of expectation, bias, and variance in design- and model-based estimation, the terminology, introduced by Cassel *et al.* (1977), uses the terms *p-expectation*, *p-bias* and *p-variance* to denote the average behaviour over all possible samples in design-based estimation whereas the *ξ-expectation*, *ξ-bias* and the *ξ-variance* denote these operators in model-based estimation.

Geostatistical techniques for estimating temporal change

Consider the following situation. We have two sets of observations, one from each of two times in a given region G of size $|G|$. On the first occasion n_1 observations denoted by $z_1(\mathbf{x}_{1,i})$ were made at locations $\mathbf{x}_{1,i}$, $i = 1, \ldots, n_1$, where each $\mathbf{x}_{1,i}$ is a vector of coordinates in the two-dimensional space. On the second occasion there were n_2 observations denoted by $z_2(\mathbf{x}_{2,i})$. In general the positions of the first, $\mathbf{x}_{1,i}$, and second sample set, $\mathbf{x}_{2,i}$, do not coincide. Let us suppose that the volume of a sample is very small in comparison with the size $|G|$ (quasi-point support). The observations of the two sample sets are considered to be drawn from a realization of two second order stationary random processes, $Z_1(\mathbf{x})$ and $Z_2(\mathbf{x})$, having ξ-expectations μ_1, μ_2, ξ-covariance functions $C_{11}(\mathbf{h})$, $C_{22}(\mathbf{h})$, and cross ξ-covariance function $C_{21}(\mathbf{h})$.

Depending on the focus of a study the temporal differences averaged over contrasting spatial scale may be important. First, predictions of *point* $\Delta(\mathbf{x}_0)$ or *local mean changes*

$\Delta(B_k)$ (Eq. 1) over a small block of land might be required to create a map showing parts of a region G with varying extent of change. The mean change over a block B_k is defined by

$$\Delta(B_k) = \frac{1}{|B_k|}\int_{B_k}\Delta(\mathbf{x})\,\mathrm{d}\mathbf{x} = \frac{1}{|B_k|}\int_{B_k}\{Z_2(\mathbf{x}) - Z_1(\mathbf{x})\}\,\mathrm{d}\mathbf{x} = Z_2(B_k) - Z_1(B_k)\,. \quad (1)$$

Second, one might be interested in the mean change over the whole region, which is denoted by the term *global mean change*. This terminology relates to the name of the corresponding geostatistical technique and is partly misleading because it implies that averages are taken over large areas. This need not be the case, and a global mean can be computed solely over a small piece of land. A mean is global if the averaging area is the same as the area over which the sampling locations are distributed. This contrasts with a local mean where the averaging area is much smaller than the sampled area. Journel & Huijbregts (1978, p.321) discuss why one must distinguish between estimating local and global means. The reasons are both theoretical and practical: First, it might be unreasonable to assume stationarity over the whole area. Second, reliable estimates of the covariance functions cannot be obtained for lag distances larger than one third or half the maximal diameter of the averaging area. Formally, the definition of the global mean is obtained from Eq. (1) by simply changing the averaging area

$$\Delta(G) = \frac{1}{|G|}\int_{G}\Delta(\mathbf{x})\,\mathrm{d}\mathbf{x} = Z_2(G) - Z_1(G)\,. \quad (2)$$

Kriging can be used to estimate local or global mean changes. Ordinary block co-kriging (Eq. 3) provides the best linear ξ-unbiased estimate of the mean change over a block B_k (Papritz & Flühler, 1993)

$$\hat{\Delta}(B_k) = \sum_{i=1}^{n_2}\lambda_2^i Z_2(\mathbf{x}_{2,i}) - \sum_{i=1}^{n_1}\lambda_1^i Z_1(\mathbf{x}_{1,i})\,, \quad (3)$$

where λ_j^i are the kriging weights. They can be determined by solving the system of linear equations

$$\begin{aligned} \sum_{v=1}^{2}(-1)^{v+u}\sum_{j=1}^{n_v}\lambda_v^j C_{vu}(\mathbf{x}_{v,j} - \mathbf{x}_{u,i}) - \psi_u &= \sum_{v=1}^{2}(-1)^{v+u}\bar{C}_{vu}(B_k, \mathbf{x}_{u,i})\,, \\ \sum_{j=1}^{n_u}\lambda_u^j &= 1\,, \end{aligned} \quad (4)$$

where $u = 1, 2$; $i = 1, \ldots, n_u$, and ψ_u are two Lagrange multipliers. As usual, $\bar{C}_{vu}(B_k, \mathbf{x}_{u,i})$ denotes the average covariance between the block B_k and the support point $\mathbf{x}_{u,i}$.

The minimal estimation variance to predict the change is given by

$$\begin{aligned} \sigma^2(B_k) &= \bar{C}_{11}(B_k, B_k) + \bar{C}_{22}(B_k, B_k) - 2\bar{C}_{21}(B_k, B_k) \\ &\quad - \sum_{u=1}^{2}\{\sum_{j=1}^{n_u}\lambda_u^j\sum_{v=1}^{2}(-1)^{v+u}\bar{C}_{vu}(B_k, \mathbf{x}_{u,j}) - \psi_u\}\,. \end{aligned} \quad (5)$$

$\bar{C}_{uv}(B_k, B_k)$ is the within-block average of the covariance functions.

The so-called *global estimation method* (Journel & Huijbregts, 1978; McBratney & Webster, 1983)—in the sequel referred to as the model-based scheme—gives an estimate of the mean change $\Delta(G)$. The region G is divided into K non-overlapping blocks B_k, and the block estimates $\hat{\Delta}(B_k)$ are computed for each of these blocks by cokriging. An estimate of the global mean is then obtained from

$$\hat{\Delta}(G) = \sum_{k=1}^{K} w_k \hat{\Delta}(B_k)\,, \tag{6}$$

where $w_k = |B_k|/|G|$ is the relative size of the block B_k. Following Journel & Huijbregts (1978) the global estimation variance $\sigma^2(G)$ is computed as a good approximation from the estimation variance of the individual blocks by

$$\sigma^2(G) \approx \sum_{k=1}^{K} w_k^2 \sigma^2(B_k)\,, \tag{7}$$

where $\sigma^2(B_k)$ is computed only from those support points lying in the block B_k.

Classical methods for estimating temporal change

The classical methods can be used to estimate local (Eq. 8) and global mean changes (Eq. 9)

$$\delta(B_k) = \frac{1}{|B_k|}\int_{B_k} \delta(\mathbf{x})\,\mathrm{d}\mathbf{x} = \frac{1}{|B_k|}\int_{B_k} \{z_2(\mathbf{x}) - z_1(\mathbf{x})\}\,\mathrm{d}\mathbf{x} = z_2(B_k) - z_1(B_k)\,, \tag{8}$$

$$\delta(G) = \frac{1}{|G|}\int_{G} \delta(\mathbf{x})\,\mathrm{d}\mathbf{x} = z_2(G) - z_1(G)\,. \tag{9}$$

Note that the variables are written in lower case to stress their non-random nature. Classical sampling theory offers a variety of schemes to estimate the mean change of a soil property over an area. The choice of a specific sampling design depends on the objectives and the available resources. If no *a priori* knowledge of the spatial variation is available then a stratified random sampling scheme with strata of equal size and selection of two observations per stratum will usually be sensible. As an alternative systematic sampling could be considered. For both schemes, the locations of the first and second sample can be selected independently (*unpaired sampling*), or the positions of the second sample can be selected at a fixed distance and direction from the locations of the first sample (*paired sampling*), i.e. $\mathbf{x}_{2,i} = \mathbf{x}_{1,i} + \mathbf{h}$. Pairing of sampling locations increases the efficiency of a design if $z_1(\mathbf{x})$ and $z_2(\mathbf{x}+\mathbf{h})$ are positively correlated.

Simple random sampling I shall illustrate the difference between the two approaches for simple random sampling. For unpaired locations we estimate the mean change over a block B_k by the difference between the means computed from the $n_{1,k}$ and $n_{2,k}$ observations that lie within the block, thus

$$\hat{\delta}_u(B_k) = \frac{1}{n_{2,k}}\sum_{i=1}^{n_{2,k}} z_2(\mathbf{X}_{2,i}) - \frac{1}{n_{1,k}}\sum_{i=1}^{n_{1,k}} z_1(\mathbf{X}_{1,i}) = \hat{z}_2(B_k) - \hat{z}_1(B_k)\,. \tag{10}$$

Note that capital $\mathbf{X}$ denotes randomly selected sampling locations. The estimators $\hat{\delta}_u$, $\hat{z}_1$, and $\hat{z}_2$—being functions of the random variables $z_j(\mathbf{X}_{j,i})$—are random quantities, too. For unpaired sampling $\hat{\delta}_u(B_k)$ is a p-unbiased estimator for $\delta(B_k)$, and its p-variance is given by

$$\mathrm{Var}_p[\hat{\delta}_u(B_k)] = \mathrm{Var}_p[\hat{z}_2(B_k) - \hat{z}_1(B_k)] = \frac{\sigma^2_{1,k}}{n_{1,k}} + \frac{\sigma^2_{2,k}}{n_{2,k}} . \tag{11}$$

The symbols $\sigma^2_{j,k}$ denote the p-variance of the populations within the block B_k. Formally they are defined by

$$\sigma^2_{j,k} = \frac{1}{|B_k|} \int_{B_k} \{z_j(\mathbf{x}) - z_j(B_k)\}^2 \, \mathrm{d}\mathbf{x} , \tag{12}$$

and a p-unbiased estimate is obtained from the sample variance

$$\hat{\sigma}^2_{j,k} = \frac{1}{n_{j,k} - 1} \sum_{i=1}^{n_{j,k}} \{z_j(\mathbf{X}_{j,i}) - \hat{z}_j(B_k)\}^2 . \tag{13}$$

Thus,

$$\widehat{\mathrm{Var}}_p[\hat{\delta}_u(B_k)] = \frac{\hat{\sigma}^2_{1,k}}{n_{1,k}} + \frac{\hat{\sigma}^2_{2,k}}{n_{2,k}} \tag{14}$$

is a p-unbiased estimator for the variance of the mean change, computed from unpaired sampling locations.

Of course, the number of observations at both times are equal for paired sampling ($n_{1,k} = n_{2,k} = n_k$). The estimator for the change can then be written as

$$\hat{\delta}_p(B_k) = \frac{1}{n_k} \sum_{i=1}^{n_k} \{z_2(\mathbf{X}_{1,i} + \mathbf{h}) - z_1(\mathbf{X}_{1,i})\} . \tag{15}$$

Unless the locations of the first and second sample are identical ($\mathbf{h} = 0$), $\hat{\delta}_p(B_k)$ is no longer a p-unbiased estimator for $\delta(B_k)$ because of the edge effect introduced by shifting the locations. The bias is negligible if $\mathbf{h}$ is small compared with the size of B_k. The p-variance of the $\hat{\delta}_p(B_k)$ equals approximately

$$\mathrm{Var}_p[\hat{\delta}_p(B_k)] = \frac{1}{n_k} \{\sigma^2_{1,k} + \sigma^2_{2,k} - 2\sigma_{12,k}\} . \tag{16}$$

The p-covariance, $\sigma_{12,k}$, is defined by

$$\sigma_{12,k} = \frac{1}{|B_k|} \int_{B_k} \{z_1(\mathbf{x}) - z_1(B_k)\}\{z_2(\mathbf{x} + \mathbf{h}) - z_2(B_{k+\mathbf{h}})\} \, \mathrm{d}\mathbf{x} , \tag{17}$$

and a sample estimate can be obtained from

$$\hat{\sigma}_{12,k} = \frac{1}{n_k - 1} \sum_{i=1}^{n_k} \{z_1(\mathbf{X}_{1,i}) - \hat{z}_1(B_k)\}\{z_2(\mathbf{X}_{1,i} + \mathbf{h}) - \hat{z}_2(B_k)\} . \tag{18}$$

The symbol $z_2(B_{k+\mathbf{h}})$ denotes the mean of $z_2(\mathbf{x})$ averaged over the domain B_k shifted by the vector $\mathbf{h}$. Comparing the p-variance of the block mean for paired and unpaired sampling shows that the increase in efficiency depends on $\sigma_{12,k}$ which in turn is a function of the vector $\mathbf{h}$. Usually, $\sigma_{12,k}$ decreases with increasing spacing between the sampling locations. This conforms with a statistical model where the cross correlation between $Z_1(\mathbf{x})$ and $Z_2(\mathbf{x})$ decreases with increasing lag distance. The sample estimate of $\text{Var}_p[\hat{\delta}_p(B_k)]$ is given by

$$\widehat{\text{Var}}_p[\hat{\delta}_p(B_k)] = \frac{1}{n_k}\{\hat{\sigma}^2_{1,k} + \hat{\sigma}^2_{2,k} - 2\hat{\sigma}_{12,k}\} . \tag{19}$$

Stratified random sampling The above results may be easily extended to stratified random sampling by noting that an estimate of the global mean is given by

$$\hat{\delta}(G) = \sum_{k=1}^{K} w_k \hat{\delta}(B_k) , \tag{20}$$

where w_k denotes again the relative size of the block B_k. Similarly, the p-variance of $\hat{\delta}(G)$ can be written

$$\text{Var}_p[\hat{\delta}(G)] = \sum_{k=1}^{K} w_k^2 \text{Var}_p[\hat{\delta}(B_k)] , \tag{21}$$

and it can be estimated from

$$\widehat{\text{Var}}_p[\hat{\delta}(G)] = \sum_{k=1}^{K} w_k^2 \widehat{\text{Var}}_p[\hat{\delta}(B_k)] . \tag{22}$$

Systematic random sampling Systematic sampling is often more precise than stratified random because the sampling points are more evenly distributed. But systematic schemes have the disadvantage that there is no general way to estimate the standard error of the mean given the observations of a single sample. If the population is considered to be a realization of a random process then there are formulae for estimating the p-variance of the mean of a systematic sample that are ξ-unbiased under the adopted model. If, for instance, $z_1(\mathbf{x})$ and $z_2(\mathbf{x})$ are realizations of two random processes that are not autocorrelated and mutually independent (*white noise*) then the classical formula to estimate the p-variance of the difference between the sample means (cf. Eq. 14) is ξ-unbiased, i.e its ξ-expectation equals the ξ-expectation of the p-variance of the difference between the means of two systematic samples.

$$\text{E}_\xi[\widehat{\text{Var}}_p[\hat{\delta}_u(G)]] = \text{E}_\xi[\frac{\hat{\sigma}_1^2}{n_1} + \frac{\hat{\sigma}_2^2}{n_2}] = \text{E}_\xi[\text{Var}_p[\hat{\delta}_{u,sys}(G)]] = \frac{C_{11}(0)}{n_1} + \frac{C_{22}(0)}{n_2} , \tag{23}$$

where $\hat{\delta}_{u,sys}(G)$ is the difference between the means of two unpaired systematic samples. If the random processes $Z_1(x)$ and $Z_2(x)$ are autocorrelated then the formula given in Eq. (14) might overestimate the p-variance of the difference. Estimating the p-variance of the mean difference by the methods of *balanced differences* is likely to give better results. Details of this method can be found in the textbook by Yates (1981) and in the review articles by Bellhouse (1988) as well as by Murthy and Rao (1988).

Choosing between geostatistical and classical estimation methods
Choosing between model-based and design-based methods might be difficult. Controversy continues between the advocates of design-based inference and those of model-based inference in the traditional fields of sampling theory (cf. Krishnaiah & Rao, 1988). De Gruijter & ter Braak (1990) extended this discussion to sampling and estimating spatially distributed variables. Provided we know the statistical model of the underlying spatial random processes, kriging is the most precise.

As an illustration, let us consider estimating the mean of a soil property over a block B_k. Assume that we repeatedly select sets of sampling locations at random with equal probability within B_k (simple random sampling). Consider the n observations of each set to be a realization of a second order stationary, ergodic random process $Z(\mathbf{x})$ with ξ-expectation μ and ξ-covariance function $C(\mathbf{h})$. For each set we compute the arithmetic sample mean $\bar{Z} = 1/n \sum_{i=1}^{n} Z(\mathbf{X}_i)$ and the ordinary block kriging estimate $\hat{Z}(B_k) = \sum_{i=1}^{n} \lambda_i Z(\mathbf{X}_i)$. The notation indicates that we regard both Z and $\mathbf{X}$ as random quantities. We can then compute the expectation of these estimators and their variances with respect to the joint distribution of Z and $\mathbf{X}$. The p-expectation corresponds to the conditional expectation given a particular realization $z(\mathbf{X})$ of the random process $\mathrm{Z}(\mathbf{X})$. Similarly, the ξ-expectation plays the role of the conditional expectation given a particular set of sampling locations $\{\mathbf{x}_i\}$, $i = 1, \ldots, n$. Thus, for the kriging estimate with optimal weights, λ_i, we may write

$$\begin{aligned} &\mathrm{E}_p[\mathrm{E}_\xi[\{\hat{Z}(B_k) - Z(B_k)\}^2]] = \\ &\quad \mathrm{E}_p[\sum_{i=1}^{n}\sum_{j=1}^{n} \lambda_i\lambda_j C(\mathbf{X}_i - \mathbf{X}_j) - 2\sum_{i=1}^{n} \lambda_i \bar{C}(\mathbf{X}_i, B_k) + \bar{C}(B_k, B_k)] \,, \end{aligned} \tag{24}$$

and for the arithmetic mean we find

$$\begin{aligned} &\mathrm{E}_\xi[\mathrm{E}_p[\{\bar{Z} - \mathrm{E}_p[\bar{Z}]\}^2]] = \mathrm{E}_p[\mathrm{E}_\xi[\{\bar{Z} - Z(B_k)\}^2]] = \\ &\quad \mathrm{E}_p[\frac{1}{n^2}\sum_{i=1}^{n}\sum_{j=1}^{n} C(\mathbf{X}_i - \mathbf{X}_j) - \frac{2}{n}\sum_{i=1}^{n} \bar{C}(\mathbf{X}_i, B_k) + \bar{C}(B_k, B_k)] \,. \end{aligned} \tag{25}$$

Clearly, equation (25) is a special case of Eq. (24) with the weights $\lambda_i = 1/n$, for all $i = 1, \ldots, n$. Now, for each set of sampling locations $\{\mathbf{x}_i\}$, the kriging variance is always smaller than or equal to $\mathrm{E}_\xi[\{\bar{Z} - Z(B_k)\}^2]$, and the difference between these expressions satisfies

$$\sum_{i=1}^{n}\sum_{j=1}^{n}\{\frac{1}{n^2} - \lambda_i\lambda_j\} C(\mathbf{x}_i - \mathbf{x}_j) - 2\sum_{i=1}^{n}\{\frac{1}{n} - \lambda_i\}\bar{C}(\mathbf{x}_i, B_k) \geq 0 \,. \tag{26}$$

Computing the p-expectation of this expression is equivalent to computing its spatial average for all possible sets of sampling locations. Therefore, the p-expectation of the kriging variance is smaller than the ξ-expectation of the p-variance of the arithmetic sample mean. Thus, if we repeatedly sample at random from an ergodic random field with autocorrelated observations then kriging should give on the average more precise results than simple random sampling.

Similar results can be derived for estimating temporal differences by cokriging. Furthermore, the extension of these results to global estimation shows that the geostatistical method will be more precise than a stratified random scheme if the statistical model is known. In practice, however, this is not the case, and one has to infer some characteristics of the unknown probability distribution (usually the second order moments) from data. The estimates of the covariance functions and of the variograms are subject to some unknown error which propagates through the kriging equations to the estimated mean. Webster & Oliver (1992) have shown that misspecification of variograms is likely to occur if the sample size from which the variograms are estimated is smaller than about 150. If the statistical model is poorly specified then kriging might still result in small estimation errors, but these might differ strongly from the fluctuations that are found by repeated estimation.

In general, it is difficult to see precisely how errors occurring in estimating variograms impair the kriging predictions. Simulating realizations of autocorrelated random fields and repeatedly drawing samples by Monte Carlo methods enables this error to be assessed and helps in choosing between design- and model-based techniques. Depending on the available resources, the method that is likely to perform best for a given population can be evaluated. The next section shows results of a simulation study to estimate short-term change of pH in the top soil on small plots in a forest.

Sampling from Simulated Random Fields

Generating two auto- and cross correlated data fields

Two large fields of auto- and cross correlated values were generated by means of a linear coregionalization model (Journel & Huijbregts, 1978). I computed only a single realization of the fields because this is similar to the situation an observer is confronted with in field studies. Due to the ergodicity assumption averaging over many realizations should be equivalent to averaging over space in one realization. Autocorrelation was imposed by computing a moving average of uncorrelated numbers over a circular averaging area. This transformation leads to circular variograms (McBratney & Webster, 1986). Both fields were squares each consisting of $560 \times 560 = 313\,600$ values. The range of the variograms was 150 grid intervals. The nugget constant and the sill of the variograms were chosen proportional to the values found in a study on short-term change of pH in a forest soil (Table 1). The means of the fields were arbitrarily set to 300 (field one) and 400 (field two) resulting in a true difference $\Delta = 100$.

If one regards a generated number as an observation measured on a soil core with a diameter of 5 cm then the two data fields represent the spatial distribution of a soil property on a small plot of 28 m $\times$ 28 m at two different times. The range then equals 7.5 m.

Table 1: Nugget constants and sills of variograms of simulated data fields.

	nugget constant	sill
variogram field 1	0.25	2.35
variogram field 2	0.8	2.5
pseudo cross variogram	0.575	2.425

Table 2: Size of samples drawn under various schemes from each simulated field.

	sample size		
name	stratified random sample	systematic sample	global estimation
very small	32	25	—
small	50	49	49
medium	98	100	101
large	200	196	—
very large	392	400	—

Sampling schemes and criteria to assess their performance
Three different sampling schemes were used to estimate the difference between the means of the two fields:

- paired stratified random samples,
- paired systematic samples, and
- model-based scheme.

The shifting vector was $\mathbf{h} = (5, 5)$. The size of samples, i.e. the number of values that were selected from each field, varied between 25 and 400 (Table 2). For the model-based scheme only two different sample sizes were tested so far.

Stratified random samples Each field was divided into square blocks (*strata*) of equal size, and two values were randomly chosen in each stratum ($n_{1,k} = n_{2,k} = 2$). The mean difference and its p-variance were estimated for each pair of samples according to Eq. (20) and (22).

Systematic samples The numbers were selected on a randomly placed square grid. For each sample, the p-variance of the mean difference was predicted by (i) the formula for

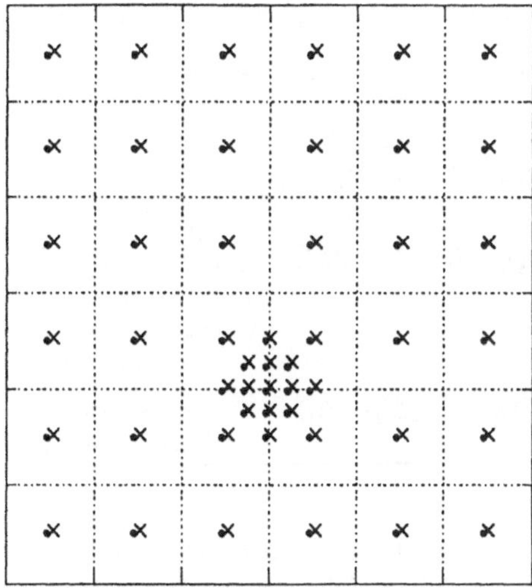

Figure 1: Spatial distribution of the sampling locations of field one (dots) and field two (crosses) for sample size 'small'.

white noise processes (cf. Eq. 23) and (ii) by the formula for balanced differences (Yates, 1981).

Model-based scheme As for the systematic samples, the values were selected on a randomly placed square grid, too. Additional values were selected in a randomly chosen part of the field to compute experimental semivariances for lag distances smaller than the spacing of the grid. This is necessary because one must know the variograms for lag distances smaller than the grid interval to compute the global estimation error by Eq. (7). Approximately one third of the total sample size was used for the denser sampling (13 of 49 and 37 of 101, respectively). This corresponds to the fraction of the resources spent for reconnaissance studies in soil survey (Webster, personal communication). As an illustration, one set of the sampling locations is shown in Fig. (1) for the smaller sample. The experimental variograms were computed from the selected data, and a linear coregionalization model was fitted by non-linear unweighted least-squares. Four different coregionalization models (circular, spherical, pentaspherical, exponential) were fitted to each set of variograms, and the model with the smallest residual sum of squares was selected for kriging.

The block differences and the corresponding estimation variances were computed according to Eq. (4) and (5), substituting the variograms for the covariance functions in the cokriging system (cf. Papritz & Flühler, 1993). The block geometry was chosen such

Table 3: Diagnostic statistics to assess performance of estimation methods.

formula	explanation
$BIAS = \hat{\Delta} - \Delta = \frac{1}{100}\sum_{i=1}^{100}\hat{\Delta}_i - \Delta$	bias of estimation method
$VARMDIFF = \frac{1}{100}\sum_{i=1}^{100}\{\hat{\Delta}_i - \bar{\hat{\Delta}}\}^2$	*var*iance of the estimated *m*ean *diff*erences
$MSE = (BIAS)^2 + VARMDIFF$	mean of squared deviation between estimated mean differences and true difference (*m*ean *s*quared *e*rror)
$AVEPVAR = \frac{1}{100}\sum_{i=1}^{100}\sigma^2(\hat{\Delta}_i)$	*ave*rage *p*redicted estimation *var*iance of the mean difference

that the support points were in the centres of the blocks (cf. Fig. 1). The support points lying in the target block and those of the adjacent blocks were used to compute the mean difference. Thus, the number of support points used in kriging the block difference varied depending on the positions of the block. For a block forming a corner of the field the support points of 3 adjacent blocks and for a block lying in the inner part of the field those of 8 neighbouring blocks were used. However, the estimation variance of a block was computed only from those support points that were in the target block. An estimate of the global mean difference was then computed by means of Eq. (6), and its estimation variance was obtained from Eq. (7).

For each sampling scheme and each sample size 100 independently selected pairs of sample sets were randomly drawn from the fields, and an estimate of the global mean difference $\hat{\Delta}_i$ and its predicted variance $\sigma^2(\hat{\Delta}_i)$ were computed for each pair.

Criteria to assess the performance of an estimation method A good method of estimation has the following properties:

- it is *unbiased*,
- it is *precise*, i.e. the fluctuations from sample to sample should be small, and
- the *predicted fluctuations* should provide a realistic description of the *fluctuations* caused by *sampling*.

To judge the performance of the tested schemes the statistics listed in Table (3) were computed. The statistic *VARMDIFF* measures the *precision* of an estimation scheme, and the statistic *BIAS* reveals whether any of the methods leads to systematic deviations. To get a realistic judgement on the significance of an estimated difference the ratio of *AVEPVAR* to *MSE* should be close to one. This is somewhat different from estimating solely the spatial mean of a soil property where we are usually satisfied to get the most efficient estimate. The exact size of the estimation variance is less important.

Results

Estimating and modelling the variograms

The population variance of field 1 (2.2117) was close to the averaged semivariance within the field (2.2481). For field 2, the agreement was equally good (2.4509 compared to 2.4180). The variance of the difference, $z_2(\mathbf{x}_i)-z_1(\mathbf{x}_i)$, was 1.1039, confirming close cross correlation between the values at the same locations (covariance of population 1.7794). The experimental variograms showed considerable deviations from the underlying variograms. Some variograms increased without reaching a sill, others showed no spatial correlation, but most seemed bounded. All model types were selected several times as 'best fit' model. On the average, the circular coregionalization model most frequently fitted the experimental variograms best. But the fitted nugget constants and sills differed in some cases strongly from the true values, the deviations being mainly due to fluctuations of the experimental variograms. A weighted least square scheme (results not shown here) did not lead to much better fit. The boxplots in Figures 2 and 3 illustrate the variation of the fitted parameters within the 100 repetitions.

BIAS of estimation methods

None of the estimation methods resulted in a large bias (Fig. 4). The ratio $BIAS^2$ to MSE was always smaller than 0.08, showing that the mean deviation from the true value was due mainly to random sampling effects.

Mean squared error (MSE) of estimation methods

Since for a given sample size the number of observations differed slightly from design to design (cf. Table 2) the mean squared error was related to a single observation by multiplying MSE with two times the sample size. I denote this quantity by *mean squared error per observation*. Figure 5 shows the MSE per observation for stratified and systematic sampling and the model-based method.

The latter resulted in a larger mean squared error than the other methods. The ratio between the MSE per observation of the model-based and the stratified estimates was 1.69 for sample size 'small' and 1.49 for sample size 'medium'. The ratios comparing the model-based with the systematic scheme were 1.92 and 1.47. Thus, the model-based method was clearly less efficient than the other schemes. The larger fluctuations were probably caused by varying influence of the support points used in block kriging. The points forming the edge of the square and those of the densely sampled part of the field contributed less to the estimate than those lying in the inner part of the grid. For three out of five sample sizes the MSE per observation was smaller for systematic than for stratified random

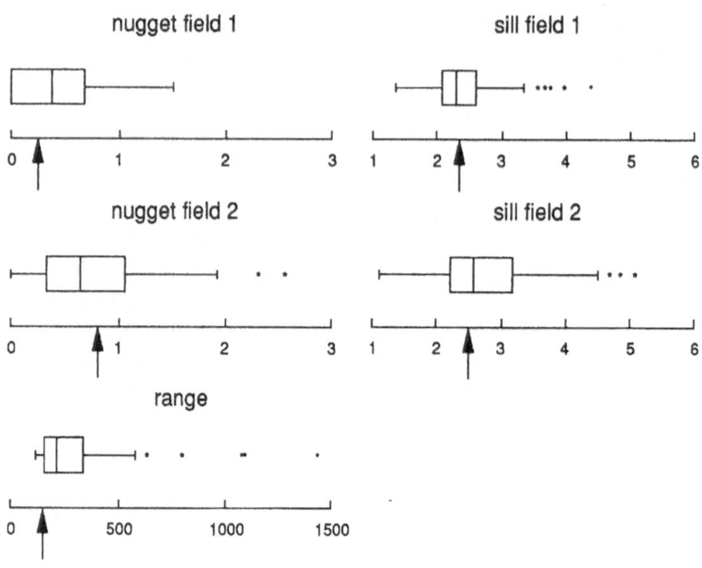

Figure 2: Fitted parameters of the variograms of field one and two for sample size 'small', arrows indicate true parameter values.

samples. For the smallest sample size, the stratified design was more efficient, and both showed approximately the same precision for the sample size 'medium'. On the average, systematic sampling seems to be more precise than stratified random sampling. This agrees with theory: Cochran (1977) has shown that for an autocorrelated superpopulation with a bounded variogram the ξ-expectation of the p-variance for systematic samples is smaller than for stratified samples.

Comparing the average predicted variance (AVEPVAR) with the mean squared error (MSE)

Predicting the variance for systematic samples Figure 6 compares the ratio of the average of the predicted variance, *AVEPVAR*, to the mean squared error, *MSE*, for systematic samples. To make deviations in either sense symmetric the logarithms of the ratio were computed. Predicting the p-variance by the formula for white noise processes overestimated the mean squared error strongly. Contrarywise, the method of 'balanced differences' predicted the sampling fluctuations fairly well. These results conform with theoretical considerations (Cochran, 1977).

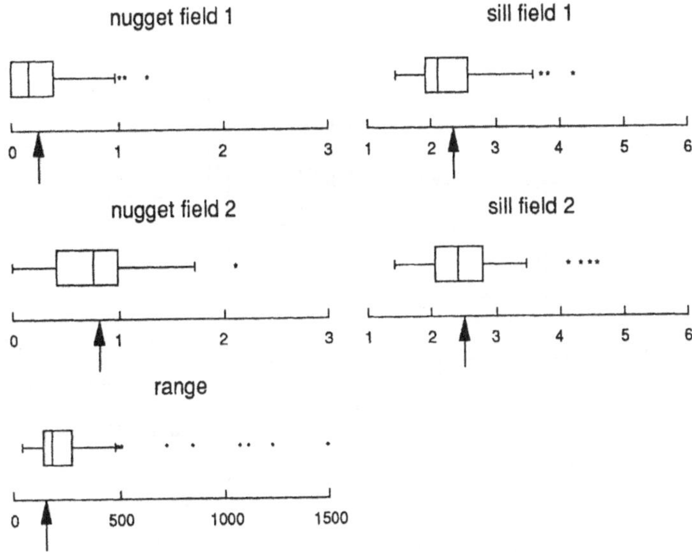

Figure 8: Fitted parameters of the variograms of field one and two for sample size 'medium', arrows indicate true parameter values.

Comparing AVEPVAR with MSE The logarithms of the ratios between *AVEPVAR* and *MSE* are compared in Fig. 7. For systematic samples, predictions by the method of 'balanced differences' were used. For this design no clear tendency is discernible. The predicted variance underestimated the mean squared error for the smallest sample size but overestimated it clearly for the largest sample size. For stratified samples the ratio *AVEPVAR/MSE* decreased with increasing sample size. It is not yet clear whether this dependence is due to a characteristic inherent in this scheme or just a random effect.

The variance predicted by the model-based scheme underestimated the mean squared error for both sample sizes. For size 'medium' the deviation was negligible, but for the smaller sample underestimating was considerable. A preliminary analysis of the results computed for the sample size 'very large' confirmed the tendency to underestimate the mean squared error ($\log AVEPVAR/MSE = -0.11$). Underestimating the fluctuations may result from errors in modelling the variograms and from the approximation used in Eq. (7). If the first cause dominates one would expect an improvement in the prediction for larger samples because the variograms should be better estimated. However, the results do not support this hypothesis. It seems that both causes contribute to the observed deviations.

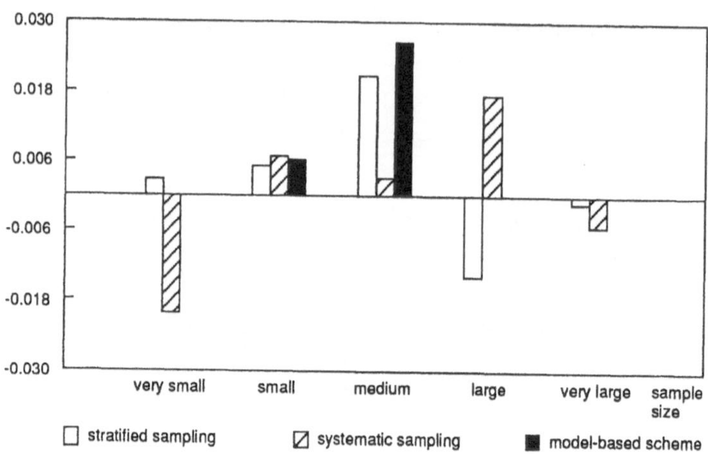

Figure 4: Bias of sampling schemes as a function of sample size.

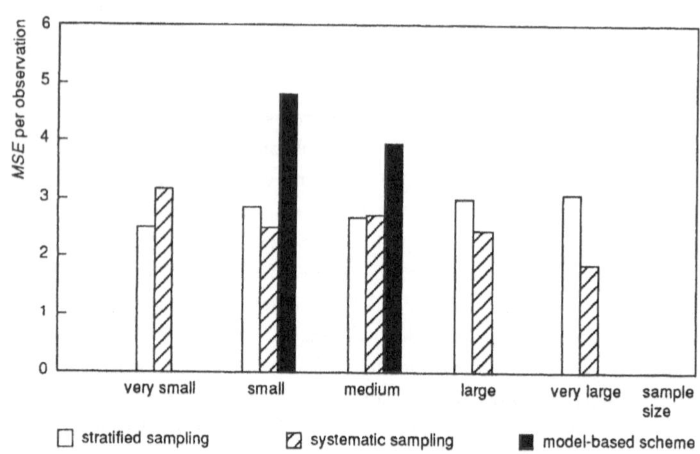

Figure 5: Mean squared error per observations of sampling designs as a function of sample size.

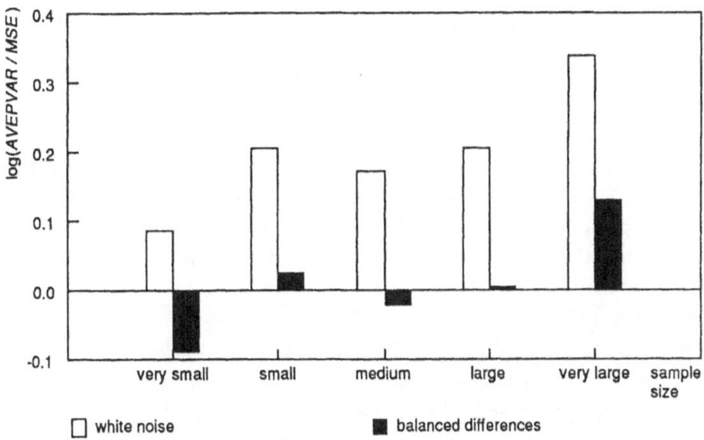

Figure 6: Common logarithm of the ratio of the average predicted variance to the mean squared error for systematic samples.

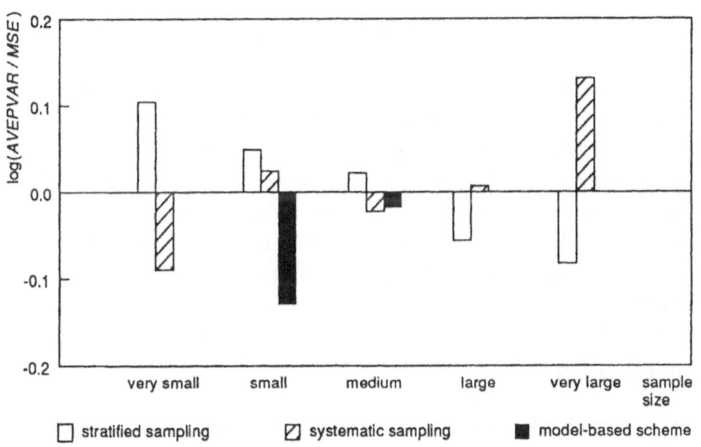

Figure 7: Common logarithm of the ratio of the average predicted variance to the mean squared error.

Summary and Conclusions

Both model-based and design-based estimation techniques can be used to estimate local mean changes over small plots. Similarly, global mean changes over a larger area can be obtained from both methods. Theory suggests that model-based estimation based on kriging should be more precise than simple or stratified random sampling plans. However, kriging demands that the true variograms are known. In practice, experimental variograms have to be computed from data, the estimates are subject to sampling error, and this influences the estimation of the mean change. It is therefore doubtful whether geostatistical global estimation methods are most efficient.

A simulation study for estimating the global mean change within an area showed that the effort necessary to make inference on the variograms rendered the model-based method less efficient than a paired stratified or paired systematic sampling scheme. The systematic scheme was slightly more precise than the stratified sampling plan. The model-based methods seemed to underestimate the mean squared error of global estimates, whereas no clear tendency was discernible for the classical schemes. For systematic samples the methods of 'balanced differences' predicted the variance well.

Although these results must be confirmed for larger sets of randomly chosen samples we can draw the preliminary conclusion that for global estimation geostatistical methods offer no advantage over classical sampling plans. They require much more computing effort, are less efficient, and seem to underestimate the estimation error. The use of stratified or systematic paired samples can be recommended to estimate the mean temporal change of soil properties.

If, however, global estimation is not the only objective of a study then the use of the geostatistical method may nevertheless be considered because this approach results in estimates of the variograms as a by-product. The knowledge of these functions can provide valuable insight into the patterns of spatial variation, and this might itself be of interest. Furthermore, knowing estimates of the variograms, a map of local estimates can be computed, showing regions with varying extent of change. Thus, these additional possibilities may compensate the smaller efficiency of the model-based approach in global estimation problems.

Acknowledgement

I thank H.R. Künsch, R. Webster, and two anonymous reviewers for their comments and contributions.

References

Bader, H.P. & Baccini, P. 1993. Monitoring and control of regional material fluxes. In *Soil Monitoring*, eds R. Schulin, A. Desaules, R. Webster, B. von Steiger, Birkhäuser Verlag, Basel, this volume.

Bellhouse, D.R. 1988. Systematic sampling. In *Sampling Handbook of Statistics Vol. 6*, eds P.R. Krishnaiah & C.R. Rao. North Holland, Amsterdam, pp. 125-145.

Cassel, C.-M., Särndal, C.-E. & Wretman, J.H. 1977. *Foundations of Inference in Survey Sampling*. John Wiley, New York.

Cochran, W.G. 1977. *Sampling Techniques*. 3rd edition. John Wiley, New York.

de Gruijter, J.J. & ter Braak, C.J.F. 1990. Model-free estimation form spatial samples: a reappraisal of classical sampling theory. *Mathematical Geology* **22**, 407-415.

Journel, A.G. & Huijbregts, C.J. 1978. *Mining Geostatistics*. Academic Press, London.

Krishnaiah, P.R. & Rao, C.R. 1988. *Sampling Handbook of Statistics Vol 6*. North Holland, Amsterdam.

Laslett, G.M. & McBratney, A.B. 1990. Estimation and implications of instrumental drift, random measurement error and nugget variance of soil attributes—a case study for soil pH. *Journal of Soil Science* **41**, 451-471.

Lischer, P. 1993. Sources, propagation and control of errors in soil monitoring. In *Soil Monitoring*, eds R. Schulin, A. Desaules, R. Webster, B. von Steiger, Birkhäuser Verlag, Basel, this volume.

McBratney, A.B. & Webster, R. 1983. How many samples are needed for regional estimation of soil properties? *Soil Science* **135**, 177-183.

McBratney, A.B. & Webster, R. 1986. Choosing functions for semi-variograms of soil properties and fitting them to sampling estimates. *Journal of Soil Science* **37**, 617-639.

Murthy, M.N. & Rao, T.J. 1988. Systematic sampling with illustrative examples. In *Sampling Handbook of Statistics Vol. 6*, eds P.R. Krishnaiah & C.R. Rao. North Holland, Amsterdam, pp. 147-185.

Papritz, A. & Flühler, H. 1993. Temporal change of spatially autocorrelated soil properties: optimal estimation by cokriging. *Geoderma*, in press.

Schulin, R. 1993. Contaminant mass balances in soil monitoring as a problem of scale. In *Soil Monitoring*, eds R. Schulin, A. Desaules, R. Webster, B. von Steiger, Birkhäuser Verlag, Basel, this volume.

von Steiger B. & Obrist, J. 1993. Available databases for the assessment of regional mass balances in agricultural land. In *Soil Monitoring*, eds R. Schulin, A. Desaules, R. Webster, B. von Steiger, Birkhäuser Verlag, Basel, this volume.

Webster, R. 1993. Dealing with spatial variation. In *Soil Monitoring*, eds R. Schulin, A. Desaules, R. Webster, B. von Steiger, Birkhäuser Verlag, Basel, this volume.

Webster, R. & Oliver, M.A. 1992. Sample adequately to estimate variograms of soil properties. *Journal of Soil Science* **43**, 177-192.

Yates, F. 1981. *Sampling Methods for Censuses and Surveys.* 4th edition, Griffin, London.

A. Papritz, Institut für terrestrische Ökologie, ETH Zürich,
Grabenstrasse 3, CH-8952 Schlieren, Switzerland

Soil Monitoring, Monte Verità, © Birkhäuser Verlag Basel

USING GIS FOR PROCESSING DATA FROM THE SURVEY AND MONITORING OF POLLUTION

A.P.J. De Roo & P.A. Burrough

Geographical Information Systems (GIS) are useful for processing many data collected during environmental and natural resource inventory and monitoring because they contain a wide range of tools for collecting, storing, retrieving, transforming, and displaying spatial data. Further, complex mathematical models, such as groundwater, surface runoff and soil erosion models, can be linked to a GIS database to give additional power to aid environmental management. In particular, a large and detailed database in a GIS linked to a mathematical model greatly benefits model predictions. A GIS can be useful for spatial analysis of specific problems, such as the identification of outliers, which might indicate point sources of pollutants. Digital elevation models can be used in GIS to aid the survey and monitoring of environmental pollution.

Geographical Information Systems

There is an increasing need for the early detection and survey of soil contamination and degradation. Because the data from such surveys contain spatial components, Geographical Information Systems (GIS) are particularly helpful for analysis. Frequently, too, data from many different sources and scales need to be used, such as existing maps, remotely sensed data from satellites or aeroplanes, or laboratory data from previous projects. Sometimes, map projections must be changed, especially when the data come from different organisations.

Burrough (1986) defined a GIS as a "powerful set of tools for collecting, storing, retrieving at will, transforming, and displaying spatial data from the real world for a particular set of purposes". However, a GIS is more than just a means of coding, storing and retrieving data about aspects of the earth's surface. Because the data in a geographical information system can be accessed, transformed, and manipulated interactively, they can serve as a test bed for studying environmental processes or for analyzing the results of trends, or for anticipating the possible results of planning decisions. In principle, it is possible for planners and decision-makers to explore a range of possible scenarios and to obtain an idea of the consequences of a course of action before irrevocable mistakes have been made.

To show how a GIS can be used in the survey and monitoring of pollution, we first give a general overview of the tools for spatial analysis that many GIS include. Then several case studies are briefly presented to illustrate the advantages of the tools.

Tools available for analysis

To obtain reliable answers to practical problems with the help of a geographical information system, the size and quality of the database should be carefully appraised. Spatial data obtained by digitizing or scanning maps, sampling and interpolation, or from remote sensors such as satellites, need to be adequate, representative, and collected at the right time with the appropriate spatial and temporal resolution. For example, it is not sensible to use coarse-grained satellite imagery to look for individual trees that are suffering from soil acidification, neither is it particularly useful to record rainfall on a weekly basis when one is interested in the storm-dependent run-off characteristics of small catchments. Those who collect data should provide information about their quality so that rational decisions can be made about whether to include any particular data in an analysis (Burrough 1992). Also, if errors in the basic data are known, the propagation of errors in spatial modelling with a GIS can be investigated (Heuvelink *et al.*, 1989; De Roo *et al.*, 1992).

The most important characteristic of a GIS is the facility with which a user can create new entities or new attribute values from existing entities and their attributes. There is an enormous range of basic operations for data analysis available to the user of a GIS, although not all will be present in a single system. The wide range of methods for obtaining new attribute values and entities can be summarised as follows (Burrough, 1992).

1. A new attribute and its values are derived from existing attributes by using exact-valued non-geographical attributes of exact objects (points, lines, polygons or stacks of grid cells).

   Example: using simple logical (Boolean) operations, e.g. 'find all mapping units that have a topsoil texture of clay loam in which the pH is 7.0 or less'. The result is a map created from maps of soil texture and pH.

   Other operations include simple arithmetical operations, weighted arithmetical operations, including complex mathematical models, and multivariate classification methods.

2. A new attribute and its values are derived from existing attributes by transformations that admit non-exact values of non-geographical attributes of exact objects.

   Example: fuzzy logical operations, using fuzzy membership functions defining how the possibility of membership varying continuously from 0 for individuals that are completely outside the set to 1 for objects that are within the central concept (Burrough *et al.*, 1992). Using the example from the class 1 operations, sites with a pH of 7.1 have a larger membership probability than sites with pH 8.0, although in a class 1 approach the final results for these two sites are not different.

Other operations include multiple regression (including error), and multivariate statistical classification and discrimination.

3. The value of an attribute at a point or at a given grid cell x is derived from the values of the same or other attributes within a defined neighbourhood surrounding that point. The attributes are attached to discrete objects.
   Example: All grid cells having more than 1 neighbouring grid cell with a clay soil (discrete) become a certain new attribute.
   Other functions include connectivity and proximity.
4. The value of an attribute at a point or at a given grid cell is derived from the values of the same or other attributes within a given neighbourhood surrounding that point. The attributes are taken to be those of a continuously varying surface.
   Example: The computation of slope gradient from the altitude values of nine grid cells: the grid cell for which slope gradient is calculated plus its eight neighbours.
   Other functions include geometric filtering, computing moving averages, edge detection, computation of first and higher order derivatives of a continuous surface (slope, aspect, slope length, curvature, shaded relief, drainage networks, saddle points, peaks, valleys, and pits), interpolation (including kriging), surface fitting, and intervisibility of points on a continuous surface.
5. An area or neighbourhood surrounding a point, line or area receives new attributes and values according to the attributes of the original spatial entity at a location.
   Example: The calculation of buffer zones along roads, factories; distance to the closest railway station, bank or shop.
   Other functions include point-in-polygon operations.
6. New spatial objects (vector space) are created from existing objects or the objects are modified.
   Example: polygon overlay of two vector maps (e.g. soil units and land use), creating a new vector map.
   Other functions include buffering in vector maps, computation of centroids, threading of contour lines, Thiessen polygons through a set of points, and line smoothing.
7. New attributes are derived from the geometrical attributes of (vector) objects or sets of objects.
   Example: measurement of the area or circumference of a map unit.
   Other functions include measurements of shape.
8. Analyses are carried out and reports are created but the geographical information system database is not increased with new attributes.
   Example: creating a histogram of the map attributes.
   Other functions include cross-tabulation, derivation of cross sections and profiles.

9. Data management operations.
   Example: change the projection of a map.
   Other functions include joining map fragments together or cutting out any part of a map to make a new file.

The operations listed above show that a GIS can contain many basic tools for environmental surveys and spatial data analysis. If a given task can be broken down into a logical series of steps, each of which can be carried out using one or more of the operations listed above, a GIS will be useful. Note that not all GIS contain functions for all the tools listed. For example, few GIS include geostatistical methods, though it is often possible to link the results of a geostatistical analysis to a GIS. Okx & Kuipers (1991) and Okx *et al.* (1992) demonstrate the use of geostatistical methods linked to a GIS for probability analysis of soil pollutants. Few GIS include complex mathematical models of crop yield, groundwater flow, surface runoff, soil erosion or pollutant leaching, but they can supply data to these models and display the results, often in the form of coloured maps.

The graphical quality of GIS output, whether resulting from data selection, simple or complex modelling, often conceals many shortcomings in both the data and the models. It is essential that all results of GIS modelling and analysis should be accompanied by some statement of quality or reliability, but this is often difficult to achieve. Uncertainties in spatial data can be quantified in terms of variances which can be derived using geostatistics. When these data are used as inputs to models the errors in data cause errors in the model results which could have important consequences for decision making. Statistical methods for following error propagation through models exist, however, and error propagation tools have been built to study how models react to errors in input data and parameter values (Heuvelink *et al.*, 1989, de Roo *et al.*, 1992). These tools permit the spatial analysis of errors and enable the user to see how error levels may be associated with particular areas or features of the terrain.

Future GIS for environmental resource analysis will include sophisticated modelling and error analysis options which will enable users to improve their estimates of the work needed to remedy pollution within given levels of tolerance (Burrough, 1992). At present we are not at this stage, however, and the following examples demonstrate the present state of the art in using GIS for monitoring and analyzing soil and water pollution.

Case study 1: Interactive exploratory analysis of spatial data

A GIS is being used in a study of polluted sediments in an estuary located in the Southwest of the Netherlands (Hazelhoff & Gunnink, 1992). In 1971 a dam was constructed at the outlet of the estuary of the rivers Rhine and Maas which resulted in the deposition of sediments from east to west: these were polluted with heavy metals and polychlorobiphenols (PCBs). Three sedimentation periods could be distinguished based on the measured thickness of the sediments in 1971, 1976, 1981 and 1987. Samples of sediment were taken at several depths and loca-

tions, and analyzed for pollutants. The concentrations of pollutants gradually decreased from 1971-1980 to the present. The upper 30 cm was analyzed separately, because this layer is considered to be active in sediment transport.
The spatial data were analyzed using the Genamap GIS, together with the Regard software (formerly called Spider) for exploratory data analysis (EDA) - (Haslett *et al.*, 1990). Using this GIS the maps can be displayed, and histograms of the pollution data can be created. Also, the spatial correlation of the pollutants was investigated by computing the variograms.
One of the strengths of linking GIS and EDA is that one can interactively select, for example the left side of a histogram of a pollutant, and at the same time in another graphic window observe the map with sampling points, on which the sampling points from the left side of the histogram are highlighted. By selecting any part or parts of the histogram one can observe the spatial distribution of the corresponding sampling points. Thus, it is very easy to locate outliers, or points exceeding critical pollution limits.
Analyzing the multivariate characteristics of the pollutants, using this GIS, revealed that all pollutants were strongly and linearly related, indicating that the pollutants are well mixed after emission from upstream sites. It was found that the first principal component, calculated from the PCB and heavy metal concentrations of the sediments, provided information on this linear relationship. The second principal component provided information on the residuals or deviations from these linear relationships, possibly indicating local emissions of pollutants. Again, using this GIS, one can select the outliers from the histogram of the scores of the second principal component and at the same time see the corresponding sampling points in the other graphic window. Thus, one can observe that the two selected points are geographically close together, indeed indicating a local emission (Figure 1).
This case study shows that a GIS linked to EDA is very useful for spatial data analysis, both in the inventory phase of a project and in an advanced phase when specific problems must be investigated which may reveal outliers that could indicate local emissions. Conventional statistical packages alone make the detection of the spatial context of outliers difficult.

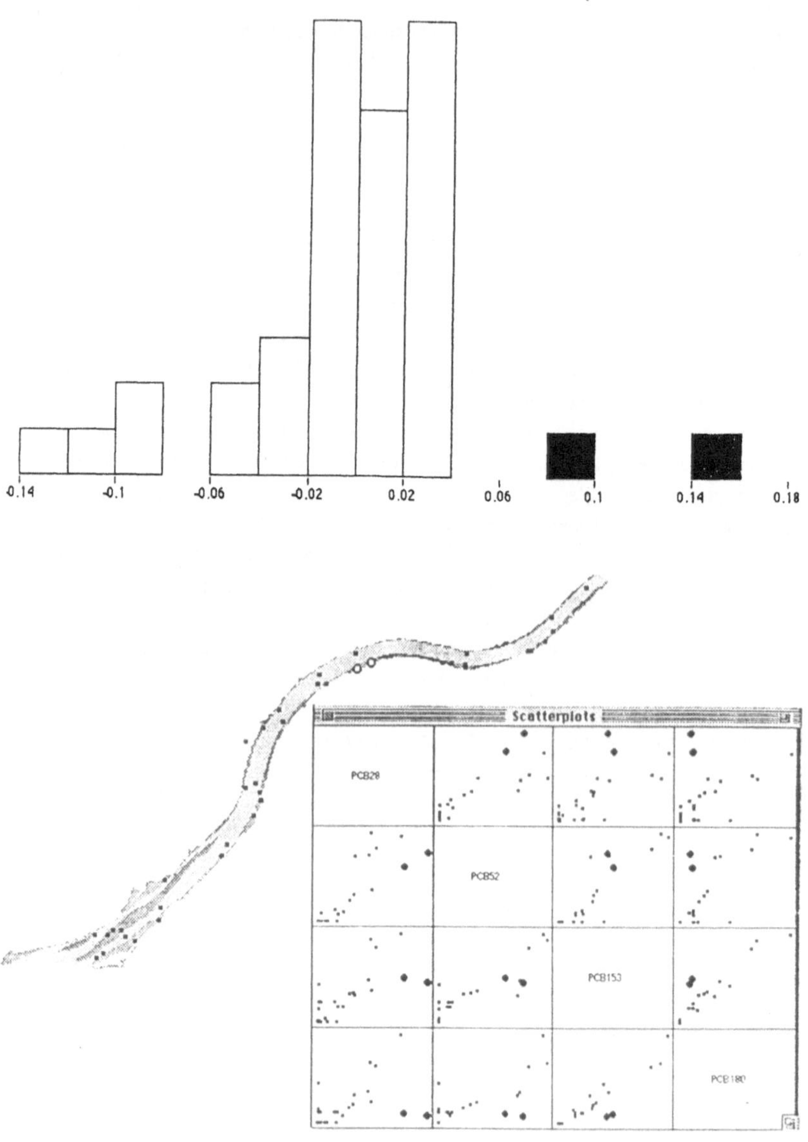

Figure 1. Histogram of the scores on the second principal component showing two outliers, the location of these points, and PCB scatterplots including these points (from Hazelhoff & Gunnink 1992).

Case study 2: Linking complex mathematical models to a GIS

The strength of a GIS in environmental studies is demonstrated by an example of a complex mathematical model linked to a GIS. Models contribute to science because the hypotheses expressed in mathematical terminology describe quantitatively chemical, biological, and hydrological processes. Pollution studies use many kinds of transport models, such as groundwater models, catchment hydrology models, river flow models, meteorological models, with or without a chemical component.

ANSWERS - an event-based runoff and erosion model
Here we report the use of the distributed and physically-based 'ANSWERS' model (Beasley *et al.*, 1980) as an example of a complex mathematical model linked to a GIS. The model simulates the hydrological response of agricultural catchments during and immediately after a rainfall event, such as a single shower or period of rain. It simulates processes such as overland flow, channel flow, soil erosion and sediment transport. In addition, the transport of pollutants attached to the sediment can be simulated easily. Sediment can also be considered a pollutant in its own right if it is washed into buildings that are flooded by excessive runoff. ANSWERS' primary application is in planning and evaluating various strategies for controlling surface runoff and sediment transport from intensively cropped land. The original version of the model has been fully integrated within a GIS so that data can be entered easily and the results can be displayed as maps and tables (De Roo *et al.*, 1989).

A catchment to be modelled is assumed to be composed of square cells. Channel flow is analyzed by a separate pattern of channel elements. Values of variables are defined for each element, e.g. slope, aspect, soil variables (porosity, moisture content, field capacity, infiltration capacity, a soil erodibility factor), crop variables (coverage, interception capacity, a management factor), surface variables (roughness and surface retention) and channel variables (width and roughness). These values may be derived from digital maps or from interpolation of point data. It is clear that several detailed maps of land attributes are needed to run the model. These maps can be stored and edited in a GIS. A rainfall event can be simulated with increments of one second up to several minutes, depending on the gridsize, thus taking into account spatial and temporal variability of rainfall. The continuity equation is used to establish the composite response of the single elements. The output of up-slope elements becomes the input of downslope elements. Several physically-based mathematical relations are used to describe water movement by interception, infiltration, surface retention, drainage, overland flow, channel flow and subsurface flow, and the detachment of sediment by rainfall and overland flow and sediment-transport by overland flow (inter-rill erosion).
Hydrological and soil erosion processes vary over the catchment as do the more permanent properties of the soil and landscape (often called 'parameters' by hydrological modellers because they are assumed to remain constant). If data about the spatial variation of these proper-

ties are stored in geographical information systems they can easily be extracted for this type of erosion modelling. Data are not equally available about all aspects of the landscape, however. At present most detailed information is available about the terrain form. To make the best possible use of distributed topographic data a catchment should be modelled by a raster-based Digital Elevation Model (DEM). In many cases this DEM will have several thousands of elements, so that entering data by hand is not feasible.

A DEM can be constructed by digitizing contour maps. The GIS computes maps of altitude, slopė and aspect, which are all input to the ANSWERS model (Burrough, 1986). Also, maps of concavity and convexity and potential stream channels can be derived from the DEM. Geostatistical interpolation techniques, incorporated in the GIS, can be used to produce maps from point observations of the soil etc., recorded during field experiments. Using a method such as block kriging, point observations are interpolated to blocks of the same size as the elements used for simulation. When there are insufficient field measurements, the distribution of a desired input variable can be derived from digitized soil or land use maps.

The University of Utrecht version of ANSWERS uses the geographical database (with the maps of the input variables at a spatial resolution appropriate for the catchment) to produce an ANSWERS input file. A specially developed program transforms the maps of input-variables to an input-file. As the input files may contain data on thousands of grid elements it is clear that distributed parameter models such as ANSWERS cannot be used without a GIS to supply data at the correct spatial resolution.

The GIS can also be used to produce modified element-files for different land use patterns or conservation measures in order to simulate alternative scenarios, and the results can be displayed as maps. The output of the models can be captured after every iteration and stored in the GIS so that dynamic maps can be produced showing the variation with time of the patterns of soil erosion, sedimentation and runoff. These maps can be compared by subtraction to yield maps indicating how erosion or sedimentation might be affected by certain control measures within the catchment. Runoff can also be displayed as an overlay on the landform surface (digital elevation model).

Applications of ANSWERS

The primary application of ANSWERS is for planning and evaluating various strategies for controlling pollution from intensively cropped land. This is demonstrated by the following example. In a catchment of 46 ha near Catsop (Stein, Limburg, the Netherlands), the impact of possible erosion control measures was assessed by simulation. The catchment was modelled by constructing 4585 elements of 10 m square (0.01 ha). A DEM was constructed by digitizing a contour map, which was rastered to cells of 10 m x 10 m, and maps of slope and aspect were derived from it. The land use map was digitized and converted to a 10 m x 10 m grid as were maps of soil, land use, management and channel descriptions. For the first simulations only two soil types were defined: soil under row-crops and soil under grassland or forest.

The resulting database described the topography, soil and actual land use of the catchment as the baseline condition. Data on rainfall and channels were also entered.
To evaluate some erosion control measures and to determine the best possible locations for them, several land-use scenarios were developed as follows:

1. *Baseline* Land use and management as in March 1987 when 88.0% of the surface was fallow with crop residues, 10.7% was grassland and 0.5% was lynchets, which are forested terrace-borders on slopes.
2. *Fallow* The grassland is converted to fallow land with crop residues, and the lynchets are removed.
3. *Grassland* All fallow land is converted to grassland.
4. *Contouring* The fallow land is ploughed on the contour.
5. *Lynchets* At several locations, lynchets (Dutch: 'graften'- cultivation terraces) are constructed as flat forested terraces.
6. *Contour-grass-strips* Strips of grass, constructed on the contour, are left unploughed between bands of cropped land.

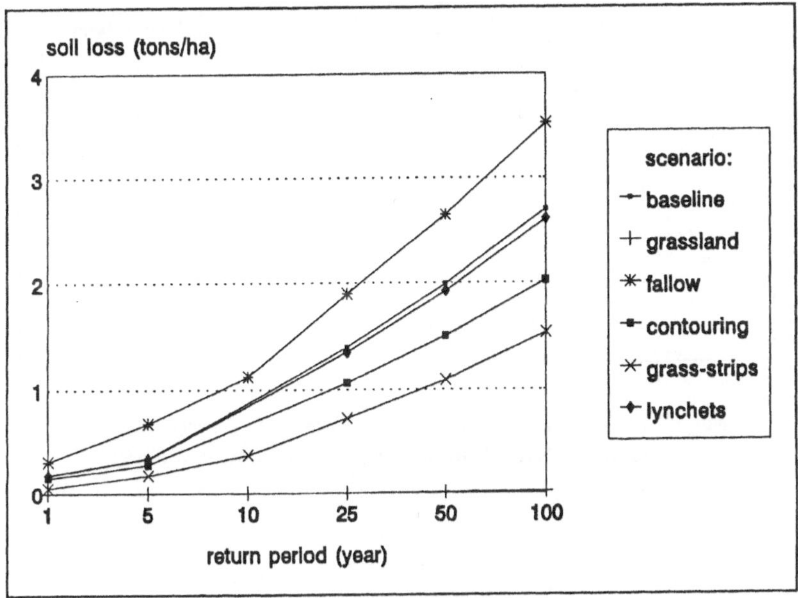

Figure 2. The effectiveness of soil erosion scenarios in the Catsop catchment (South-Limburg, The Netherlands) for soil erosion.

ANSWERS was run several times with rainfall events of different magnitude, with different return periods. By simulating these events it can be determined how the catchment might possibly react under conditions of normal and extreme precipitation. Figures 2 and 3 show the results of the scenarios which suggest that the *Contouring* scenario is effective in reducing soil erosion. The *Contour-grass-strips* scenario is effective both for runoff and soil erosion, and the maximum rate of runoff decreases spectacularly. Under conditions of extreme rainfall the effectiveness of this scenario decreases. The *Grassland* scenario reduces soil erosion to zero and minimizes the surface runoff.

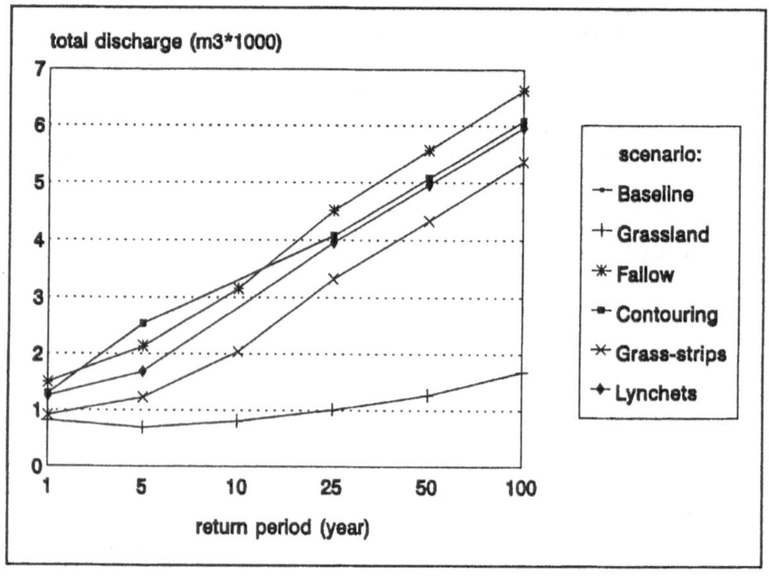

Figure 3. The effectiveness of soil erosion scenarios in the Catsop catchment (South-Limburg, The Netherlands) for total catchment discharge.

Maps of soil erosion and sedimentation of the scenarios were compared by subtraction, as described above. These simulations indicated where possible control measures would have the greatest positive and negative consequences.

Planners can decide to combine the positive elements from several scenarios and leave out the negative elements to develop new scenarios. The advantage of using ANSWERS in combination with the GIS is that the analysis can be done very quickly because combining maps in a GIS is a standard operation. The database can easily be updated with new data (such as field and laboratory measurements), necessitating new interpolations to produce maps.

To validate this model the spatial predictions of soil erosion were compared with measured values of the ^{137}Cs in the soil, which was used as a tracer of soil erosion (De Roo, 1991). The simulated soil erosion map and the interpolated Cs map were compared and the original point Cs data were compared with the estimates from the soil erosion model for the corresponding grid cells (Figure 4).

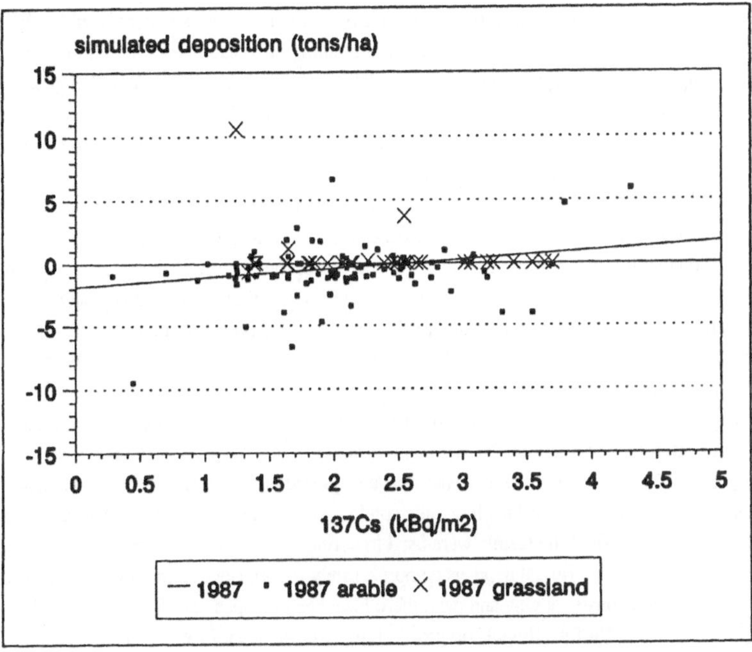

Figure 4. *Comparison of measured ^{137}Cs data, indicating soil erosion and deposition, and simulated soil erosion in the Etzenrade catchment (The Netherlands).*

Figure 4 shows that there is a weak positive correlation between ^{137}Cs (large values indicate deposition, small values indicate erosion) and the simulation results (positive values represent deposition, negative values represent soil erosion) with a correlation coefficient, *r*, of 0.27 (significant at a 95% confidence interval $p<0.002$). The large dispersion of the points on Figure 4 demonstrates that techniques for modelling and monitoring soil erosion still need to be improved for successful, routine prediction.

This case study shows that models simulating pollution can greatly benefit from the linkage with a GIS. The quality of the estimates can be seen through comparisons with measurements which are also stored in the GIS database. Different scenarios can be set up quickly for evaluation.

Case study 3: Using detailed Quaternary geological information stored in a GIS for groundwater protection in the Netherlands

Another example of the strength of a GIS for environmental studies is demonstrated by the following case study in which a large database containing information on soil texture is used to provide input values for a groundwater model (Weerts & Bierkens, 1992). Detailed geohydrological information for groundwater protection is becoming increasingly important in the Netherlands because of pollution of drinking water wells. It is common practice to assign parameter values of, for example permeability and porosity to groundwater models based on a limited number of pumping tests and samples of the subsoil. Protection zones around wells are derived from these models. Hydrological parameters such as permeability and porosity are strongly related to the texture and structure properties of the subsoil. If this relation is known, qualitative geological information can be translated into quantitative geohydrological information. Thus, the results of groundwater models that predict pollution-plumes and residence times, can be improved. However, detailed geological information is often ignored because it is too complex to handle, but this can be remedied by a GIS.

A study currently being carried out at the University of Utrecht concerns a fluvial area near Velddriel (Bommellerwaard, the Netherlands) from which drinking water is abstracted. A well pumps water from an aquifer that is covered by an aquitard of clayey Holocene fluvial deposits. Pumping tests have established the hydraulic conductivity of the aquifer and the hydraulic resistance of these deposits. The results were used to define a groundwater protection area around the well. The Holocene fluvial deposits comprise sandy stream ridges and crevasse splays, and backswamps that consist of clay and peat: these have been mapped and the detailed geological map was digitised. Various fluvial soil profile-types within the Holocene aquitard were distinguished and the hydraulic conductivity, porosity and grain-size distribution of the profile-types of each unit were measured, and were used to create a detailed geohydrological map of the confining Holocene layer in the GIS database.

Linking the GIS to the MODFLOW groundwater model (McDonald & Harbaugh 1988) revealed the infiltration spots contributing to the drinking water well. It was found that the total contribution of water seeping through the confining layer is small. Most of the contributing spots appeared to be outside the groundwater protection area. The conclusion from this study is that a large data base in a GIS linked to a mathematical model improves spatial resolution and greatly benefits prediction.

Case study 4: Using elevation data to map of soil pollution on floodplains efficiently

Geomorphological information including elevation is a large and readily available source that is often not used for environmental studies. Leenaers *et al.* (1990) showed that knowledge of geomorphological processes can often improve the interpolation of properties that are expensive to measure, such as heavy metals. Here we summarize their results.

Along the riverbanks of the River Geul, a tributary of the River Maas which crosses the Belgium - Netherlands border, there are large heaps of reject material and tailings of metal mining left over from the 19th century. Erosion of these locally highly contaminated streambank deposits provides a source of heavy metals. Further, since the techniques for processing the ore were inefficient, the resulting effluent, which was discharged directly into the river, had a large concentration of ore particles and metal-rich spoil.

This pollution has contaminated the flood deposits of the entire downstream stretch of the River Geul for 30 km to its junction with the River Maas. The concentrations of heavy metals in the flood deposits decline with distance from the source as a result of dilution with relatively clean materials provided by the erosion of tributary river banks and from hillslope processes such as soil erosion (De Roo, 1991). Leenaers *et al.* (1990) observed local deviations from this pattern when they mapped the floodplain in detail. They noted that the surface form of the floodplain controlled both the relation between inflow and outflow of water and the sedimentary conditions during a flood which led to a large spatial variation in heavy metal concentrations over the floodplain.

A Digital Elevation Model (DEM) of a part of the Geul valley was constructed from readily available elevation data. By using a simple model of generalized flood water levels flood hazard zones were delineated on the DEM. Soil in areas most frequently flooded had significantly larger metal concentrations than areas bordering the valley side, where flooding is rare. Classifying measurements of heavy metal concentration according to flood hazard zone and performing an analysis of variance yielded an explained variance fraction of 65 per cent. This clearly demonstrates the importance of the relation between the degree of pollution and the elevation relative to the river bed.

The relation between elevation and heavy metal concentration on the floodplain was further investigated by co-kriging, and it was demonstrated that in this situation the use of relative elevation as a co-variable in a co-kriging procedure to estimate heavy metal concentrations improved the predictions still further.

Conclusions

The examples presented show that GIS are useful for many applications in environmental inventory and analysis. Numerous tools for collecting, storing, retrieving, transforming, and displaying spatial data are available in GIS. Further, linking complex mathematical models, such as groundwater and surface runoff and soil erosion models to a GIS database, provides extra power for environmental management. The availability of large data bases in a GIS, such as one with Quaternary geological information, linked to a mathematical model, greatly benefits model predictions. GIS (e.g. Genamap) linked to EDA (Regard) is very useful for analysing spatial data, both in the inventory phase of a project, and also in an advanced phase for revealing outliers which could be local emissions. Data from DEMs can be used in many ways for inventory and monitoring of environmental pollution.

Acknowledgements

The authors thank L. Hazelhoff, J. Gunnink, H. Weerts, M. Bierkens, and H. Leenaers for making their work available as examples of using GIS for pollution studies.

References

Beasley, D.B., Huggins, L.F. & Monke, E.J. 1980. ANSWERS: A Model for Watershed Planning. *Transactions of the ASAE* **23**, 938-944.

Burrough, P.A. 1986. *Principles of Geographical Information Systems for Land Resources Assessment.* Clarendon Press, Oxford.

Burrough, P.A. 1992. Development of intelligent geographical information systems. *International Journal of Geographical Information Systems (IJGIS)* **6**, 1-11.

Burrough, P.A., MacMillan, R.A. & Van Deursen, W. 1992. Fuzzy classification methods for determining land suitability from soil profile observations and topography. *Journal of Soil Science* **43**, 193-210.

De Roo, A.P.J. 1991. The use of ^{137}Cs as a tracer in an erosion study in South-Limburg (The Netherlands) and the influence of Chernobyl fallout. *Hydrological Processes* **5**, 215-227.

De Roo, A.P.J., Hazelhoff, L. & Burrough, P.A. 1989. Soil erosion modelling using 'ANSWERS' and Geographical Information Systems. *Earth Surface Processes and Landforms* **14**, 517-532.

De Roo, A.P.J., Hazelhoff, L. & Heuvelink, G.B.M. 1992. The use of Monte Carlo simulations to estimate the effects of spatial variability of infiltration on the output of a distributed hydrological and erosion model. *Hydrological Processes* **6**, 127-143.

Haslett, J., Wills, G. & Unwin, A.R. 1990. SPIDER - An interactive statistical tool for the analysis of spatial data. *International Journal of Geographical Information Systems (IJGIS)* **4**, 285-296.

Hazelhoff, L. & Gunnink, J.L. 1992. Linking tools for exploratory analysis of spatial data with GIS. In *Proceedings of the Second European Conference on Geographical Information Systems*, ed. J-J Harts, H.F.L. Ottens and H.J. Scholten. EGIS Foundation, Utrecht.

Heuvelink, G.B.M., Burrough, P.A. & Stein, A. 1989. Propagation of errors in spatial modelling with GIS. *International Journal of Geographical Information Systems (IJGIS)* **3**, 303-322.

Leenaers, H., Okx, J.P. & Burrough, P.A. 1990. Employing elevation data for efficient mapping of soil pollution on floodplains. *Soil use and management* **6**, 105-114.

McDonald, M.G. & Harbaugh, A.W. 1988. A modular three-dimensional finite-difference groundwater flow model. *Techniques Water Resources Investigations Book 6, Chapter A1: Modeling techniques*, US Geological Survey.

Okx, J.P. & Kuipers, B.R. 1991. 3-Dimensional Probability Mapping for Soil Remediation Purposes. In *Proceedings of the Second European Conference on Geographical Information Systems*, ed. J-J Harts, H.F.L. Ottens and H.J. Scholten. EGIS Foundation, Utrecht, pp. 765-774.

Okx, J.P., Leenaers, H. & Krzanowski, R.M. 1992. Probability kriging as a decision support tool for local soil pollution problems. In *Geostatistics Tróia '92*, Vol. 2, ed. A. Soares, Kluwer Academic Publishers, Dordrecht, pp. 673-683.

Weerts, H.J.T. & Bierkens, M.F.P. 1992. Using detailed Quaternary geological information for groundwater protection in the Netherlands. In *Proceedings of the first Dutch Earth Sciences Conference*, Veldhoven, the Netherlands.

A.P.J. De Roo & P.A. Burrough
Department of Physical Geography
University of Utrecht
Heidelberglaan 2
P.O.Box 80.115
3508 TC Utrecht
The Netherlands.

Discussion

Question by S. van der Zee
Many people use models for applications at different scales from that originally intended. Is that a methodological problem in GIS?
Answer by A. de Roo
It is not a methodological problem in a GIS, but it is a fundamental problem in distributed modelling. All physically-based models use equations developed at small scales, and are run on larger scale elements/pixels. As far as I know, this problem is recognized (e.g. Beven, K.J. (1989), Changing ideas in hydrology - the case of physically-based models. Journal of Hydrology, **105**, 157-172) but not solved yet. Another problem arises when going from point data to the size of a raster or vector element. This can be partially solved using geostatistics (the change of support).
Question by M. Flury
You have shown the advantages of a GIS. What are the disadvantages? Are there any?
Answer by A. de Roo
The predictions and results based on a GIS analysis depend on the data available and also on the underlying models used. If the relation between model resolution and database resolution is not correct then the results produced by a GIS that uses that model on these data can be misleading. Also erroneous algorithms can lead to inaccurate results.
Question by J.-P. Dubois
You have pointed out the differences between raster and vector information. Is there commercial software that can use simultaneously these two types of information in the same GIS?
Answer by A. de Roo
At present there are several GIS that I know of which combine vector and raster, namely GENAMAP, the new version of ARC-INFO, Intergraph, GRASS, SPANS, and the PC-package called IDRISI.

Soil Monitoring, Monte Verità, © Birkhäuser Verlag Basel

TREND DETECTION IN SOIL MONITORING — LOOKING FOR THE NEEDLE IN THE HAYSTACK

M. A. Oliver

Introduction

Soil monitoring is concerned to detect change (*the needle*) in soil properties in space or time or both that signify stress in part of or in the whole of the environmental system, and that will trigger action. This includes the soil's response to changes in its environment or its treatment, and the effects that such changes might have on plants and animals that it supports and on other components of the environment such as water supply and the atmosphere. The changes will often be small, and they might be masked by the background variation (*the haystack*) which comprises measurement error and natural random fluctuation in space and time. These latter factors limit our ability to identify and quantify change, and we must take them into account when we use data for decisions to ameliorate the land or to impose restrictions on industry or the population at large.

We must know what we should do to take account of such limitations in our data both when planning a monitoring scheme and when it is reappraised.

Measurement error

Errors arising from measurement are nuisance variation, and are the easiest to quantify and eliminate. They arise from poor laboratory practice, lack of standardization, poor calibration, and inadequate instrumentation. Such variation can be diminished by replicating analyses, establishing protocols for standardizing laboratory and field procedures and adhering to them, and by improving instruments.

Temporal change

Change over time has to be detected from short and long term fluctuations, and if sampling is destructive we have the additional complication of the change in time being confounded with local spatial variation. Both of these sources of variation might be much larger than the change itself.

Temporal fluctuation can be dealt with by sampling regularly at suitable time intervals. The interval could quite sensibly be a year if we do not want to detect any seasonal effects. For some pollutants the change might have a strong seasonal element, and then we should sample twice a year at appropriate times, such as summer and winter. Sampling needs to be for a long enough period overall. For certain heavy metals sampling intervals of more than thirty years may be required to find significant changes. This requires guaranteed funding and safeguards for the sites.

Temporal change must also be separated from spatial variation. This can be achieved by choosing sites with care. The plots should be homogeneous in a statistical sense. Unusually variable terrain should be avoided. Under these constraints the size of the plots will in general be fairly small, typically no more than half a hectare. The sites should be in areas that are known to respond rapidly to stress, for instance where the soil is sandy or with a small buffering capacity, or sites that receive water from surrounding catchments.

Samples from the sites should be bulked to give an average value for plot; bulking smoothes local spatial variation, and it also overcomes the difficulty arising from destructive sampling.

Information from a few well chosen sites with good temporal resolution should be the aim.

Moving averages can be used to detect temporal change, using time periods of different lengths to filter out the long and short term fluctuations. With a long enough series of measurements time series analysis would allow some prediction of the likely magnitude of changes in the future and can also summarize the overall pattern of temporal change.

Spatial change

The points considered for detecting temporal change apply here as well. Spatial changes can occur at the regional scale or more locally related to specific point source, such as an industrial installation or road. They will require different kinds of sampling design and analysis.

To detect spatial change of soil properties without any prior information is difficult. It requires regional sampling on a grid at an intensity that is adequate to detect the underlying spatial variation and any likely change in the extent of pollutants. The major problem is likely to be the cost of sampling at a reasonable intensity for geostatistical analyses. Block kriging of data for a series of years would be the most effective way of detecting spatial change from the local spatial variation. With prior knowledge Bayesian kriging should be more effective. Trend Surface Analysis might be useful in certain regional situations if there is a known physical model to describe the process before analysis.

For point sources the sampling design could be a series of transects radiating out from the source to positions beyond the present extent of contamination. To monitor the change in spatial extent of pollution from roads transects perpendicular to the road are needed. For such deterministic situations spatial change can be modelled and regression analysis used to separate change from natural variation.

Situations might arise where a combination of spatial and temporal analysis is desirable. There are now developments in geostatistics to satisfy this need for single properties and for multivariate data.

Properties to measure

The *properties* monitored are important. Some will be prescribed officially and monitored because of their relevance given by law or other norms, but others may be chosen because they are sensitive indicators of change, e.g. species composition of the soil microbial population, or because they correlate strongly with many others. Quantitative data from such properties will enable us to identify small changes. However, qualitative information, such as changes in land use, agricultural methods, social conditions (increased car ownership), policy (limits set by Governments or other statutory bodies), and the perceptions and awareness of policy makers and the general public arising from scientific developments are also relevant because they might guide the selection of quantitative properties and, perhaps more importantly, the choice of monitoring sites.

Conclusion

The needle is undoubtedly difficult to find. Nevertheless, it is important that it is found if it exists, and to choose methods that have a good chance of detecting it. These will be different depending on whether we want to detect change in time or in space. A few well located monitoring sites with regular sampling and measurement in time will be the best way of detecting temporal change. For detecting regional spatial change overall grid sampling is needed, but the time between samplings can be greater. In planning new monitoring schemes or reappraising established ones the following points for identifying change from the background variation should be considered.

1) Specimens from long term experimental monitoring sites should be banked to allow their history to be re-examined. It is a safeguard against insensitive technique at the time when the monitoring scheme is established, and also the failure to identify some of the problems or questions that become important as science and policy develop.

2) It is essential that policy makers guarantee funding and protection of sites from development for a minimum period of 30 years and ideally 50 years if we are to make sen-

sible statements about temporal and spatial changes. Any scheme should also be designed with scope to deal with changing knowledge and demands from society.

3) The ability to detect temporal and spatial changes depends on choosing significant variables, on adequate sampling and sound design, on care in locating sampling sites in the field and adhering to established protocols there, on reliable standard methods in the laboratory, on the time interval between sampling, and on proper statistical analysis of the resulting data.

M. A. Oliver
School of Geography
The University, Edgbaston
Birmingham B15 2TT
Great Britain

SYNTHESIS

Soil Monitoring, Monte Verità, © Birkhäuser Verlag Basel

SYNTHESIS

H. Flühler, P. Baccini, R.Webster and R.Schulin

Actors and interests in soil monitoring

If there was a point on which all participants of the workshop could certainly agree it was the notion that the term *soil monitoring* embraces a broad spectrum of definitions, motivations, operations, approaches and concepts in applied environmental protection as well as basic soil research. It became clear that there are two main interests in soil monitoring (Fig. 1). One may be labelled as the search for concepts, the other as the operational realization of soil monitoring.

Because of the diversity of interests and motivations, the various actors in soil monitoring tend to form themselves into corresponding groups. People whose aims are to elucidate the nature of soil and to find strategies to ensure the long-term sustainable use of soil resources tend to be oriented towards *conceptual soil monitoring*. They include primarily members of research organizations and environmental pressure groups. There are others who are occupied in one way or another with legislation and codes of practice to protect soil, and these people tend to focus on *operational soil monitoring*. This group consists of politicians, their professional advisers, the executives of governments and the entrepreneurs who put the instructions into practice.

These groups tend to create barriers between themselves. The barriers are imaginary, of course. Nevertheless, they are also unfortunate and inappropriate because they polarize where integration is needed most. None of the groups is homogeneous, nor well-defined. The actors in the field of soil monitoring constitute a social continuum with a wide range of overlapping interests. Some actors might not even be aware of their involvement, e.g. scientists of other disciplines who do not realize that they are contributing to improve knowledge in soil monitoring. Depending on their position in the continuum, the various actors are confronted either mainly with operational or more with conceptual soil monitoring problems.

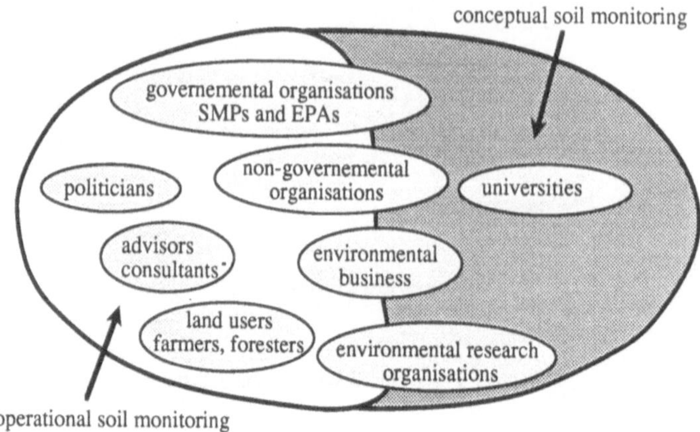

Figure 1. The social continuum of actors in soil monitoring, SMP = Soil Monitoring Programme, EPA = Environmental Protection Agency.

Tasks of soil monitoring

The workshop identified three main areas or tasks of soil monitoring.

The first task is the *anticipation* and *early detection* of potentially hazardous impacts on soil ecosystems. Hence, by definition we think of *foreseeable* events or developments. This is the nature of prediction. The entire scientific community (not only the soil scientists) has a duty to improve our predictive capabilities. Good predictions are based on sound models, correct assumptions, and comprehensive knowledge (data and understanding of processes), which further depend on support from professional tools such as computing systems and analytical instruments. Together these represent the *scientific base* of soil monitoring.

The second task is the *anticipation of unthinkable events and developments*. New, yet unknown problems will appear, and it is our duty to be prepared for them. Soil monitoring programmes should save as much information as possible about the present status of the soil for future soil monitoring. This includes *archiving* soil specimens (pedothek, soil sample bank) for future investigation as well as *documenting* data and methods using comprehensive, precise and understandable formats so that they can be used by our yet unborn successors.

The third task is the *diagnosis of the current status and developmental trend of the soil ecosystem*. Although soil monitoring is primarily a prospective tool to forecasting, it must also look backwards to identify the causes of harm such as soil degradation by overloading with pollutants. Many soils are ecologically overstressed today and need protection to prevent degradation. Others are already degraded and require remedial treatment to restore their ecological multifunctionality. The *judgement base* for long-term regulatory or political action must be obtained from soil monitoring. For this purpose it must include scientific long-term studies of reference soils. An improper diagnosis of soil malfunctioning will enhance the risks that countermeasures remain ineffective or may even worsen the situation. Experimental evidence of on-going or past contamination and degradation not only aids the dialogue between scientists and policy-makers but also provides for a better scientific understanding.

In addition we recognize a fourth task. This is the *surveillance* of the state of the soil after either remedial treatment of the soil or control of emissions of contaminants or other pollution abatement.

Time scales in the control of soil contamination and degradation

Man has been spoiling the environment for millenia. However, in the current context of soil monitoring we can represent environmental degradation and protection as four stages. They begin with the scientific perception, which is not meant to be a privilege of scientists but rather the way certain phenomena are discovered; this stage is followed by political perception, which leads to the political will and legislation. There is a final stage which is maintenance of the desired quality once it has been achieved. These stages are shown as points on the time axis in Fig.2, with the first three separated from one another by a time constant Δt. For example, water protection was first appreciated by scientists in the early 1950s (say $t_0 \approx 1950$). It took 15 years for politicians to become convinced, and a further 15 years for them to act. The recent developments in air pollution control suggest that Δt is of the same magnitude, but the political process began approximately 20 years later ($t_0 \approx 1970$). If the time constant Δt is controlled by the cognitive process of society then we may postulate that recognition of soil contamination and degradation will lead to successful governmental action by the year 2010 ($t_0 \approx 1980$). The quantity Δt separates again the time when successful governmental action starts from the time when the severity of the problem starts to decrease.

The preceding speculations depend on the assumption that the length Δt of the various stages is equal and remains constant over time. However, this is highly questionable. In the figure the reaction time to legal, political and technical actions (implementation time t_i) is in this graph set equal to Δt, but it could be significantly larger depending on the responsiveness of the system (overshooting). The soil is likely to respond more slowly to treatment than aquatic systems, and so t_i may be considerably larger for soil than for the water and air pollution legislation and control. The 'severity curve' need not to be symmetric. Degradation and recovery might span different lengths of time.

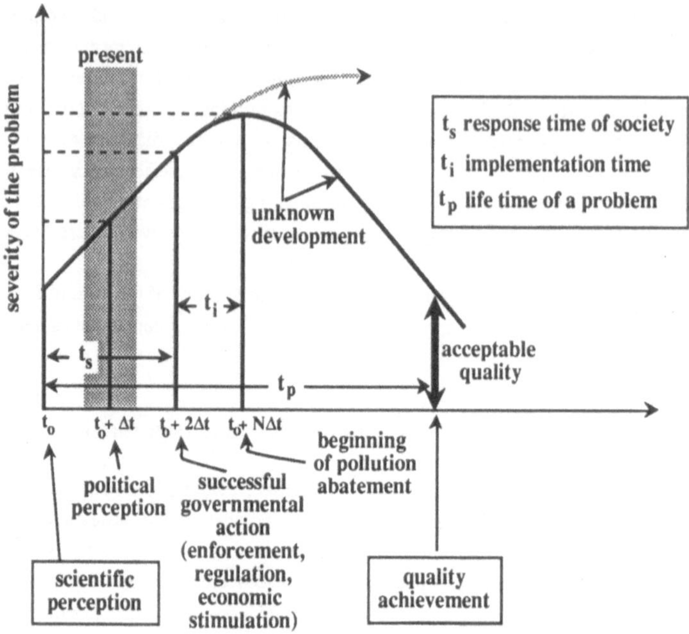

Figure 2. Part of the life cycle of man-made damage to the environment

The minimum relevant time scale for soil monitoring must encompass the entire problem life time t_p, from t_o when the problem is perceived, to t_o+t_p when it has been solved. Monitoring does not stop at this point. We do not want to go through that cycle a second time, and society will want to ensure that the quality it has taken pains to achieve is maintained. Monitoring will be required to ensure this.

What is soil monitoring?

Because of the diverse views, motives, and aims of the above actors a generally accepted definition of soil monitoring has not been found yet. We therefore present our personal view as follows.

Soil monitoring is often used in the narrow and purely *operational* sense of quantifying contaminant accumulation in soil by means of national or regional *soil observation networks*, as they have been established now in several countries. Some of the networks are defined as organizationally independent programmes and others are implemented in parallel networks such as forest inventories, meteorological, hydrological or ecological observation programmes, as well as government statistics on fertilizer and pesticide use.

Monitoring the soil is not only data gathering, however. The term 'monitoring' embraces a broad spectrum of scientific actions comprising *observation, measurement, estimation* and most importantly also *identification* of relevant features of the soil ecosystem. The fact that we concentrate today on the feature of contaminant accumulation or changes of 'easily observable' properties reveals our problem. Ultimately we ought to anticipate failures of soil processes or soil functions.

'Soil functions' is an abbreviation for all the interlaced processes that guarantee the sustainable functioning of the soil (i) to serve as a production base for many renewable goods not only for men but also for nature, (ii) to buffer heat (energy), water and element exchange between atmosphere and geosphere, (iii) to filter material fluxes, and (iv) to serve as a reactor for decomposing man-made or natural components.

Whoever monitors the soil must adopt a scientific approach because proper scientific methods and valid scientific understanding are needed to obtain testable forecasts of future developments. However, soil monitoring is not a synonym for soil ecosystem research. Soil monitoring is focused on wide-range and long-term changes of the multifunctionality of soils with respect to potential counteractions.

To sum up, a reasonable working definition of soil monitoring could be as follows.

Soil monitoring is the scientifically based continuous observation of the soil ecosystem, the aims of which are to detect change, to assess and predict the impacts of man's activities, and to forestall deterioration or restore the quality of the soil.

Putting this definition into practice itself needs political will and organization. It must win public support and funding in the long term. It needs a body of scientists and technicians of sufficient size to embrace all the expertise required for its implementation, and for education and publicity. The proponents will need to clarify their aims, and convert general objectives into more specific ones that can be written in terms of operations. They will have to maintain dialogue with politicians, executives and research scientists, for communication will be vital for success.

Key issues and gaps in knowledge

The discussions at the workshop gave an informative picture of agreement and disagreement in the appraisal of both what we do know and what we ought to know.

Recognizing that the relevance of the different issues and fields with insufficient knowledge may vary from region to region, from discipline to discipline and especially from one soil problem to another, we do not intend here to impose a ranking on the scientific issues, on the political or information needs, or on the gaps in knowledge, because this could be misinterpreted as a tactical move to shift attention and resources. Nevertheless, we shall try to structure the issues and gaps.

To be prepared for potentially hazardous developments of the soil ecosystem in the long term it is essential to detect unfavourable changes as early as possible. In particular we need to observe the *material balance* of the soil, asking whether (i) undesirable compounds accumulate, (ii) naturally occurring compounds accumulate to an unfavourable degree, or (iii) the stock of essential compounds is becoming depleted.

Accumulation and depletion can be quantified only in terms of a well defined *compartment*. Choosing proper compartments is one of the most controversial aspects in setting up mass balances. Clearly each compartment should be defined in such a way that the relevant fluxes across its boundaries can be measured reliably. In practice however, the *choice of boundaries* is often based on traditions rather than on the reliability of flux estimation. In many instances (mini-)catchments or farms as management units or regional compartments could provide a better framework for material balances than small isolated sites. Natural lysimeters are examples of minicatchments, with an impervious bedrock serving as a well defined flux boundary below the root zone and making the flume at the catchment exit an optimal flux boundary for solute export.

Not only do accumulation and depletion need increased attention but even more so does the *deterioration* of the soil as a habitat for the entire spectrum of soil organisms. These are vital to the whole functioning of the soil, and they deserve even more attention than they have received in the past. We should then ask what (easily) observable variables we should use to assess the *biological performance* of the soil and to guide those who need to define the various legal intervention levels. We should concentrate our attention on the susceptibility of soil organisms to man-made stresses, and we should devise methods of biomonitoring to measure what is happening.

The soil is highly variable in space, whereas most soil properties, especially the concentrations of contaminants, change only slowly. Detecting *temporal change* against the background of *spatial variation* can demand a great deal of sampling, and there is a need to work out efficient designs. Fortunately the spatial variation is structured; there are distinct types of soil the distributions of which can be mapped and which respond in their particular ways to human impact. In these circumstances it should pay to recognize and record the variation, and to use the soil classes as strata to monitor each major soil type separately. In other circumstances the structure may be represented geostatistically, and other kinds of sampling can take advantage of this. A few soil properties, such as the microbial activity and the pH, may fluctuate rapidly, and for these the long term trend must be perceived against this fluctuation. Again, substantial sampling is likely to be needed.

Although we can economize on sampling by taking into account the structured variation in soil the very fact of *heterogeneity* makes the measurement of temporal change difficult. To detect small changes with limited resources will always demand difficult decisions. Thus, should we:

(i) choose a few sites and then sample frequently to measure all properties of interest?

(ii) choose many sites and measure the soil properties there infrequently?

(iii) sample differently for different properties according to the kind of variation expected?

There is no easy answer, but almost certainly the same sampling strategy applied for monitoring all pollutants on all kinds of soil in all circumstances will not be cost effective. This may also be called the *sampling dilemma*.

In fact the cost of measurement must be taken into account. Therefore, we recommend a *hierarchy* of monitoring intensity. We suggest a network of numerous sites where we record all properties that we can measure cheaply. We should choose a subset at which we record the really diagnostic properties at greater expense: From these we should select even fewer sites where we can analyse the ecological functioning of the soil on an experimental basis. This does not mean that we abandon existing networks. Rather, we ask that they be used to structure enquiry more purposefully so that resources are used to best advantage and the need for extra resources clearly demonstrated.

Another suggestion to deal with the problem of cost-effectiveness in soil sampling is to use more easily measurable *surrogate* or *indicator variables*. These are variables that are not themselves of interest but which are so closely related to the primary variables of interest that they can substitute for them reliably. Recording soil indicators can be compared to measuring the body temperature in human medicine. The interpretative value of indicator variables depends on the depth of understanding of the related processes.

The soil is variable at all scales. Therefore we shall find it difficult to keep any state variable of the soil at every location within the same legally defined limits. In addition, there will always remain a risk that undesired conditions occur at unsampled locations. In monitoring, as in other fields, all actors need to learn to think in terms of probabilities of tolerable and intolerable conditions and in terms of site specific risk factors. Besides a deeper understanding of the processes, this requires that they learn to communicate and to accept these uncertainties.

Conclusions

Many aspects of soil monitoring are still vague. In particular, a wide gap exists between formulations of the general objectives of soil monitoring and their realization by operational instructions. The main efforts in the next decade should be on the conceptual work of soil monitoring programmes. We think that too many current programmes are self-reinforcing. Because of conceptual gaps the data collected are not sufficient and, as a consequence, more and very costly means for additional field measurements are demanded. As an alternative we recommend a combination of the on-going programmes with studies of material fluxes into and out of defined regions, based on data from existing programmes, and specific soil investigations. Stipulating an operational strategy for soil monitoring will be an iterative process of successive approximations, which requires continuous discussions between all the involved groups of actors, decision-makers, executives, as well as researchers.

Acknowledgment

The above represents our views and conclusions at the end of the workshop. We are grateful to Margaret Oliver for her many suggestions for improvement and to Thomas Joller from whose critique we have drawn.

H. Flühler, R. Webster, and R. Schulin
Institut für terrestrische Ökologie
ETH Zürich
Grabenstrasse 3/11a
CH-8952 Schlieren
Switzerland

P. Baccini
Lehrstuhl für Stoffhaushalt und Entsorgungstechnik
ETH Zürich
CH-8600 Dübendorf
Switzerland

LIST OF PARTICIPANTS

Prof. Peter Baccini, Lehrstuhl für Stoffhaushalt und Entsorgungstechnik, ETH Zürich, 8600 Dübendorf, Switzerland

Dr Hans-Peter Bader, EAWAG, Abt. AWS, 8600 Dübendorf, Switzerland

Giovanni Bidoglio, CEC Joint Research Centre Ispra, Environment Institute, 21020 Ispra (Va), Italy

Prof. Peter Blaha, Kosice University, c/o UNI Lausanne, BFSH 2, 1015 Lausanne, Switzerland

Dr Peter Blaser, Eidgenössische Forschungsanstalt für Wald, Schnee und Landschaft, Zürcherstr. 111, 8903 Birmensdorf, Switzerland

Dr Lage Bringmark, Swedish University of Agricultural Sciences, Center for Environmental Monitoring, P. O. Box 7050, 75007 Uppsala, Sweden

Dr Philip C. Brookes, Rothamsted Experimental Station, Soil Science Department, Harpenden, Herts AL5 2JQ, Great Britain

Toni Candinas, Eidgenössische Forschungsanstalt für Agrikulturchemie und Umwelthygiene, 3097 Liebefeld-Bern, Switzerland

Jean-Pierre Clément, Office fédéral de l'environnement, forêt et paysage, Hallwylstr. 4, 3003 Bern, Switzerland

Prof. Jeffrey Collett, University of Illinois, Institute for Environmental Studies, 1101 West Peabody Drive, Urbana, IL 61801, USA

Dr Eros Crivelli, Cantone Ticino, Sezione protezione aria e acqua, Via Salvioni 2, 6500 Bellinzona, Switzerland

Ruedi Dahinden, Eidgenössische Forschungsanstalt für Agrikulturchemie und Umwelthygiene, Schwarzenburgstr., 3097 Liebefeld-Bern, Switzerland

Dr Ad de Roo, Unversity of Utrecht, Dept. of Physical Geography, P.O. Box 80115, 3508 TC Utrecht, The Netherlands

Dr André Desaules, Eidgenössische Forschungsanstalt für Agrikulturchemie und Umwelthygiene, Schwarzenburgstr., 3097 Liebefeld-Bern, Switzerland

Jean-Pascal Dubois, EPF Lausanne, IATE-Pédologie, GR-Ecublens, 1015 Lausanne, Switzerland

Urs Eggenberger, Universität Bern, Mineralogisch-Petrographisches Institut, Balzerstr. 1, 3012 Bern,

Peter Federer, ETH Zürich, Institut für terrestrische Oekologie, Bodenchemie, Grabenstrasse 3, 8952 Schlieren, Switzerland

Prof. Peter Fitze, Universität Zürich, Geograph. Institut, Winterthurerstr. 190, 8057 Zürich, Switzerland

Prof. Hannes Flühler, ETH Zürich, Institut für terrestrische Oekologie, Bodenphysik, Grabenstrasse 3, 8952 Schlieren, Switzerland

Markus Flury, ETH Zürich, Institut für terrestrische Oekologie, Bodenphysik, Grabenstrasse 3, 8952 Schlieren, Switzerland

Irene Forrer, ETH Zürich, Institut für terrestrische Oekologie, Bodenphysik, Grabenstrasse 3, 8952 Schlieren, Switzerland

Roger Frossard, Eidgenössische Forschungsanstalt für Agrikulturchemie und Umwelthygiene, Schwarzenburgstr., 3097 Liebefeld-Bern, Switzerland

Patricia Fry, ETH Zürich, Institut für terrestrische Oekologie, Bodenschutz, Grabenstrasse 3, 8952 Schlieren, Switzerland

Dr Gerhard Furrer, ETH Zürich, Institut für terrestrische Oekologie, Bodenschutz, Grabenstrasse 3, 8952 Schlieren, Switzerland

Gabi Geiger, ETH Zürich, Institut für terrestrische Oekologie, Bodenschutz, Grabenstrasse 3, 8952 Schlieren, Switzerland

Emmanuel Glenck, SKUB/EAWAG, Abt. AWS, 8600 Dübendorf, Switzerland

Dr Matthias Grottker, EAWAG, Ingenieurabteilung, 8600 Dübendorf, Switzerland

Rolf Gsponer, AGW Kt. Zürich, Fachstelle Bodenschutz, Walchetor, 8090 Zürich, Switzerland

Dr Rudolf Häberli, Swiss Priority Program Environment, Längassstrasse 23, 3012 Bern, Switzerland

Dr Uli Hoins, ETH Zürich, Institut für terrestrische Oekologie, Bodenschutz, Grabenstrasse 3, 8952 Schlieren, Switzerland

Dr Thomas Joller, Kant. Luzern, Amt für Umweltschutz, Klosterstrasse 31, 6002 Luzern, Switzerland

Prof. William Jury, University of California, Dept. of Soil and Environmental Sciences, Riverside, CA 92521, USA

Dr Catherine Keller, EPFL - IGR, IATE-Pédologie, GR-Ecublens, 1015 Lausanne, Switzerland

Raimund Kohl, Landesamt für Umweltschutz Baden-Württemberg, Referat Bodenschutz, Hertzstrasse 173, 7500 Karlsruhe, Germany

Dr Gerald Kuhnt, Universität Kiel, Geographisches Institut, 2300 Kiel, Germany

Prof. H.R. Künsch, ETH Zürich, Seminar für Statistik, ETH-Zentrum, 8092 Zürich, Switzerland

Dr Piet Lagas, National institute of public health and environmental protection, Laboratory for soil and groundwater research, P.O. Box 1, 3720 BA Bilthoven, The Netherlands

Dr Christian Lajaunie, Ecole des Mines de Paris, Centre de Géostatistique, 35, rue St Honoré, 77305 Fontainebleau Cedex, France

Dr Peter Lischer, Eidgenössische Forschungsanstalt für Agrikulturchemie und Umwelthygiene, Schwarzenburgstr., 3097 Liebefeld-Bern, Switzerland

Prof. M. Maignan, UNI Lausanne, Institut de Minéralogie, BFSH 2, 1015 Lausanne, Switzerland

Dr Mari Marinussen, Agricultural University Wageningen, Soil Science and Plant Nutrition, Dreijenplein 10, 6703 HB Wageningen, The Netherlands

Prof. Egbert Matzner, Universität Bayreuth, Lehrstuhl für Bodenökologie, Postfach 101251, 8580 Bayreuth, Germany

Prof. Robert Mayer, Gesamthochschule Kassel, Landscape Ecology, Postfach 101380, 3500 Kassel, Germany

Dr Harald Menzi, Eidgenössische Forschungsanstalt für Agrikulturchemie und Umwelthygiene, Schwarzenburgstr., 3097 Liebefeld-Bern, Switzerland

Reto Meuli, ETH Zürich, Institut für terrestrische Oekologie, Bodenschutz, Grabenstrasse 3, 8952 Schlieren, Switzerland

Konrad Meyer, WWF Schweiz, Postfach, 8037 Zürich, Switzerland

Dr Markus Müller, Eidgenössische Forschungsanstalt für Obst-, Wein- und Gartenbau, Schloss, 8820 Wädenswil, Switzerland

Jürg Obrist, ETH Zürich, Institut für terrestrische Oekologie, Bodenschutz, Grabenstrasse 3, 8952 Schlieren, Switzerland

Dr Margaret Oliver, Birmingham University, School of Geography, The University, Edgbaston, Birmingham B15 2TT, Great Britain

Yvan Pannatier, UNI Lausanne, Institut de Minéralogie, BFSH 2, 1015 Lausanne, Switzerland

Andreas Papritz, ETH Zürich, Institut für terrestrische Oekologie, Bodenphysik, Grabenstrasse 3, 8952 Schlieren, Switzerland

Anja Pepels, ETH Zürich, Institut für terrestrische Oekologie, Bodenschutz, Grabenstrasse 3, 8952 Schlieren, Switzerland

Prof. Hans-Rudolf Pfeifer, UNI Lausanne, Center d'analyse minérale, BFSH 2, 1015 Lausanne, Switzerland

Prof. Harald Puchelt, Universität Karlsruhe, Institut für Petrographie und Geochemie, Kaiserstr. 12, 7500 Karlsruhe, Germany

John Quinton, Cranfield Institute of Technology, Silsoe College, Silsoe, Bedford MK45 4DT, Great Britain

Germano Righetti, Cantone Ticino, Sezione protezione aria e acqua, Via Salvioni 2, 6500 Bellinzona, Switzerland

Werner Rohr, ETH Zürich, Institut für terrestrische Oekologie, Bodenchemie, Grabenstrasse 3, 8952 Schlieren, Switzerland

Dr G. Schädlich, Inst. für Geographie und Geoökologie, Georgi-Dimitroff-Platz 1, 7010 Leipzig, Germany

Dr Thomas Scheurer, Schweiz. Akadamie der Naturwissenschaften, Bergacker 128, 3304 Zuzwil, Switzerland

Dr Bernd Schilling, Bayerisches Geologisches Landesamt, Hessstr. 128, 8000 München 40, Germany

Prof. Rainer Schulin, ETH Zürich, Institut für terrestrische Oekologie, Bodenschutz, Grabenstrasse 3, 8952 Schlieren, Switzerland

Christian Skark, Institut für Wasserforschung, Zum Kellerbach 46, 5840 Schwerte, Germany

Dr Erich Stamm, Ciba-Geigy Basel, Ökochemie, PP 2.54, 4002 Basel, Switzerland

Prof. Zygmunt Strzyszcz, Polish Academy of Sciences, Institute of Environmental Engineering, Skodowskiej-Curie 34, 41-800 Zabrze, Poland

Konrad Studer, Eidgenössische Forschungsanstalt für Agrikulturchemie und Umwelthygiene, Schwarzenburgstr., 3097 Liebefeld-Bern, Switzerland

Dr S.E.A.T.M. van der Zee, Agricultural University, Soil Science and Plant Nutrition, PO Box 8005, 6700 HB Wageningen, The Netherlands

Berchtold von Steiger, ETH Zürich, Institut für terrestrische Oekologie, Bodenschutz, Grabenstrasse 3, 8952 Schlieren, Switzerland

Prof. Richard Webster, ETH Zürich, Institut für terrestrische Oekologie, Grabenstrasse 3, 8952 Schlieren, Switzerland

Stefan Zimmermann, Eidgenössische Forschungsanstalt für Wald, Schnee und Landschaft, Zürcherstr. 111, 8903 Birmensdorf, Switzerland

Dr Martin Zysset, Eidgenössische Forschungsanstalt für Wald, Schnee und Landschaft, Zürcherstr. 111, 8903 Birmensdorf, Switzerland